Introduction to Semiconductor Devices

For Computing and Telecommunications Applications

From semiconductor fundamentals to state-of-the-art semiconductor devices used in the telecommunications and computing industries, this book provides a solid grounding in the most important devices used in the hottest areas of electronic engineering today. The book includes coverage of future approaches to computing hardware and RF power amplifiers, and explains how emerging trends and system demands of computing and telecommunications systems influence the choice, design, and operation of semiconductor devices.

The book begins with a discussion of the fundamental properties of semiconductors. Next, state-of-the-art field effect devices are described, including MODFETs and MOSFETs. Short channel effects and the challenges faced by continuing miniaturization are then addressed. The rest of the book discusses the structure, behavior, and operating requirements of semiconductor devices used in lightwave and wireless telecommunications systems.

This is both an excellent senior/graduate text, and a valuable reference for engineers and researchers in the field.

Kevin Brennan (1956–2003) was the recipient of a National Science Foundation Presidential Young Investigator Award. He was named School of ECE Distinguished Professor at Georgia Tech in 2002, and awarded a special commendation from the Vice Provost for Research in recognition of his contributions to graduate-level education in 2002. In 2003, he received the highest honor that a Georgia Tech faculty member can attain: the Class of 1934 Distinguished Professor Award. He also served as an IEEE Electron Device Society Distinguished Lecturer.

Introduction to Semiconductor Devices

For Computing and Telecommunications Applications

KEVIN F. BRENNAN

CAMBRIDGE
UNIVERSITY PRESS

PUBLISHED BY THE PRESS SYNDICATE OF THE UNIVERSITY OF CAMBRIDGE
The Pitt Building, Trumpington Street, Cambridge, United Kingdom

CAMBRIDGE UNIVERSITY PRESS
The Edinburgh Building, Cambridge, CB2 2RU, UK
40 West 20th Street, New York, NY 10011–4211, USA
477 Williamstown Road, Port Melbourne, VIC 3207, Australia
Ruiz de Alarcón 13, 28014 Madrid, Spain
Dock House, The Waterfront, Cape Town 8001, South Africa

http://www.cambridge.org

First published 2005

Printed in the United Kingdom at the University Press, Cambridge

Typefaces Times 10.25/13 pt. and Antique Olive   *System* LATEX 2$_\varepsilon$ [TB]

A catalog record for this book is available from the British Library

ISBN 0 521 83150 4 hardback

To my family, Lea, Casper, and Jack

Contents

Preface   *page* xi
List of physical constants   xv
List of materials parameters for important semiconductors,
   Si and GaAs   xvi

1 Semiconductor fundamentals   **1**
1.1 Definition of a semiconductor   2
1.2 Equilibrium carrier concentrations and intrinsic material   7
1.3 Extrinsic material   16
    Problems   21

2 Carrier action   **23**
2.1 Drift and diffusion   23
2.2 Generation–recombination   28
2.3 Continuity equation and its solution   33
    Problems   36

3 Junctions   **38**
3.1 p–n homojunction in equilibrium   38
3.2 p–n homojunctions under bias   47
3.3 Deviations from ideal diode behavior   57
3.4 Carrier injection, extraction, charge control analysis,
    and capacitance   61
3.5 Schottky barriers   68
    Problems   75

4 Bipolar junction transistors   **78**
4.1 BJT operation   78
4.2 Secondary effects in BJTs   92
4.2.1 Drift in the base region   92
4.2.2 Base narrowing or the Early Effect   94
4.2.3 Avalanche breakdown   95
4.3 High frequency operation of a BJT   97
    Problems   99

5   JFETs and MESFETs   **101**

5.1   JFET operation   101
5.2   MESFET and MODFET operation   104
5.3   Quantitative description of JFETs and MESFETs   112
5.4   Small signal model for a JFET   121
      Problems   124

**6   Metal–insulator–semiconductor structures
      and MOSFETs**   **127**

6.1   MIS systems in equilibrium   127
6.2   MIS systems under bias   133
6.3   Basic theory of MOSFET operation   144
6.4   Small signal operation of MESFETs and MOSFETs   155
6.5   CMOS circuits   160
      Problems   165

7   Short-channel effects and challenges to CMOS   **169**

7.1   Short-channel effects   169
7.2   Scaling theory   176
7.3   Processing challenges to further CMOS miniaturization   183
      Problems   186

8   Beyond CMOS   **188**

8.1   Evolutionary advances beyond CMOS   188
8.2   Carbon nanotubes   195
8.3   Conventional vs. tactile computing, molecular and
      biological computing   197
8.4   Moletronics – molecular diodes and diode–diode logic   201
8.5   Defect tolerant computing   206
8.6   Quantum dot cellular automata   210
      Problems   219

9   Telecommunications systems – an overview   **220**

9.1   Fiber transmission   220
9.2   Amplifiers and repeaters   223
9.3   Mobile cellular telecommunications systems   225
9.4   Device types for cellular systems   228

**10   Optoelectronic devices – emitters, light amplifiers, and
       detectors**   **230**

10.1   LEDs   230
10.2   Stimulated emission   238
10.3   Laser operation   244

10.4	Types of semiconductor lasers	248
10.5	EDFAs	255
10.6	SOAs	258
10.7	p–i–n photodetectors	260
10.8	Avalanche photodiodes	265
	Problems	273

11 Transistors for high frequency, high power amplifiers for wireless systems **275**

11.1	Transistor figures of merit for wireless systems	275
11.2	Heterostructures	281
11.3	MODFET devices	286
11.4	HBTs	290
11.5	Wide band gap semiconductors	294
	Problems	298

| | References | 300 |
| | Index | 303 |

Preface

At the time of this writing the microelectronics industry is poised at the threshold of a major turning point. For nearly fifty years, the industry has grown from the initial invention of the integrated circuit through the continued refinement and miniaturization of silicon based transistors. Along with the development of complementary metal oxide semiconductor circuitry, miniaturization of semiconductor devices created what has been called the information revolution. Each new generation of devices leads to improved performance of memory and microprocessor chips at ever reduced cost, thus fueling the expansion and development of computing technology. The growth rate in integrated circuit technology, a doubling in chip complexity every eighteen months or so, is known as Moore's First Law. Interestingly, the semiconductor industry has been able to keep pace with Moore's First Law and at times exceed it over the past forty years. However, now at the beginning of the twenty-first century doubts are being raised as to just how much longer the industry can follow Moore's First Law. There are many difficult challenges that confront CMOS technology as device dimensions scale down below 0.1 μm. Many people have predicted that several of these challenges will be so difficult and expensive to overcome that continued growth in CMOS development will be threatened. Further improvement in device technology will then require a disruptive, revolutionary technology.

One might first wonder why is it important to continue to improve microprocessor speed and memory storage much beyond current levels? Part of the answer to this question comes from the simultaneous development of the telecommunications industry. Both lightwave communications and cellular communications systems have grown rapidly. Over just the past ten years, the cellular telephone industry has increased exponentially, making it one of the fastest growing industries in the world. The expansion of cellular telephony to the transmission of data, internet connections, and video information is already beginning. Cellular transmission of video information will require much higher bandwidth operation and greater sophistication than is currently available in cellular systems. Lightwave systems already handle video and internet communications and are pressed to improve bandwidth for faster operation. Though improvement in software and algorithms has been highly instrumental in improving telecommunications system capacity, hardware improvements are equally as important to maintain growth in these systems. Therefore, there is an acute need for faster electronics with a concurrent memory enhancement to improve telecommunications systems, thus further fueling the information revolution.

It is my opinion that the microelectronics industry will necessarily continue to grow to meet the demands of future computing and telecommunications systems.

However, this growth may not be confined to silicon CMOS but may extend into several other technologies as well. The goal of this book is to present an introductory discussion to undergraduate students of the basic workings of current semiconductor devices used in computing and telecommunications systems and to present some of the emerging revolutionary approaches that microelectronics could take in the near future. Throughout the book, the applications and operating requirements imposed on semiconductor hardware by computing and telecommunications applications are used to describe the important figures of merit of each device. In this way, the student can clearly see what fundamental properties a particular device must have to meet the system application requirements for which it is designed.

One might wonder why yet another book is needed on semiconductor devices for undergraduate education. This question is particularly relevant in that several universities have recently decided to abandon requiring an undergraduate course in semiconductor devices, making it solely an elective instead. Given that there are several excellent texts, such as Streetman and Banerjee *Solid State Electronic Devices* (2000) or Pierret *Semiconductor Device Fundamentals* (1996), one might wonder why another undergraduate book is needed especially in light of the fact that the need for undergraduate books is apparently decreasing. Though the above mentioned books are unquestionably excellent, they do not provide a discussion of the future of microelectronics and how it relates to the greatest existing growth industries of computing and telecommunications. It is the primary purpose of this book to provide the context, namely computing and telecommunications, in which semiconductor devices play their most important and ubiquitous role. Further, the present book provides a look at not only the state-of-the-art devices but also future approaches that go beyond current technology. In this way, a new, refreshing, up-to-date approach to teaching semiconductor devices and exciting the students about the future of the field is provided. It is my opinion that through an enlightened approach the negative trend of the removal of microelectronics courses from undergraduate curriculums can be reversed. Ironically, I believe that microelectronics is poised for its greatest surge. Thus rather than abandoning teaching microelectronics, it should be more widely presented and the approach should be more interdisciplinary at least addressing possibilities in molecular and biological systems for future computing hardware. This book presents a first cut at such an interdisciplinary approach.

This book has grown out of notes used for an undergraduate course I teach in the School of Electrical and Computer Engineering at Georgia Tech. The course is one semester long and follows a required course in circuit theory that includes some of the basics of semiconductor devices. However, the book does not draw on the student's knowledge of circuits and can thus be used as a first course in semiconductor devices. Given that the presentation is a bit briefer than most semiconductor device texts on the fundamentals, the book is probably better suited for either a second level course, as is done at Georgia Tech, or a first level course for more advanced students. As for scientific and mathematical background, the book requires knowledge of calculus and differential equations. However, no knowledge of quantum mechanics, solid state physics or statistical mechanics is required. Computer based assignments have not been included in the text. The main reasons for their exclusion is that we are preparing a computer based exercise book for use in all of our undergraduate level microelectronics courses. The proposed book will have computer exercises that follow the present book providing another path for learning.

The present book is organized as follows. It begins with a presentation of the essential fundamentals of semiconductors. The second chapter discusses carrier action. The third chapter focuses on junctions including p–n homojunctions, Schottky barriers, and ohmic contacts. In the fourth chapter, bipolar junction transistors are presented. JFETs and MESFETs are discussed in Chapter 5, including ac models. Chapter 6 presents a discussion of metal insulator semiconductor systems particularly MOS devices, long channel MOSFETs, and CMOS circuits. Short channel devices, scaling and challenges to further improvement of CMOS devices are discussed in Chapter 7. Chapter 8 presents a discussion of several different technical approaches that go beyond CMOS. The topics in Chapter 8 are limited to those that do not require knowledge of quantum mechanics. These topics are included in the graduate level textbook *Theory of Modern Electronic Semiconductor Devices* (2002) by Kevin F. Brennan and April S. Brown. The balance of the book focuses on device use in lightwave and cellular telecommunications systems. Chapter 9 gives an overview of telecommunications systems, both wired and wireless. In Chapter 10 a discussion of optoelectronic devices used in lightwave communications systems such as LEDs, lasers, erbium doped fiber amplifiers, semiconductor optical amplifiers and photodetectors is presented. The book concludes with a discussion of transistors used in high frequency, high power amplifiers such as MODFETs and HBTs in Chapter 11.

This book is designed to be the first in a series of texts written by the current author. It provides an introduction to semiconductor devices using only for the most part classical physics. Some limited discussion about spatial quantization is included, however. Thus the present book is well suited to the typical junior or senior level undergraduate student. After completing a course that utilizes the present book, the student is prepared for graduate level study. At Georgia Tech graduate students in microelectronics begin their study, following an undergraduate course at the level of the present book, with the basic science of quantum mechanics, statistical mechanics, and solid state physics covered in *The Physics of Semiconductors with Applications to Optoelectronic Devices* (1999), by Kevin F. Brennan. This material is covered in a first semester graduate level course that is followed by a second semester graduate level course on modern electronic devices. The textbook for the second semester graduate level course at Georgia Tech is *Theory of Modern Electronic Semiconductor Devices* (2002) by Kevin F. Brennan and April S. Brown.

Pedagogically, the undergraduate course this book has been developed from is taught three times a year at Georgia Tech. This course is a second level course in semiconductor devices that follows a required course that contains both circuit theory and elementary semiconductor material. Since the book is used at Georgia Tech for a second level course, we typically quickly cover the topics in Chapters 1–3 in about 2–3 weeks. Depending upon the student's preparation, the fourth chapter can be skipped, substituting a brief review instead. The course gets "down to business" beginning with

Chapter 5 and goes through the remaining chapters for the balance of the semester. Often we skip the section on CMOS (this is covered in the circuits level course) as well as Chapter 9 which is generally just assigned reading. Homework problems are typically selected from those at the back of the chapters. Two in-class quizzes and a final examination are given. Instructors can obtain a solutions manual for the problems on-line at www.ece.gatech.edu/research/labs/comp_elec. The solutions manual can be downloaded and is password protected. Instructors only are given access to the solutions. Please follow the directions at the web site to obtain the necessary password.

The author would like to thank his many colleagues and students at Georgia Tech that have provided constructive criticism in the writing of this book. Specifically, the author is thankful to Mike Weber for his help on some of the figures and for assisting in creating the book web site. Thanks go to Professor Wolfgang Porod of Notre Dame University and to Dr. Phaedron Avouris at IBM for granting permission to reproduce some of their work.

Finally, I would like to thank my family and friends for their enduring support and patience.

Postscript

Professor Kevin Brennan, my colleague at Georgia Tech, and one of my best friends, passed away on August 2, 2003. After he became ill, he continued to work on this text during the last year of his life, and had essentially completed it at the time of his death. I became involved at the copy-editing stage, and would like to express my appreciation to his wife, Lea McLees, for allowing me to assist in bringing this text to conclusion. I would like to acknowledge the effort of Ms. Maureen Storey, whose meticulous attention to detail was essential to the completion of the project. Most of my corrections and additions were reactive to her questions and comments. Eric Willner at Cambridge University Press showed considerable patience in both coaxing us and allowing us time to polish the text. The Chair of the School of Electrical and Computer Engineering at Georgia Tech, Dr. Roger Webb, provided both emotional and tangible support during this difficult period. Professor Christiana Honsberg and Professor Tom Gaylord at Georgia Tech provided answers to questions from me at critical junctures. Kevin Brennan was a superb teacher, accomplished researcher, and prolific author. I am appreciative of the fact that he is able to teach us one last time.

Atlanta, GA   **W. Russell Callen**
March, 2004

Physical constants

Avogadro's constant	N_{AVO}	6.022×10^{23}	Mol^{-1}
Boltzmann's constant	k_B	1.38×10^{-23}	J/K
		8.62×10^{-5}	eV/K
Electron charge	q	1.6×10^{-19}	C
Electron rest mass	m_0	0.511×10^6	eV/C^2
		9.11×10^{-31}	kg
Permeability – free space	μ_0	1.2566×10^{-8}	H/cm
Permittivity – free space	ε_0	8.85×10^{-14}	F/cm
Planck's constant	h	4.14×10^{-15}	eV s
		6.63×10^{-34}	J s
Reduced Planck's constant	$\hbar$	6.58×10^{-16}	eV s
		1.055×10^{-34}	J s
Speed of light	c	3.0×10^{10}	cm/s
Thermal voltage – 0300 K	$k_B T/q$	0.0259	V

Material parameters for important semiconductors, Si and GaAs

Bulk material parameters for silicon

Lattice constant (Å)	$a = 5.43$
Dielectric constant	11.9
Intrinsic carrier concentration (cm^{-3})	1.0×10^{10}
Energy band gap (eV)	1.12
Sound velocity (cm/s)	9.04×10^{5}
Density (g cm^{-3})	2.33
Effective mass along X (m^*/m_0) – transverse	0.19
Effective mass along X (m^*/m_0) – longitudinal	0.916
Effective mass along L (m^*/m_0) – transverse	0.12
Effective mass along L (m^*/m_0) – longitudinal	1.59
Heavy hole mass (m^*/m_0)	0.537
Electron mobility at 300 K (cm^2/(V s))	1450
Hole mobility at 300 K (cm^2/(V s))	500
Thermal conductivity at 300 K (W/(cm °C))	1.5
Effective density of states in conduction band (cm^{-3})	2.8×10^{19}
Effective density of states in valence band (cm^{-3})	1.04×10^{19}
Nonparabolicity along X (eV^{-1})	0.5
Intravalley acoustic deformation potential (eV)	9.5
Optical phonon energy at Γ (eV)	0.062
Intervalley separation energy, X–L (eV)	1.17

Bulk material parameters for GaAs

Lattice constant (Å)	$a = 5.65$
Low frequency dielectric constant	12.90
High frequency dielectric constant	10.92
Energy band gap at 300 K (eV)	1.425
Intrinsic carrier concentration (cm^{-3})	2.1×10^{6}
Electron mobility at 300 K (cm^2/(V s))	8500
Hole mobility at 300 K (cm^2/(V s))	400
Longitudinal sound velocity (cm/s) along (100) direction	4.73×10^{5}
Density (g/cm^3)	5.36

Effective mass at Γ (m^*/m_0)	0.067
Effective mass along L (m^*/m_0)	0.56
Effective mass along X (m^*/m_0)	0.85
Heavy hole mass (m^*/m_0)	0.62
Effective density of states conduction band (cm^{-3})	4.7×10^{17}
Effective density of states valence band (cm^{-3})	7.0×10^{18}
Thermal conductivity at 300 K (W/(cm °C))	0.46
Nonparabolicity at Γ (eV^{-1})	0.690
Intravalley acoustic deformation potential (eV)	8.0
Optical phonon energy at Γ (eV)	0.035
Intervalley separation energy, Γ–L (eV)	0.284
Intervalley separation energy, Γ–X (eV)	0.476

Note: Γ designates a point in k-space; X and L designate directions in k-space. Γ refers to the $k = 0$ point at the center of the Brillouin zone. X refers to the {100} directions and L to the {111} directions.

1

Semiconductor fundamentals

In this chapter, we review the basic fundamentals of semiconductors that will be used throughout the text. Only the fundamental issues that we will need to begin our study of semiconductor devices utilized in computing and telecommunications systems are discussed.

Before we begin our study it is useful to point out how semiconductor devices are instrumental in many applications. In this book we will mainly examine the application of semiconductor devices to computing and telecommunications systems. Specifically, we will examine the primary device used in integrated circuits for digital systems, the metal oxide semiconductor field effect transistor, MOSFET. The discussion will focus on state-of-the-art MOSFET devices and future approaches that extend conventional MOSFETs and revolutionary approaches that go well beyond MOSFETs. It is expected that computing hardware will continue to improve, providing faster and more powerful computers in the future using either some or all of the techniques discussed here or perhaps using completely new technologies. In any event, there is almost certainly going to be a large growth in computing hardware in order to maintain the pace of computer development and this book will help introduce the student to emerging technologies that may play a role in future computing platforms.

The second major topic of this book involves discussion of semiconductor devices for telecommunications applications. We will examine devices of use in lightwave communications as well as wireless communications networks. Among these devices are emitters, detectors, amplifiers, and repeaters.

Some mention should be made of the various commercial products that are and will be greatly impacted by semiconductor devices. The development of blue and blue-green light emitting diodes (LEDs) and lasers foments the evolution of new, highly efficient, rugged, ultra-long-life illumination elements. White light emitters using LEDs are now becoming commercially available. These emitters are far more efficient than incandescent bulbs, cost about the same or less, have lifetimes measured in years rather than months, are rugged and durable. It is expected that replacing incandescent lighting by LEDs worldwide can result in a substantial energy savings and potentially reduce consumption of fossil fuels. Perhaps this will lead to a reduction in greenhouse gas emission and help combat global warming and environmental decay in general. Blue lasers enable the development of very small compact discs for data storage, video and audio systems thus greatly expanding the storage capacity of CDs.

New semiconductor materials, such as gallium nitride (GaN) and silicon carbide (SiC), are emerging that are far more tolerant of high temperatures, and operate at significantly higher current densities and frequencies than existing devices. Devices made from these materials are highly attractive for high power, high frequency, and high temperature operation. Specific applications are as power amplifiers for base stations in wireless telecommunications systems, hybrid electric vehicles, switching elements for electric power grids, and high power amplification for radar and satellite communications. Thus GaN and SiC may emerge as important semiconductor materials for many important applications.

1.1 Definition of a semiconductor

The first question one might raise is why are semiconductor materials important in electrical engineering? To answer this question let us first consider a useful characterization scheme for solids based on their electrical properties, specifically their electrical conductivity. Generally, all crystalline solids can be classified into one of four categories. These categories, arranged from highest electrical conductivity to lowest, are metals, semimetals, semiconductors, and insulators. The distinction among these four categories is of course, somewhat vague. For instance, some materials can be either metallic or semimetallic depending upon the form into which they crystallize. Additionally, the distinction between semiconductors and insulators can often become blurred for the wide band gap materials. Nevertheless, we will find it convenient to classify solids into one of these four categories.

Of the four classes of materials, semiconductors are arguably the most important in electrical engineering. The principal reason underlying the importance of semiconductors is the fact that their electrical properties can be readily engineered. Semiconductors are unique in that their conductivity can be significantly altered in several different ways. For the other three types of solids, metals, semimetals, and insulators, their conductivity cannot be readily and significantly altered making them far less attractive for electrical engineering.

There are numerous ways in which the conductivity of a semiconductor can be altered. In this book, we will address most of these approaches and how they can be utilized to make useful semiconductor devices. Before we outline the approaches to manipulating the electrical conductivity of a semiconductor, we should first review what a semiconductor is.

The most commonly used semiconductors are the elemental semiconductors silicon and germanium, and the compound semiconductors, consisting of compound materials. There are numerous compound semiconductors but they are generally formed from two, three, or four different elements and are referred to as binary, ternary, and quaternary compounds respectively. The most important compound semiconductors are based on Column IIIA and Column VA elements in the Periodic Table. For this reason, these compounds are called the III–V compound semiconductors or III–Vs. Examples of the III–V compounds are gallium arsenide (GaAs), indium phosphide (InP), aluminum arsenide (AlAs), indium arsenide (InAs), etc. Notice that in each case the cation is a Column III element while the anion is a Column V element. Ternary compounds can be formed using three elements such as $Al_xGa_{1-x}As$, where the subscript x represents the mole fraction of aluminum present in the compound. Similarly, quaternary compounds can be formed in which four elements are combined. An example of a quaternary compound semiconductor is $In_xGa_{1-x}As_yP_{1-y}$.

How though can we identify which materials are semiconductors? To answer this question we must first consider a fundamental result in the physics of solids. Every crystalline solid has translational symmetry. A system is said to have translational symmetry if it can be broken into a set of identical basic unit cells such that when the system is translated by a distance equal to the length of one unit cell it remains invariant. An obvious example is that of a uniform brick wall. If one translates a row of bricks by a length equal to that of a single brick, the wall looks precisely the same as before. The wall is said to be invariant under a linear translation. A similar situation holds for a crystalline solid. The arrangements of atoms forming a crystalline solid are like the bricks of a uniform wall. The atoms, much like the bricks, are arranged in periodic intervals. Therefore, when the system is translated by a distance equal to the separation between two adjacent atom centers, called the lattice constant, the system remains the same and is said to be invariant. Since the arrangement of the positions of the atoms in a crystalline solid is periodic, the electrostatic potential corresponding to the atoms is also periodic. The potential of the solid is thus also translationally symmetric. The fact that all crystalline solids have a periodic potential is extremely important. There is a fundamental result from quantum mechanics that applies to any system with a periodic potential. This result (Brennan, 1999, Chapter 8) states that for a system with a translationally symmetric potential, the electron energy levels are arranged in bands. These bands can either be conducting or forbidden. As the name implies a conduction band is one in which the electrons can propagate or conduct. Conversely, a forbidden band is one in which no conducting states exist. Electrons cannot be placed into a forbidden band.

In addition to the formation of energy bands, the presence of a periodic potential introduces energy gaps in the allowed energy spectrum. These gaps are called forbidden gaps. Forbidden gaps correspond to energy ranges wherein no allowed electronic states exist. A typical diagram showing a valence band, forbidden energy band and conduction band is shown in Fig. 1.1. As can be seen from the figure, allowed energy states exist only within the conduction and valence energy bands. As mentioned above, electrons within the conduction band can propagate through the crystal and thus carry a current. Electrons cannot be located within the forbidden band. In the valence band, electronic states exist but these states are not free. In other words, electrons within the valence band are localized into bound states that are formed by the molecular bonds between the constituent host atoms of the crystal.

A completely empty band cannot conduct a current. This should be obvious since an empty band has no carriers within it and thus there is nothing to carry the current. A less obvious fact is that a completely filled energy band also cannot conduct a current. This follows from the fact that no two electrons can simultaneously occupy the same quantum state. The general formal statement of this is the Pauli Principle, which applies to the class of particles called fermions, and includes electrons, protons, and neutrons. The Pauli Principle plays a strong role in the formation of atoms. As the reader is aware from fundamental chemistry, each atom in the Periodic Table is formed by

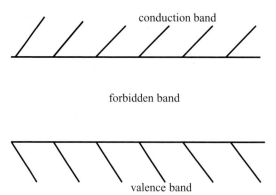

conduction band

forbidden band

valence band

Figure 1.1 Sketch of the conduction, forbidden, and valence bands within a semiconductor. Electrons within the conduction band can freely propagate through the crystal and thus can carry a current. Electrons within the valence band are localized into bound electronic states formed by the molecular bonding of the constituent atoms of the crystal. In the forbidden band, no electronic states exist and thus electrons cannot exist within the forbidden band. The forbidden band is also called the energy gap.

progressively adding an electron and proton (and possibly neutrons) to each previous atom starting with hydrogen. In the case of hydrogen the only charged particles present are one electron and one proton. The electron is placed into the lowest lying energy state of the atom. The next element is helium which comprises two electrons and two protons as well as two neutrons. The additional electron cannot be added to the same quantum state as the first electron and is placed into the first level, 1s, but with a different spin state. The 1s level is completely filled by two electrons. Thus for the next element, lithium with three electrons and three protons plus neutrons, the third electron in lithium must go into a higher energy state than that of the first two electrons, the 2s orbital. Thus ever larger atoms containing more electrons and protons are configured such that the additional electrons enter higher energy states. If electrons did not obey the Pauli Principle, then all of the electrons in an atom, no matter how many electrons are present, would be put into the lowest energy, 1s state. As a result, chemistry would be very different from what is observed.

According to the Pauli Principle, an electron cannot move into an already occupied state. This situation is similar to that of parking automobiles in a parking lot. No two cars can be put into the same parking spot simultaneously. Obviously, a parking spot must initially be unoccupied in order to place a car into it. Electrons behave in much the same way. In the case of electrons, quantum states assume the same role as parking spaces do for cars. It is important to further recognize that a filled parking lot cannot accept any more cars without removing one and similarly a filled energy band cannot accept any more electrons without removing one. Now we can understand why a filled energy band does not conduct a current. For a current to flow, electrons must move from one state to another. In a filled band there are no vacancies into which the electrons can move since all possible states are filled. Hence, no current can flow.

The distinction among each of the four categories of solids can now be made based on the energy bands in the material. An insulator is a material in which the highest occupied band, called the valence band, is completely filled and the next available energy band, called the conduction band, is completely empty. The energy separation between the filled and empty bands is called the energy gap. In order for a material to be insulating, it is also necessary that the energy gap be very high such that carriers cannot be readily promoted from the valence band into the conduction band. Therefore in an insulator, the valence band is completely filled and the conduction band is completely empty and no current can flow in the material. Conversely, a metal is a highly conductive material. Metals are solids in which the conduction band is only partially filled. The conduction band consists then of many electrons and many empty states. A large current can be supported within a metal since most of the electrons within the conduction band can contribute to the current conduction since there exist many vacancies into which the electrons can move under the action of a driving field. Consequently, metals have a very high electrical conductivity. The other two categories of materials, semimetals and semiconductors, are somewhat intermediate between metals and insulators. Semimetals are materials like insulators in that the conduction band is unoccupied and the valence band is fully occupied at zero temperature. However, in semimetals the energy gap vanishes in part such that the conduction and valence bands intersect. Electrons from the valence band can be readily accelerated into the conduction band at the point or points of intersection of the two bands and the material can thus support a current. In this way, semimetals exhibit a relatively high conductivity but not as high as that of a metal. Finally, a semiconductor is something like an insulator but with a relatively small energy gap separating the conduction and valence bands. At absolute zero temperature within a semiconductor the conduction band is completely empty and the valence band is completely filled. However, as the temperature is raised to room temperature, the energy gap is sufficiently small that some measurable population of the conduction band occurs. Therefore, a semiconductor will conduct a current at room temperature but with a much higher resistance than that of a metal.

The electrical resistance of a crystal is a function of the electron concentration in the conduction band. In a metal, the electron concentration within the conduction band is extremely high, on the order of $\sim 10^{23}$ cm^{-3}. In a semiconductor the electron concentration within the conduction band is many orders of magnitude lower. Therefore, the conductivity of a semiconductor is much less than that of a metal. To quantify the conductivity it is essential to determine the electron concentration. In the next section the technique used to determine the electron concentration within a semiconductor will be discussed.

Before we end this section, it is useful to discuss the shape of the energy bands in a crystal. One of the basic concepts of quantum mechanics is that fundamental particles have a wave-particle duality. This implies that a fundamental particle like an electron for example sometimes manifests itself as a wave and sometimes as a particle, but never simultaneously. Therefore, an electron has a wavelength associated with it, called the de Broglie wavelength, that accounts for its wavelike behavior. The momentum of an electron can be described using its wavelike behavior as

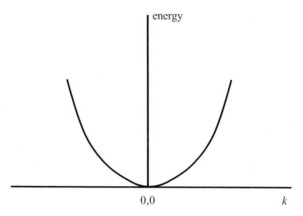

Figure 1.2 Sketch of the energy vs. k relationship for free electrons. Energy bands that obey this relationship are called parabolic energy bands. To a good approximation the energy bands within a semiconductor, at least near the band edge (bottom of the conduction band and top of the valence band), are parabolic.

$$p = \hbar k \tag{1.1}$$

where $\hbar$ is Planck's constant divided by 2π and k is defined as

$$k = \frac{2\pi}{\lambda} \tag{1.2}$$

λ is the electron wavelength and k is called the electron wavevector. A free electron has only kinetic energy given by

$$E = \frac{p^2}{2m} \tag{1.3}$$

Substituting into (1.3) for p the expression given by (1.1) obtains

$$E = \frac{\hbar^2 k^2}{2m} \tag{1.4}$$

The energy of the electron varies quadratically with the wavevector, k. The relationship between E and k given by (1.4) is called a parabolic energy vs. k relationship and is sketched in Fig. 1.2. Notice that the energy vs. k diagram shown in Fig. 1.2 is a parabola with vertex at $E = 0, k = 0$. Since the electron energy varies with respect to the electron wavevector, the $E(k)$ relationship is very important in semiconductors. The behavior of the electron as a function of k is referred to as the electron motion in k-space. In general, the wavevector $\vec{k}$ for an electron in a crystal is a three-dimensional vector. In free space, we can replace the vector $\vec{k}$ by its one-dimensional scalar magnitude, k. We can also often use this scalar one-dimensional model to gain insight into the behaviour of an actual semiconductor.

Typically, the mass that appears in the denominator of (1.4) is quite different from the free space mass and is referred to as the effective mass, usually written as m^*. The effective mass is usually less than the free space mass and takes into account the motion of the electron within the crystalline lattice. The electron effective mass is defined as

$$\frac{1}{m^*} = \frac{1}{\hbar^2} \frac{d^2 E}{dk^2}$$

(1.5)

Notice that (1.5) implies that the curvature of the $E(k)$ relationship determines the effective mass of the electron. If the curvature is high, meaning that E changes greatly with a small change in k, then the effective mass of the electron is small. Conversely, if the curvature is low, implying that the energy E changes slowly with change in k, then the effective mass of the electron is large. In the limit of a horizontal line in the $E(k)$ relationship, the effective mass is infinite; the energy never changes for any change in momentum or k.

The energy bands within most semiconductors deviate from the simple parabolic energy relationship given by (1.4) at high energy, defined as several kT above the conduction band minimum or edge or several kT below the valence band edge.[†] The valence band edge is the point of minimum hole energy within the valence band and typically lies at $k = 0$ in k-space. The energy band structure in general is very complicated in most semiconductors, yet can have a profound effect on device operation, as will be seen in later chapters.

1.2 **Equilibrium carrier concentrations and intrinsic material**

It is important first to understand the concept of equilibrium. A full discussion of equilibrium can be found in the book by Brennan (1999). The most exacting definition of a system in equilibrium is that a closed system, isolated from the external environment, if left to itself over time will evolve towards equilibrium. Under equilibrium conditions there are no external agents, i.e., external voltages, external fields, radiative excitations, or any other external perturbation acting on the system. The system is completely isolated from the external world and as such is unperturbed. There is an important difference between equilibrium and steady-state. In steady-state the system does not change with time, but it is not isolated from the external world. In equilibrium the system is completely isolated from the external world and thus does not change with time but also has no net current flow. A system in steady-state though it does not change with time still has a net current flow. One simple way to view the difference between equilibrium and steady-state is to imagine a partially filled sink. In equilibrium the water level does not change and remains constant. Additionally, there is no net current flow. There is no input or output of water from the sink, the faucet is off and the drain is closed. For a sink in steady-state the water level also doesn't change. However, there is a net current flow. The faucet is on and the drain is open such that the input matches the output and thus the water level does not change. However, the

---

[†] Here, $k = k_B$, Boltzmann's constant. It is usually multiplied by T, the absolute temperature. The factor kT appears in the Fermi–Dirac distribution function, discussed in Section 1.2.

system interacts with its external environment and thus is not in equilibrium but in steady-state instead.

In order to calculate the electron concentration within the conduction band of a semiconductor in equilibrium it is useful to again draw an analogy to parking spaces and cars. In order to park one's car two conditions must be met. First, there must be a parking space. One cannot park one's car, at least legally and safely, in the middle of the road. There must be a parking space. However, the mere presence of a parking space does not ensure that one can park one's car. The second condition is that the space must be unoccupied. The obvious statement that one must have a vacant parking space available to park one's car has an analogy for electrons. In order to put an electron into an energy state, a similar set of two conditions must exist. These are that there must exist a state matching the energy of the electron into which it can be put and this state must be unoccupied. The total number of electrons in the conduction band depends upon the number of available states at a given energy multiplied by the probability that each state is occupied. Summing this product over all possible energies will give the total number of electrons within the conduction band. Mathematically, we can determine the electron concentration in the conduction band by integrating the product of the function that describes the number of available states at a given energy, called the density of states, $D(E)$, and the function that gives the probability that a state at that energy will be occupied, called the distribution function, $f(E)$. The electron concentration, n, is given then as

$$n = \int D(E) f(E) dE \tag{1.6}$$

where the integration is taken over the full range of energy values. In order to evaluate this expression it is necessary to determine both $D(E)$ and $f(E)$. The density of states function $D(E)$ for a three-dimensional system is given as (Brennan, 1999, Section 5.1),

$$D(E) = \frac{1}{2\pi^2} \left(\frac{2m}{\hbar^2} \right)^{\frac{3}{2}} \sqrt{E} \tag{1.7}$$

where $\hbar$ is the reduced Planck constant, $h/2\pi$.

The probability distribution function, $f(E)$, depends upon whether the system is in equilibrium or not. What then is the form of the equilibrium probability distribution function for electrons? To answer this question let us consider Fig. 1.3. Figure 1.3 shows a collection of bins, arranged in ascending energy into which one can place an electron. Let each bin represent an allowed energy state. It is important to recall that no two electrons can occupy the same quantum state simultaneously in accordance with the Pauli Principle. Therefore, once an electron has been placed into a bin, no additional electrons can be put into that bin. To attain the minimum energy configuration of the system the first electron must be put into the first bin. The next electron must then be placed into the second bin, the third electron into the third bin and so forth. This process continues until all of the electrons are placed into a bin. For example, in Fig. 1.3, if only six electrons are present they are placed into the first six bins as shown in the diagram. The normalized probability of each of the first six

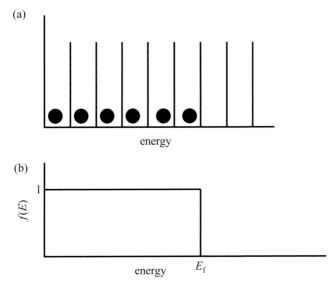

Figure 1.3 (a) Collection of energy bins representing energy states arranged in ascending energy. Into each bin only one electron can be placed in accordance with the Pauli Principle. Each circle represents an electron. The figure shows the minimum energy configuration of an arrangement of six electrons. (b) Corresponding probability distribution function, $f(E)$.

bins being occupied is thus 1. All bins above the sixth bin are empty in the example since no additional electrons are present. Hence the normalized probability of the bins higher than six being occupied is zero. The resulting probability distribution function is shown in Fig. 1.3(b). Note that the probability distribution shown in Fig. 1.3(b) holds for $T = 0$ K. Clearly, the probability distribution function reflects the physical situation, each of the first six states or bins is occupied, while those above six are empty. Inspection of Fig. 1.3 shows that the distribution has the value of 1 until an energy, E_f, is reached. This energy is called the Fermi level and is related to the number of electrons present in the system. For the present example, the energy corresponding to the Fermi level lies at the energy corresponding to the sixth bin.

What happens though at temperatures greater than absolute zero? Temperature is a measure of the internal energy of the system. At temperatures greater than zero, the total energy of the system must be greater than that corresponding to $T = 0$ K. Let us again consider a system with only six electrons. For simplicity let us set the energy of each bin to be an integer multiple of E. Thus for the system shown in Fig. 1.3(a), the total energy is given as the sum of the occupied bins as $E + 2E + 3E + 4E + 5E + 6E = 21E$. The next highest energy configuration, or higher temperature of the system is obtained by promoting the sixth electron into the seventh bin. The corresponding energy of the resulting configuration is then equal to $E + 2E + 3E + 4E + 5E + 7E = 22E$, which is obviously higher than that of the $T = 0$ K configuration. Higher temperature configurations are similarly achieved, i.e. by promoting electrons from the lower energy states into higher energy states. An example system is shown

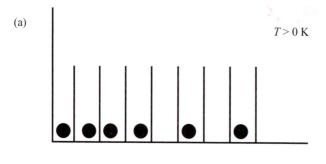

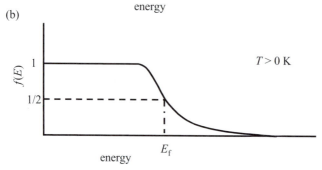

Figure 1.4 (a) Distribution of a collection of six electrons in energy bins corresponding to a collective energy or temperature greater than 0 K. (b) Corresponding probability distribution function for a $T > 0$ K distribution.

in Fig. 1.4 along with the corresponding probability distribution function. As can be seen from the figure, at higher energies there exists a tail in the distribution function. This implies that there is a nonzero probability of a state with energy greater than E_f being occupied and correspondingly a nonzero probability that states below E_f are unoccupied.

The equilibrium probability distribution function, $f(E)$, called the Fermi–Dirac distribution can be expressed mathematically. Its derivation is rather complicated and will not be repeated here. The interested reader is referred to the books by Brennan (1999, Chapter 5) or Kittel and Kroemer (1980). The Fermi–Dirac distribution is given as

$$f(E) = \frac{1}{1 + e^{\frac{(E-E_f)}{kT}}} \tag{1.8}$$

where k is Boltzmann's constant.

It is instructive to examine how $f(E)$ behaves and to show it replicates the distributions shown in Figs. 1.3 and 1.4. Consider first its behavior at $T = 0$ K. There are two conditions, $E < E_f$ and $E > E_f$. For $E < E_f$, the exponent in (1.8) is negative infinity (due to the division by zero), and exp of negative infinity is zero. Thus $f(E)$ for $E < E_f$, is $1/(1 + 0)$ or simply 1. This is of course exactly what is expected; for energies less than the Fermi level, at $T = 0$ K, $f(E) = 1$. The second case, $E > E_f$ at $T = 0$ K leads to the following. Notice that in this case, the exponent is now positive infinity, and exp of positive infinity is infinity. Thus the denominator of (1.8) becomes infinity and $f(E) = 1/\infty = 0$. Again this is consistent with Fig. 1.3; for energies greater than E_f at $T = 0$ K, $f(E) = 0$. At energy $E = E_f$, the Fermi–Dirac function has value $\frac{1}{2}$ as is readily seen from (1.8).

For temperatures greater than 0 K, $f(E)$ is no longer a simple step function and has a tail as shown in Fig. 1.4. If we consider the situation where E is large, such that $e^{(E-E_f)/kT} \gg 1$, then $f(E)$ can be approximated as

$$f(E) = \frac{1}{1 + e^{\frac{(E-E_f)}{kT}}} \sim \frac{1}{e^{\frac{(E-E_f)}{kT}}} \sim e^{-\frac{(E-E_f)}{kT}} \tag{1.9}$$

Under this condition, the Fermi–Dirac distribution behaves as the Maxwell–Boltzmann distribution, and clearly the occupation probability of a state of energy E decreases exponentially with increasing energy. This is as it should be since few electrons, if any, will occupy very high energy states.

At this point, we can now determine the equilibrium electron concentration in a semiconductor using (1.6) by substituting in for $D(E)$ and $f(E)$ the expressions given by (1.7) and (1.8). The general expression, valid for all possibilities, involves using the Fermi–Dirac distribution for $f(E)$. However, this choice of $f(E)$ necessitates that the integral in (1.6) be solved numerically. Alternatively a closed form solution can be obtained if the distribution function $f(E)$ is approximated by the Maxwell–Boltzmann distribution as

$$f(E) \sim e^{-\frac{(E-E_f)}{kT}} \tag{1.10}$$

Usage of the Maxwell–Boltzmann distribution for $f(E)$ holds when the Fermi level lies within the forbidden gap about $3kT$ below the conduction band edge or $3kT$ above the valence band edge. When this condition is valid, the semiconductor is said to be nondegenerate. If the Fermi level lies close to or within the conduction or valence bands the material is said to be degenerate and the full Fermi–Dirac formulation for $f(E)$ should be used. A degenerate material is produced by heavily doping the semiconductor. The ranges in which a semiconductor is nondegenerate and degenerate are shown in Fig. 1.5. Using the Maxwell–Boltzmann distribution the electron concentration within the conduction band can be determined from

$$n = \int_0^\infty \frac{8\pi m^{\frac{3}{2}}\sqrt{2E}}{h^3} e^{-\frac{(E-E_f)}{kT}} \, dE \tag{1.11}$$

In (1.11) the lower bound on the integral is set to zero since we assume that the minimum energy is the conduction band edge. Of course, the upper limit on the energy in a realistic energy band is not infinity but the upper bound can be extended to infinity with little error since the probability distribution decreases exponentially with increasing energy. Therefore, the error introduced by integrating $n(E)$ to infinity is exceedingly small. Equation (1.11) can be evaluated using

$$\int_0^\infty \sqrt{x}\,e^{-ax}\,dx = \frac{\sqrt{\pi}}{2a\sqrt{a}} \tag{1.12}$$

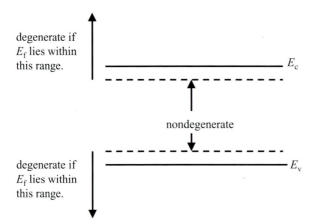

degenerate if E_f lies within this range.

E_c

nondegenerate

degenerate if E_f lies within this range.

E_v

Figure 1.5 Sketch of the energy bands of a semiconductor illustrating the conditions for degenerate and nondegenerate doping. Notice that a degenerate material is highly doped such that the Fermi level lies either near or above the conduction band in n-type material or near or below the valence band in p-type material.

to be

$$n = 2\left(\frac{2\pi m_e^* kT}{h^2}\right)^{\frac{3}{2}} e^{\frac{E_f}{kT}} \tag{1.13}$$

where m_e^* is the electron effective mass. The mass of an electron within the semiconductor is not the same as in free space. Instead, the electron behaves within the crystal as if it has a different mass, called the reduced mass. The reduced mass arises from the motion of the electrons in the periodic potential of the ions forming the crystalline lattice. If we call the bottom of the conduction band E_c instead of zero, then (1.13) becomes

$$n = 2\left(\frac{2\pi m_e^* kT}{h^2}\right)^{\frac{3}{2}} e^{\frac{-(E_c - E_f)}{kT}} \tag{1.14}$$

Defining the effective density of states, N_c, as

$$N_c = 2\left(\frac{2\pi m_e^* kT}{h^2}\right)^{\frac{3}{2}} \tag{1.15}$$

the value of n can be written as

$$n = N_c e^{-\frac{(E_c - E_f)}{kT}} \tag{1.16}$$

A similar expression holds for the equilibrium hole concentration within the valence band as

$$p = N_v e^{-\frac{(E_f - E_v)}{kT}} \tag{1.17}$$

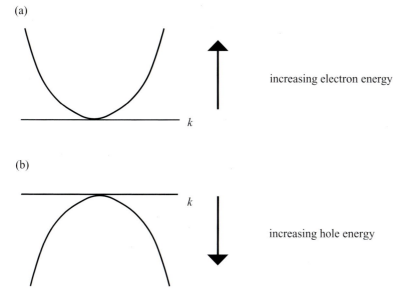

Figure 1.6 Sketch of (a) the conduction band and (b) the valence band showing the direction of increasing electron and hole energy.

where N_v is the same as (1.15) with the hole mass, m_h^*, used in place of the electron mass, m_e^*, yielding

$$N_v = 2 \left(\frac{2\pi m_h^* kT}{h^2} \right)^{\frac{3}{2}} \tag{1.18}$$

Equations (1.16) and (1.17) apply for the electron and hole concentrations for a semiconductor in equilibrium.

An intrinsic semiconductor has no intentionally added impurities. Conversely, an extrinsic semiconductor has intentionally added impurities called dopants. In an intrinsic semiconductor promoting electrons from the valence band produces carriers within the conduction band. The vacancies produced within the valence band from the promotion of electrons are called holes. The two most salient features of holes necessary for our discussion are that holes are positively charged and that hole energy increases downwards in the energy band diagram as opposed to electron energy which increases upwards as shown in Fig. 1.6. A hole is a vacancy within an otherwise filled band. Thus when an electron is promoted from the valence band into the conduction band a vacancy or hole is left behind in the valence band. As a result, the conduction band is no longer empty and can now conduct a current. This is obvious since there is now an electron within the band and it can move between different unoccupied energy states under the action of an applied field. Similarly, the valence band is no longer completely filled and thus it too can conduct a current. Given that there are now vacancies within the valence band, the electrons within the valence band can move from one

the fermi level in an intrinsic semiconductor lies in the middle of the forbidden gap, $EF_{intrinsic} = \dfrac{Ec + Ev}{2}$

vacancy to the next. This movement of electrons within the nearly filled valence band is equivalent to the movement of the corresponding number of holes within an empty band, provided that the holes have opposite sign to that of the electrons. For example, if there are $N - 1$ electrons within the valence band equivalently there is one hole. The current carrying species within the conduction band is of course the free electron that has been promoted from the valence band. The current carrying species within the valence band is the hole.

What though is the current carried by the hole? If the band is completely filled, then no current flows. The corresponding current density is given by summing over all of the electron velocities as

$$\vec{j} = -q \sum_{i=1}^{N} \vec{v}_i \tag{1.19}$$

For a completely filled band, as mentioned above the net current is zero. This implies that there are on average, an equal number of electrons crossing a Gaussian surface moving to the left as are moving to the right. Therefore, the net flux of electrons across the surface on average is zero. Consequently, the net average velocity of the carriers must also be zero; for every electron motion there is another electron with an opposite but equal motion. Thus the net velocity of the entire system vanishes as well as the current density. The current density corresponding to the motions of the electrons within a partially filled band can be related to the motion of the vacancies by recognizing that

$$\vec{j}_{\text{filled}} - \vec{j}_{\text{occupied}} = \vec{j}_{\text{vacancies}} \tag{1.20}$$

But the current density due to the filled band is zero. Consequently, the current density produced by the motion of the vacancies must be the exact negative of that produced by the motion of the electrons. The current density due to the motion of the holes (vacancies) is then

$$\vec{j} = +q \sum_{i=1}^{N} \vec{v}_i \tag{1.21}$$

A hole thus behaves like a positively charged particle. Hence, holes move in the opposite direction from electrons under the action of an applied field.

Within an intrinsic material the electron concentration is equal to the hole concentration, $n = p$. The intrinsic carrier concentration in equilibrium is called n_i. The Fermi level in intrinsic material is referred to as the intrinsic level, E_i. The position of the intrinsic level can be determined as follows. Since $n = p$, the equilibrium electron and hole concentrations in intrinsic material can be related as

$$N_c e^{\frac{(E_i - E_c)}{kT}} = N_v e^{\frac{(E_v - E_i)}{kT}} \tag{1.22}$$

where E_i has been inserted in place of E_f in (1.16) and (1.17). Solving for E_i in (1.22) obtains

$$\frac{N_v}{N_c} = e^{\frac{(E_i - E_c - E_v + E_i)}{kT}}$$

$$E_i = \frac{(E_c + E_v)}{2} + \frac{kT}{2} \ln\left(\frac{N_v}{N_c}\right) \tag{1.23}$$

Substituting into (1.23) the relationships for N_v and N_c given by (1.15) and (1.18) obtains

$$E_i = \frac{(E_c + E_v)}{2} + \frac{3kT}{4} \ln\left(\frac{m_h^*}{m_e^*}\right) \tag{1.24}$$

In some materials the effective masses of the electrons and holes are roughly equal. In this case, the intrinsic level lies at midgap, halfway between the conduction and valence bands. Even if the effective masses are substantially different it is a reasonable assumption to set the intrinsic level equal to the midgap energy. For example, in GaAs the electron and hole effective masses are 0.067 and 0.62 respectively. The last term in (1.24) is equal to 0.043 eV. The midgap energy is 0.71, so we see that the correction due to the difference in the effective masses, even when they are substantially different, is small.

The intrinsic concentration, n_i, can be obtained as follows. For an intrinsic semiconductor the electron and hole concentrations are equal:

$$n = p = n_i \tag{1.25}$$

As we discussed above, the Fermi level, E_f can be replaced by E_i the intrinsic level. Using the above results, n_i can be written as

$$n_i = N_c e^{\frac{(E_i - E_c)}{kT}}; \quad n_i = N_v e^{\frac{(E_v - E_i)}{kT}} \tag{1.26}$$

Rearranging the terms in (1.26) n_i can be expressed as

$$N_c e^{-\frac{E_c}{kT}} = n_i e^{-\frac{E_i}{kT}} \tag{1.27}$$

The electron concentration, n, can now be expressed in terms of the intrinsic concentration using (1.27). Starting with (1.16)

$$n = N_c e^{-\frac{E_c}{kT}} e^{\frac{E_f}{kT}} \tag{1.28}$$

Substituting (1.27) into (1.28) obtains

$$n = n_i e^{\frac{(E_f - E_i)}{kT}} \tag{1.29}$$

Similarly, the hole concentration can be written as

$$p = n_i e^{\frac{(E_i - E_f)}{kT}} \tag{1.30}$$

Equations (1.29) and (1.30) hold for a nondegenerate semiconductor in equilibrium in which approximating the Fermi–Dirac distribution by the Maxwell–Boltzmann distribution is valid.

Consider the product of n and p for an intrinsic semiconductor. Using (1.29) and (1.30) the np product is given as

$$np = n_i^2 e^{\frac{(E_f - E_i)}{kT}} e^{\frac{(E_i - E_f)}{kT}} \tag{1.31}$$

which is simply

$$np = n_i^2 \tag{1.32}$$

Equation (1.32) is called the Law of Mass Action. The Law of Mass Action applies only in equilibrium but it is true for any semiconductor either intrinsic or extrinsic. The np product can be written in an alternative manner using (1.16) and (1.17) as

$$n = N_c e^{-\frac{E_c}{kT}} e^{\frac{E_f}{kT}} \qquad p = N_v e^{\frac{E_v}{kT}} e^{-\frac{E_f}{kT}} \tag{1.33}$$

Taking the product of n and p given by (1.33) obtains,

$$np = N_c N_v e^{-\frac{(E_c - E_v)}{kT}} = N_c N_v e^{-\frac{(E_g)}{kT}} = n_i^2 \tag{1.34}$$

where $E_g = E_c - E_v$, the energy difference between the top of the valence band and the bottom of the conduction band, the "gap energy" of the semiconductor. Substituting in for N_c and N_v we finally obtain for n_i

$$n_i = \frac{2}{h^3} (2\pi kT)^{\frac{3}{2}} (m_e^* m_h^*)^{\frac{3}{4}} e^{-\frac{E_g}{2kT}} \tag{1.35}$$

which is a constant.

1.3 **Extrinsic material**

A semiconductor into which impurities, called dopants, are intentionally added in order to alter the conductivity of the sample is said to be extrinsic. There are two dopant types. These are n-type dopants called donors and p-type dopants called acceptors. In a semiconductor doped with donors the equilibrium electron concentration becomes larger than the equilibrium hole concentration and the semiconductor is said to be n-type. Similarly, if the semiconductor is doped with acceptors, the equilibrium hole concentration is greater than the equilibrium electron concentration and the semiconductor is said to be p-type.

An example donor atom in silicon is phosphorus. Phosphorus is a Column VA element while silicon is a Column IVA element. Therefore, phosphorus has an outer valence of five while silicon has an outer valence of four. Silicon crystallizes such that each silicon atom forms bonds with four other silicon atoms fully accommodating all four outer valence electrons. If a phosphorus atom is substituted for a silicon atom in an otherwise silicon lattice, then one of the outer valence electrons within the phosphorus atom is not bound to a neighboring silicon atom as shown in Fig. 1.7(a). The phosphorus atom then only weakly holds the unbound electron. The other four valence electrons in phosphorus are chemically bound to four neighboring silicon

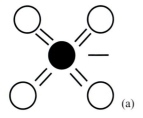

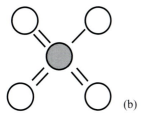

Figure 1.7 (a) A two-dimensional representation of a donor atom, phosphorus, shown in black, within a silicon matrix. Each silicon atom is represented by an open circle. Each line represents an outer valence electron. Notice that the extra electron in the outer shell of the phosphorus atom is unbound. (b) A two-dimensional representation of an acceptor atom, boron, shown in gray, within a silicon matrix. Each silicon atom is again represented by an open circle. Notice that one of the bonds is not filled yielding a hole.

atoms. The unbound electron can be readily ionized since it is not chemically bound. Once ionized, the unbound electron can move freely through the crystal and thus lies within the conduction band.

p-type doping can be achieved by adding an atom with fewer electrons in the outer shell than silicon. An example of a p-type dopant is aluminum. Aluminum is a Column IIIA element and as such has only three electrons in its outermost atomic orbital. If an atom of aluminum substitutes for an atom of silicon within the silicon matrix, one of the four bonds to the nearest neighbor atoms is unfilled as shown in Fig. 1.7(b). The vacant bond is called a vacancy or a hole. The vacancy can propagate through the lattice as a result of electrons jumping from one occupied state into another, each time leaving a vacancy behind.

It is useful to picture the donor and acceptor states in an energy level diagram. The key to understanding the energy level diagrams for donors and acceptors is to recognize that the donor and acceptor states, being impurity states, lie somewhere between the conduction and valence bands of the host semiconductor material. Donor and acceptor atoms are special types of impurities in that they introduce levels near the conduction and valence band edges respectively as shown in Fig. 1.8. Deep levels formed by impurity atoms added to the host semiconductor cannot easily be ionized. As a result these levels act as traps. The energy levels in this case lie near midgap and thus require extensive energy in order to be ionized.

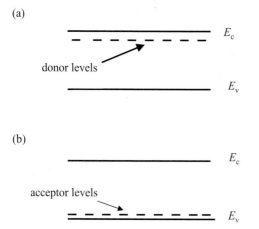

Figure 1.8 Sketch of the energy band diagrams of a semiconductor doped with (a) donors and (b) acceptors. Notice that the donor and acceptor atoms lie near the conduction and valence band edges respectively. Therefore, only a small amount of energy is needed to ionize either dopant.

In Section 1.2 we found that the electron and hole concentrations can be written in terms of the intrinsic level and Fermi level as

$$n = n_i e^{\frac{(E_f - E_i)}{kT}} \qquad p = n_i e^{\frac{(E_i - E_f)}{kT}} \tag{1.36}$$

The position of the Fermi level with respect to the intrinsic level determines the doping type in the semiconductor. Notice that the product of n and p once again yields the Law of Mass Action.

To decide if a semiconductor is n- or p-type it is necessary to compare the electron and hole concentrations with the intrinsic concentration. To do this the charge neutrality condition must be invoked. The charge neutrality condition is given as

$$0 = p - n + N_d^+ - N_a^- \tag{1.37}$$

Equation (1.37) implies that the net charge within the semiconductor is zero. The net positive charge contributed by the holes and ionized donors is balanced by the net negative charge contributed by the electrons and ionized acceptors. In most situations, only one ionized dopant atom dominates, either the donors or acceptors. Consider first an intrinsic semiconductor. In an intrinsic semiconductor the donor and acceptor concentrations are zero. Thus (1.37) becomes

$$0 = p - n \tag{1.38}$$

which of course is simply $p = n$. Alternatively for extrinsic material if $N_d \gg N_a$ then the material is n-type. Under this assumption, the acceptor concentration can be neglected with respect to the donor concentration. The electron concentration can be determined as follows. The charge neutrality condition becomes

$$0 = p - n + N_d^+ \tag{1.39}$$

The hole concentration can be expressed in terms of n using the Law of Mass Action (equilibrium conditions) as

$$p = \frac{n_i^2}{n} \tag{1.40}$$

Substituting (1.40) into (1.39) obtains

$$\frac{n_i^2}{n} - n + N_d = 0 \tag{1.41}$$

Equation (1.41) is simply a quadratic equation in n given as

$$n^2 - N_d n - n_i^2 = 0 \tag{1.42}$$

It can be solved as

$$n = \frac{N_d \pm \sqrt{N_d^2 + 4n_i^2}}{2} \tag{1.43}$$

Notice that if $n_i \ll N_d$ then (1.43) simplifies to

$$n = N_d \tag{1.44}$$

Similarly, if $N_a > N_d$ and $n_i \ll N_a$ then we obtain

$$p = N_a \tag{1.45}$$

and of course the material is p-type.

If the condition given by (1.44) holds then the Fermi level can be calculated as follows. The electron concentration in n-type material is given as

$$n = N_d = n_i e^{\frac{(E_f - E_i)}{kT}} \tag{1.46}$$

Solving for E_f obtains

$$E_f = E_i + kT \ln\left(\frac{N_d}{n_i}\right) \tag{1.47}$$

for n-type material. Notice that the Fermi level in n-type material is greater than the intrinsic level. Similarly, for p-type material the Fermi level is given as

$$E_f = E_i - kT \ln\left(\frac{N_a}{n_i}\right) \tag{1.48}$$

In this case the Fermi level lies below the intrinsic level. Thus for n-type material the Fermi level lies above the intrinsic level and in p-type material the Fermi level lies below the intrinsic level as shown in Fig. 1.9.

Example Problem 1.1

Consider a silicon sample doped with donors at 1×10^{17} cm^{-3}. If the intrinsic concentration within silicon is 1.0×10^{10} cm^{-3}, determine the location of the Fermi level relative to the valence band.

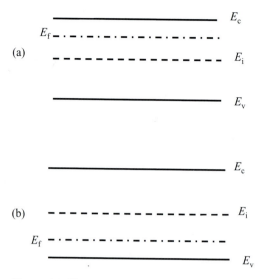

Figure 1.9 Sketch of the band structure showing the intrinsic level. In (a) the material is n-type since the Fermi level lies above the intrinsic level. In (b) the material is p-type since the Fermi level lies below the intrinsic level.

Since the donor doping concentration is very much larger than the intrinsic concentration, using the approximation that the electron concentration is equal to the donor concentration is valid. Therefore, the Fermi level relative to the intrinsic level is given as

$$E_f = E_i + kT \ln \left(\frac{N_d}{n_i} \right)$$

Substituting in for each term, $kT = 0.0259$ eV at 300 K, the Fermi level becomes

$$E_f = E_i + 0.417 \text{ eV}$$

By evaluating E_i, the position of the intrinsic level relative to the valence band can be determined. E_i is given in the text as

$$E_i = \frac{(E_c + E_v)}{2} + \frac{3kT}{4} \ln \left(\frac{m_h^*}{m_e^*} \right)$$

The effective masses in silicon for holes and electrons are

$$m_e^* = 0.328 \qquad m_h^* = 0.55$$

With these values for the effective masses the intrinsic level relative to the valence band edge is

$$E_i = \frac{1.12}{2} \text{ eV} + 0.01 \text{ eV} = 0.57 \text{ eV}$$

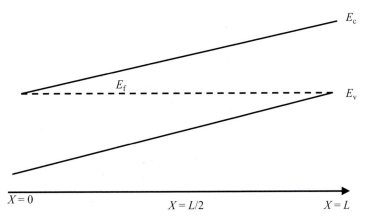

E_c

E_f

E_v

$X = 0$ $X = L/2$ $X = L$

Figure 1.10 Energy band diagram as a function of position. L is the total length of the sample. E_f corresponds to the flat dashed line in the figure.

Thus the Fermi level is located at 0.57 eV + 0.417 eV or 0.987 eV above the valence band edge.

Problems

1.1 Show that the probability that a state ΔE above the Fermi level, E_f, is filled is equal to the probability that a state ΔE below the Fermi level, E_f, is empty.

1.2 Determine the position of the Fermi level relative to the intrinsic level in silicon at 300 K if the electron concentration is 1×10^{16} cm^{-3}. Use $n_i = 1.0 \times 10^{10}$ cm^{-3}.

1.3 (a) Show that the minimum conductivity of a semiconductor sample occurs when

$$n_0 = n_i \sqrt{\frac{\mu_p}{\mu_n}}$$

   (b) What is the expression for the minimum conductivity, σ_{min}?

   (c) Calculate σ_{min} for Si at 300 K and compare with the intrinsic conductivity.

   $\mu_n = 1350$ cm^2/(V s) $\mu_p = 480$ cm^2/(V s) $n_i = 1.0 \times 10^{10}$ cm^{-3}.

1.4 What are the equilibrium concentrations of electrons and holes at $T = 300$ K in:

   (a) Si doped with $N_d = 3 \times 10^{14}$ cm^{-3};

   (b) Ge doped with $N_a = 3 \times 10^{14}$ cm^{-3};

   $n_i(\text{Si}) = 1.0 \times 10^{10}$ cm^{-3} $n_i(\text{Ge}) = 2.4 \times 10^{13}$ cm^{-3}

1.5 An intrinsic semiconductor sample has a resistance of 5 Ω at 360 K and a resistance of 50 Ω at 330 K. Assume that the only factor that changes the resistance between the two cases is the change in the intrinsic carrier concentration. Determine an expression relating the energy gaps at 360 K and 330 K.

1.6 A silicon sample has a length of 1 cm, a height of 0.01 cm and a width of 0.1 cm. The temperature is assumed to be 300 K. The electron mobility is given as

1450 cm^2/(V s) and the hole mobility is given as 500 cm^2/(V s). Determine the resistance of the sample if the doping concentration is given as

(a) intrinsic

(b) donor doping of 10^{16} cm^{-3}

(c) acceptor doping of 10^{15} cm^{-3}

1.7 A semiconductor is characterized by the energy band diagram sketched in Fig. 1.10. The system is doped such that there is a band bending as shown in the figure. Given $E_g = 1.12$ eV, $n_i = 10^{10}$ cm^{-3}, and $kT = 0.0259$ eV determine the following.

(a) Determine n and p at $X = L/2$

(b) Determine n at $X = L/4$

(c) If $L = 1$ cm, what is the magnitude of the electric field in the semiconductor?

(d) In the region, $L/2 < X < L$ is the material n- or p-type? Explain why.

1.8 Consider a semiconductor sample doped n-type with 5.0×10^{12} cm^{-3} donors. If the intrinsic concentration of the semiconductor is 10^{13} cm^{-3} what is the electron concentration in the semiconductor? Neglect the acceptor concentration.

1.9 Consider a collection of electrons at a temperature of 300 K. If the Fermi energy is equal to 1.2 eV, determine the probability that a state is NOT occupied at an energy of 1.25 eV. Assume kT is 0.0259 eV.

1.10 Determine the energy band gap of a semiconductor at 300 K if the electron and hole masses are 0.067 and 0.50 times the free electron mass respectively. The intrinsic carrier concentration of the semiconductor is 1.5×10^7 cm^{-3}.

2

Carrier action

In this chapter we examine the dynamics of carriers in semiconductors. We consider three general types of dynamics: drift, diffusion, and generation–recombination. In the first section, we discuss both drift and diffusion, which govern electron transport dynamics in semiconductors. The next section is devoted to the study of generation–recombination mechanisms active in semiconductors. Finally, we conclude with a discussion of the carrier continuity equation and its solution.

2.1 Drift and diffusion

The two major mechanisms that govern current flow in a semiconductor are drift and diffusion. Drift is charged particle motion in response to an applied electric field. The carrier drifts under the action of the applied electric field $\mathcal{E}$ as

$$F = q\mathcal{E} \tag{2.1}$$

where F is the force acting on the electron.[†] For an electron q is negative while for a hole q is positive. Notice that for an electron the force acts in the opposite direction from the field. For a hole the force and field point in the same direction. The work done on the carrier from a constant electric field is given as

$$\Delta E = q\mathcal{E}\Delta x \tag{2.2}$$

where ΔE is the change in energy of the carrier. The carrier cannot be accelerated continuously; otherwise its energy would "run away" and approach infinity, which is of course not observed. The electron suffers an occasional scattering with the lattice leading to energy transfer from the electron to the lattice. In this way, the electron energy is reduced. Ultimately a balance is achieved between the gain of energy from the field and the loss of energy via lattice collisions. Once this balance is achieved, the electron has no net energy gain; the time rate of change of the energy of the carrier is zero. This condition is steady-state. Once steady-state is reached the system no longer changes with time. Under steady-state conditions the energy gained from the field must be equal to the energy lost to the lattice through scatterings which can be expressed as

$$\Delta E(\text{gain from the field}) = \Delta E(\text{lost to the lattice}) \tag{2.3}$$

[†] In this chapter, although electric field and current density are vector quantities, we consider only a one-dimensional approach.

The electron motion under the application of an applied field is directed. If the field is shut off the carriers relax through lattice scattering back to their equilibrium distribution. As a result there is no net current flow in any direction.

The drift current density can be expressed as

$$J = qnv_d \tag{2.4}$$

where v_d is the drift velocity. The current density is simply I/A, where I is the current and A the area. Thus the electron drift current can be written as

$$I = -qnv_d A \tag{2.5}$$

while the hole drift current is given as

$$I = qnv_d A \tag{2.6}$$

There exists a linear relationship between the drift velocity and the field. This relationship is

$$v_d = \mu \mathcal{E} \tag{2.7}$$

where μ is the mobility. The mobility is a measure of how readily a carrier can move through the crystal. The drift velocity is then

$$v_d = -\mu_m \mathcal{E} \tag{2.8}$$

for electrons and

$$v_d = \mu_p \mathcal{E} \tag{2.9}$$

for holes. Therefore, the hole and electron drift current densities become

$$J_p(\text{drift}) = q\mu_p p \mathcal{E} \qquad J_n(\text{drift}) = q\mu_n n \mathcal{E} \tag{2.10}$$

The electrical conductivity in general is equal to the ratio of the current density to the electric field. Thus the conductivity for n-type material is

$$\sigma_n = q\mu_n n \tag{2.11}$$

The corresponding electrical conductivity for p-type material is

$$\sigma_p = q\mu_p p \tag{2.12}$$

The mobility itself can be expressed (for electrons) as

$$\mu_n = \frac{q\tau}{m_e^*} \qquad \qquad Scattering\ rate = 1/\tau \tag{2.13}$$

where τ is the mean free time between collisions. Equation (2.13) shows how the mobility varies with both the effective mass and mean time between scatterings, τ. Notice that as the mean time between scatterings increases, implying that the scattering rate (which is inversely proportional to τ) decreases, the mobility increases. This is as expected since if the scattering rate is decreased then there are fewer impeding collisions. As a result, the carrier more easily moves through the crystal and hence has a higher mobility. The mobility is also a function of the effective mass. As the effective mass increases, the mobility decreases. An increased effective mass produces a higher inertia for the carrier and thus it is less easily moved. At high doping levels the mobility is reduced due to the enhanced scattering rate caused by ionized impurity scattering. The mobility varies also with temperature. For a low doped material, the mobility decreases with increasing temperature. This is because the lattice scattering increases with increasing temperature reducing the mobility (the mean time between collisions is lowered).

When both electrons and holes are present in the material the total current density due to drift, $J_{total}(drift)$, is equal to the sum of the electron and hole drift current densities:

$$J_{total}(drift) = J_n + J_p \tag{2.14}$$

Substituting in the expressions for J_n and J_p given by (2.10) $J_{total}(drift)$ becomes

$$J_{total}(drift) = qn\mu_n\varepsilon + qp\mu_p\varepsilon \tag{2.15}$$

which simplifies to

$$J_{total}(drift) = q(n\mu_n + p\mu_p)\varepsilon \tag{2.16}$$

The electrical conductivity is defined as the ratio of the current density to the field so when both carriers are present the electrical conductivity is given as

$$\sigma = \frac{J_{total}(drift)}{\varepsilon} = q(n\mu_n + p\mu_p) \tag{2.17}$$

Let us next consider the form of the conductivity for three different cases. Case 1 is intrinsic material. In this situation we have

$$n = p = n_i$$
$$\sigma = qn_i(\mu_n + \mu_p) \tag{2.18}$$
$$\rho = \frac{1}{\sigma} = \frac{1}{qn_i(\mu_n + \mu_p)}$$

where ρ is the resistivity. Case 2 is for n-type material. In this case in equilibrium,

$$n = N_d \qquad p = \frac{n_i^2}{N_d} \qquad p \ll n \tag{2.19}$$

Under these conditions the electrical conductivity and resistivity are given as

$$\sigma = qN_d\mu_n$$
$$\rho = \frac{1}{\sigma} = \frac{1}{qN_d\mu_n} \tag{2.20}$$

Finally, in the third case the material is p-type. In this case in equilibrium the electron and hole concentrations are given as

$$p = N_a \qquad n = \frac{n_i^2}{N_a} \qquad n \ll p \tag{2.21}$$

negative potential

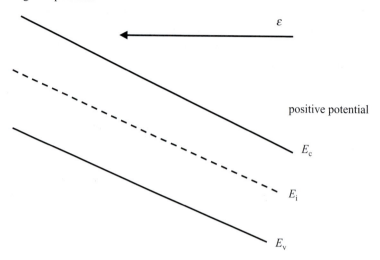

positive potential

E_c

E_i

E_v

Figure 2.1 Energy bands under the application of an applied field.

The resulting conductivity and resistivity are

$$\sigma = q N_a \mu_p$$

$$\rho = \frac{1}{\sigma} = \frac{1}{q N_a \mu_p} \tag{2.22}$$

Consider next band bending and its effect on electron and hole motion. Energy band diagrams depict electron energy not potential. Recall that the electron energy is lowered near a positive potential and raised near a negative potential. The band bending is shown in Fig. 2.1. Notice that the potential energy is positive at the lowest point and negative for the highest potential. An electron will be attracted by the positive potential and repelled by the negative potential. Thus an electron will "roll downhill" in the diagram towards the positive potential. This is a general result. Electrons "roll downhill" in potential energy diagrams. Let us examine the electric field. The electric field can be found from the potential as the negative gradient of the potential. This is given as

$$\mathcal{E} = -\frac{dV}{dx} \tag{2.23}$$

But $V(x)$ is given as $E_i/(-q)$. With this definition the electric field becomes

$$\mathcal{E} = -\frac{dV}{dx} = -\frac{d}{dx}\left(\frac{E_i}{-q}\right) = \frac{1}{q}\frac{dE_i}{dx} \tag{2.24}$$

Aside from drift motion a carrier can move via diffusion. Carriers will diffuse from regions of high concentration to regions of low concentration until the concentration gradient is zero. A vivid example of diffusion is an opened perfume bottle in a closed room. Initially all of the perfume is in the bottle and no vapor has spread. However, after some time the perfume molecules will diffuse throughout the room until the concentration gradient vanishes. Under these conditions the concentration of perfume molecules is everywhere the same. Diffusion is governed by Fick's Law. The particle density current is proportional to the concentration gradient:

$$J' = -D\frac{dn}{dx} \tag{2.25}$$

where D is the diffusion constant. Notice that the particle density current flows from the region of high concentration to the region of low concentration which produces the negative sign. For electrons the diffusion current is given as

$$J_n(\text{diff}) = (-q)\left(-D_n\frac{dn}{dx}\right) = qD_n\frac{dn}{dx} \tag{2.26}$$

Similarly, for holes the diffusion current is given as

$$J_p(\text{diff}) = (q)\left(-D_p\frac{dp}{dx}\right) = -qD_p\frac{dp}{dx} \tag{2.27}$$

Using (2.26) and (2.27) the total current densities can be determined. The total current densities may be expressed as the sum of the drift and diffusion current densities:

$$\vec{J}_p = q\mu_p p\vec{\varepsilon} - qD_p\vec{\nabla}p \tag{2.28}$$

for holes and

$$\vec{J}_n = q\mu_n n\vec{\varepsilon} + qD_n\vec{\nabla}n \tag{2.29}$$

for electrons.

In equilibrium no net current flows by drift and diffusion. The current due to the drift of carriers in the applied electric field must exactly balance on average the current due to diffusion. When the current vanishes the gradient of the Fermi level is zero

$$\frac{dE_f}{dx} = 0 \tag{2.30}$$

Hence in equilibrium, the Fermi level is flat. The electron carrier concentration in equilibrium can be written as

$$n = n_i e^{\frac{(E_f - E_i)}{kT}} \tag{2.31}$$

The derivative of n with respect to x is

$$\frac{dn}{dx} = n_i e^{\frac{(E_f - E_i)}{kT}}\frac{1}{kT}\left(\frac{dE_f}{dx} - \frac{dE_i}{dx}\right) \tag{2.32}$$

But dE_f/dx is zero. With this substitution dn/dx becomes

$$\frac{dn}{dx} = -\frac{1}{kT}n\frac{dE_i}{dx} \tag{2.33}$$

but

$$\mathcal{E} = \frac{1}{q}\frac{dE_i}{dx} \tag{2.34}$$

Combining (2.33) and (2.34) obtains

$$\frac{dn}{dx} = -\frac{q}{kT}n\mathcal{E} \tag{2.35}$$

In equilibrium the electron current density vanishes. Thus,

$$J = 0 = q\mu_n n\mathcal{E} + qD_n\frac{dn}{dx} \tag{2.36}$$

Substituting (2.35) into (2.36) yields

$$q\mu_n n\mathcal{E} + qD_n\left(-\frac{q}{kT}n\mathcal{E}\right) = 0 \tag{2.37}$$

Simplifying, (2.37) becomes

$$\mu_n = \frac{q}{kT}D_n \tag{2.38}$$

which is known as the Einstein relation. The Einstein relation relates the diffusion constant to the mobility but it holds rigorously only in equilibrium. A similar expression exists for holes.

2.2 Generation–recombination

Generation and recombination events change the electron and hole concentrations. A generation event creates free electrons and holes while a recombination event removes free electrons and holes. There are two general types of generation–recombination events. These are band-to-band and band-to-bound transitions. Band-to-band transitions are between the valence and conduction bands. A band-to-band generation event occurs when an electron within the valence band is promoted into the conduction band as shown in Fig. 2.2(a). A band-to-band recombination event occurs when an electron within the conduction band recombines with a vacancy in the valence band as shown in Fig. 2.2(b).

Conversely, a band-to-bound event occurs between a band state, either the conduction or valence band, and an impurity state located within the band gap. Examples of band-to-bound events are shown in Fig. 2.3. In Fig. 2.3(a) an electron is generated from an impurity state located within the energy gap. As can be seen from the figure, the impurity state lies near midgap. Similarly, an electron can be captured by an impurity state and thus be removed from the conduction band as shown in Fig. 2.3(b). This is a band-to-bound recombination event.

A net electron-hole pair can be either generated or recombined through band-to-bound transitions. First the electron is trapped by the bound impurity state which lies near midgap. Then after some time the electron can be emitted from the impurity state and recombine with a vacancy in the valence band. As a result an electron–hole pair has recombined. Alternatively, an electron–hole pair can be generated by the action of

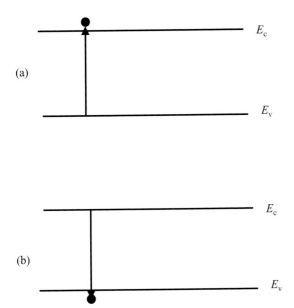

Figure 2.2 Band-to-band generation–recombination processes: (a) band-to-band generation event; (b) band-to-band recombination event. The solid circle represents the final state of the electron.

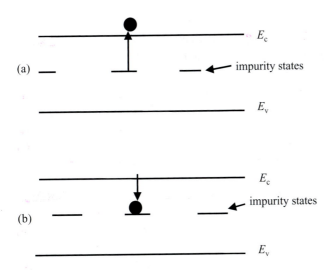

Figure 2.3 Band-to-bound generation–recombination events. In (a) an electron is generated from an impurity state within the energy gap. In (b) an electron is captured by a impurity state leading to an electron recombination event. The solid circle represents the final state of the electron in the process.

the impurity state. An electron from the valence band jumps up to a trap state leaving behind a hole within the valence band. After some time, the electron can be emitted from the trap state into the conduction band creating an electron–hole pair.

There are three different mechanisms by which either a band-to-band or band-to-bound generation–recombination event can proceed. These are: thermal, radiative, and Auger. A thermal generation–recombination event occurs by the emission or absorption of the quantum of lattice vibrations, a phonon. Phonons are similar to photons. A photon is the quantum of an electromagnetic vibration (light). A phonon is the quantum of a lattice vibration. In a phonon emission event a phonon is given up by the electron to the lattice. Thus the electron energy is lowered by the energy of a phonon. In a phonon absorption event a phonon is absorbed by the electron thus increasing the electron's energy by the amount given by the phonon.

A radiative generation–recombination event occurs with either the absorption or the emission of a photon. A photon absorption event leads to generation while a photon emission event leads to recombination. In order for a band-to-band radiative generation event to occur, the incident photon must have energy equal to or greater than the energy band gap. Conversely, a band-to-band radiative recombination event emits a photon with an energy equal to or greater then the energy band gap.

Finally, the last mechanism is Auger generation or recombination. An Auger generation–recombination event occurs via energy transfer between two carriers, either two electrons or two holes. A band-to-band Auger generation event occurs when a high energy carrier, either an electron or a hole, makes a collision with the lattice and transfers its excess kinetic energy to the lattice to produce an electron–hole pair. A band-to-band Auger generation event is a threshold process since the initial carrier must have kinetic energy equal to or greater than the energy gap in order to produce an electron–hole pair. Band-to-band Auger generation is often referred to as impact ionization. Band-to-band Auger recombination occurs when an electron–hole pair recombines and transfers its excess kinetic energy to either a free electron or hole. The free carrier is then promoted to a high energy within either the conduction band (electron) or the valence band (hole).

Let us next consider generation–recombination quantitatively. Define n_0 and p_0 as the equilibrium electron and hole concentrations respectively; n and p are the general carrier concentrations within the material. The excess concentrations are given as

$$\delta n = n - n_0 \qquad \delta p = p - p_0 \tag{2.39}$$

The time rate of change of the carrier concentration due to generation–recombination is written as

$$\left.\frac{\partial n}{\partial t}\right|_{R-G} \qquad \left.\frac{\partial p}{\partial t}\right|_{R-G} \tag{2.40}$$

Radiative transitions depend to some extent upon the energy band structure of the semiconductor. The simplest energy relationship for a semiconductor assumes completely free carriers. Under this condition the energy is completely kinetic and is

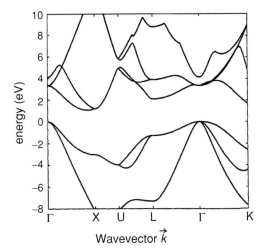

Figure 2.4 Electronic energy band structure of bulk Si. Notice that the conduction band minimum occurs at the X point, the (100) point, in k-space. Since the maximum hole energy occurs at the Γ point, the material is indirect.

given by (1.4) as

$$E = \frac{\hbar^2 k^2}{2m} \tag{2.41}$$

The energy is a quadratic function of the wavevector, k; $E(k)$ is parabolic. Under this condition, the energy bands assume their simplest form. Generally, the energy band structure of a realistic semiconductor departs from this simple relationship. The band structures of Si and GaAs are shown in Figs. 2.4 and 2.5 respectively. Notice that in both diagrams there exist multiple bands for the valence bands. These are the heavy and light hole bands. Only a single conduction band is shown for both materials in the diagram. Hole energy increases downwards in the diagram while electron energy increases upwards. At $k = (000)$, the valence bands reach their minimum energy value in both materials. However, the minimum electron energy occurs at different points in k-space for GaAs and Si. As can be seen from the figures, the minimum electron energy within the conduction band occurs at $k = (000)$ for GaAs but at $k = (100)$ for Si. Since the minimum electron and hole energies occur at the same location in k-space in GaAs, it is called a direct gap semiconductor. Alternatively, since the minimum electron and hole energies occur at different points in k-space in Si the material is called an indirect semiconductor. A direct gap semiconductor like GaAs is a far more efficient absorber and emitter of radiation than an indirect gap semiconductor like Si. As we will see in Chapter 10, most optoelectronic devices are made using direct gap semiconductors.

For a constant radiative generation rate, G_L, the time rate of change of the carrier distributions is given as

$$\left.\frac{\partial n}{\partial t}\right|_{R-G} = \left.\frac{\partial p}{\partial t}\right|_{R-G} = G_L \tag{2.42}$$

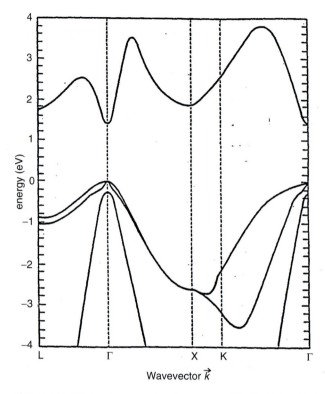

Figure 2.5 Electronic energy band structure of bulk GaAs. Notice that the conduction and valence band minimums occur at the same point in k-space, the Γ point. For this reason the material is said to be direct.

Radiative absorption requires that the incident photon have energy equal to or greater than the energy band gap. Let $h\nu > E_g$ and $I(x)$ be the intensity of light incident onto the semiconductor sample. The rate of change of the intensity is proportional to the intensity:

$$\frac{dI(x)}{dx} = -\alpha I(x) \tag{2.43}$$

which can be readily solved for $I(x)$ as

$$I(x) = I_0 e^{-\alpha x} \tag{2.44}$$

where α is called the absorption coefficient. The intensity of the light a distance l into the sample is given then as

$$I(l) = I_0 e^{-\alpha l} \tag{2.45}$$

If an external agent, such as illumination, acts on a semiconductor to generate carriers the semiconductor is driven out of equilibrium. As a result, the excess carrier concentration is increased above the equilibrium concentration. To relax back to equilibrium once the illumination is removed, there must be a net recombination rate to return the concentration back to its equilibrium concentration. The full details as to the quantitative behavior of band-to-band and band-to-bound transitions are discussed in Brennan (1999). Here we only quote the salient details and leave the interested reader to examine the details in Brennan (1999). The change in the electron and hole concentrations from generation/recombination can be written as

$$\frac{\partial n}{\partial t}\bigg|_{R-G} = -\frac{\delta n}{\tau_n} \qquad \frac{\partial p}{\partial t}\bigg|_{R-G} = -\frac{\delta p}{\tau_p} \tag{2.46}$$

where τ_n and τ_p are the electron and hole recombination lifetimes respectively.

Example Problem 2.1 Light absorption in a semiconductor

Given:

incident intensity, $I_0 = 5$ mW; $\alpha = 5 \times 10^4$ cm^{-1}; thickness, $l = 0.50 \times 10^{-4}$ cm, photon energy, $h\nu = 2.0$ eV.

(a) Find the total amount of power absorbed.

The transmitted intensity is given as

$$I(l) = I_0 e^{-\alpha l} = 5\,\text{mW} \times 0.082 = 0.41\,\text{mW}$$

Therefore, the absorbed power is simply, $(5.0 - 0.41)$ mW $= 4.59$ mW.

(b) Assume perfect quantum efficiency and determine the number of electron–hole pairs per second produced.

First calculate the number of photons per second absorbed. This is given by number of photons per second = energy absorbed per second/energy of each photon

$$\frac{4.59 \times 10^{-3}\,\text{J/s}}{(1.6 \times 10^{-19}\,\text{J/eV})(2\,\text{eV/photon})} = 1.43 \times 10^{16}\ \text{photons/s}$$

To find the number of electron–hole pairs produced we need to use the quantum efficiency. The quantum efficiency is defined as the number of electron–hole pairs produced per photon. If the quantum efficiency is 100% this means that for every incident photon, one electron–hole pair is produced. So we see in this example that 1.43×10^{16} photons are produced per second. Thus the number of electron–hole pairs per second produced is the same, 1.43×10^{16}.

2.3 **Continuity equation and its solution**

The continuity equation is an expression of charge conservation. In its most basic form it is given for electrons as

$$\frac{\partial n}{\partial t} = \frac{\partial n}{\partial t}\bigg|_{\text{drift}} + \frac{\partial n}{\partial t}\bigg|_{\text{diffusion}} + \frac{\partial n}{\partial t}\bigg|_{\text{thermal G–R}} + \frac{\partial n}{\partial t}\bigg|_{\text{other G–R}} \tag{2.47}$$

and for holes as

$$\frac{\partial p}{\partial t} = \frac{\partial p}{\partial t}\bigg|_{\text{drift}} + \frac{\partial p}{\partial t}\bigg|_{\text{diffusion}} + \frac{\partial p}{\partial t}\bigg|_{\text{thermal G–R}} + \frac{\partial p}{\partial t}\bigg|_{\text{other G–R}} \tag{2.48}$$

The sum of the time rate of change of n and p with respect to drift and diffusion can be rewritten as

$$\frac{1}{q}\vec{\nabla} \circ \vec{J}_n = \frac{\partial n}{\partial t}\bigg|_{\text{drift}} + \frac{\partial n}{\partial t}\bigg|_{\text{diffusion}} \tag{2.49}$$

and

$$-\frac{1}{q}\vec{\nabla} \circ \vec{J}_p = \frac{\partial p}{\partial t}\bigg|_{\text{drift}} + \frac{\partial p}{\partial t}\bigg|_{\text{diffusion}} \tag{2.50}$$

Substituting (2.49) and (2.50) into (2.47) and (2.48) respectively obtains

$$\frac{\partial n}{\partial t} = \frac{1}{q}\vec{\nabla} \circ \vec{J}_n + \frac{\partial n}{\partial t}\bigg|_{\text{thermal R--G}} + \frac{\partial n}{\partial t}\bigg|_{\text{other R--G}} \tag{2.51}$$

and

$$\frac{\partial p}{\partial t} = -\frac{1}{q}\vec{\nabla} \circ \vec{J}_p + \frac{\partial p}{\partial t}\bigg|_{\text{thermal R--G}} + \frac{\partial p}{\partial t}\bigg|_{\text{other R--G}} \tag{2.52}$$

Typically, $(\partial n/\partial t)(\text{other})$ and $(\partial p/\partial t)(\text{other})$ are due to radiative generation, G_L.

Let us consider now the solution of the continuity equation. We make the following simplifying assumptions:

(i) the system is one-dimensional;
(ii) there is no applied field present so that the drift term vanishes;
(iii) the equilibrium concentration is constant and uniform for both electrons and holes;
(iv) there is low level injection;
(v) the time rate of change of the carrier concentrations due to other generation/recombination mechanisms is equal to G_L.

With these assumptions, the divergence of J_n becomes

$$\frac{1}{q}\vec{\nabla} \circ \vec{J}_n = \frac{1}{q}\frac{\partial J_n}{\partial x} = \frac{1}{q}\frac{\partial}{\partial x}\left(q D_n \frac{\partial n}{\partial x}\right) = D_n \frac{\partial^2 n}{\partial x^2} \tag{2.53}$$

where we have neglected the drift current. The electron concentration can be written as the sum of the equilibrium concentration and the excess concentration as

$$n = n_0 + \delta n \tag{2.54}$$

Since the equilibrium concentration is uniform and constant the second derivative of n becomes

$$\frac{\partial^2 n}{\partial x^2} = \frac{\partial^2 \delta n}{\partial x^2} \tag{2.55}$$

Thus,

$$\frac{1}{q}\vec{\nabla} \circ \vec{J}_\text{n} = D_\text{n}\frac{\partial^2 \delta n}{\partial x^2} \tag{2.56}$$

Recognizing that the last two terms on the right hand side of (2.51) are

$$\left.\frac{\partial n}{\partial t}\right|_\text{R-G} = -\frac{\delta n}{\tau_\text{n}} \qquad \left.\frac{\partial n}{\partial t}\right|_\text{other} = G_\text{L} \tag{2.57}$$

the continuity equations then become

$$\frac{\partial \delta n}{\partial t} = D_\text{n}\frac{\partial^2 \delta n}{\partial x^2} - \frac{\delta n}{\tau_\text{n}} + G_\text{L} \tag{2.58}$$

for electrons and

$$\frac{\partial \delta p}{\partial t} = D_\text{p}\frac{\partial^2 \delta p}{\partial x^2} - \frac{\delta p}{\tau_\text{p}} + G_\text{L} \tag{2.59}$$

for holes. Let us consider the sample solutions of these equations given by Examples 2.2 and 2.3.

Example Problem 2.2 The continuity equation in steady-state for p-type material

Solve the continuity equation in steady-state for a p-type material.

Since the material is p-type the minority carriers are electrons. We will thus solve for the excess minority carrier concentration. We assume that the radiative generation rate, G_L, is zero, and the system is in steady-state. These assumptions imply

$$\frac{\partial \delta n}{\partial t} = 0 \qquad G_\text{L} = 0$$

The electron continuity equation becomes then

$$0 = D_\text{n}\frac{\partial^2 \delta n}{\partial x^2} - \frac{\delta n}{\tau_\text{n}}$$

Simplifying yields

$$0 = \frac{\partial^2 \delta n}{\partial x^2} - \frac{\delta n}{D_\text{n}\tau_\text{n}}$$

Let L_n be defined as

$$L_\text{n} = \sqrt{\tau_\text{n}D_\text{n}}$$

L_n is called the diffusion length, and physically represents the mean distance a carrier diffuses before recombining. This leads to the following general solution for δn:

$$\delta n = Ae^{-x/L_\text{n}} + Be^{x/L_\text{n}}$$

Example Problem 2.3 Solution of the continuity equation with no concentration gradient

Solve the continuity equation for the case in which there is no concentration gradient.

Consider the solution of the continuity equation when $G_L = 0$ and when there is no concentration gradient. A vanishing concentration gradient implies that the second derivative with respect to x is zero. Thus,

$$\frac{\partial^2 \delta n}{\partial x^2} = 0$$

The continuity equation becomes then

$$\frac{\partial \delta n}{\partial t} = -\frac{\delta n}{\tau_n}$$

The solution of this differential equation is simple and is given as

$$\delta n(t) = \delta n(0) e^{-t/\tau_n}$$

We see that the excess electron concentration decays exponentially in time.

Problems

2.1 A hole current of 10^{-4} A/cm^2 is injected into the side $(x = 0)$ of a long n-type silicon bar. Assume there is no electric field and hence no drift current and that as x approaches infinity the excess carrier concentration decays to zero. Use a one-dimensional solution. Given: $\mu_p = 400$ cm^2/(V s), $kT = 0.0259$ eV, $\tau_p = 25$ μs $= 25 \times 10^{-6}$ s.

Determine the steady-state excess hole concentration at $x = 0$.

2.2 An electric field of 10 V/cm is applied to an intrinsic silicon sample. If the carriers drift 1 cm in 100 μs, determine at $T = 300$ K $(kT = 0.0259$ eV) the following:
(a) the drift velocity;
(b) the diffusion constant;
(c) the conductivity.
Assume that the electron and hole mobilities are equal and that $n_i = 10^{10}$ cm^{-3}.

2.3 Determine an expression for the steady-state minority hole concentration as a function of the one-dimensional position x in a semiconductor sample of length, L. Assume that no generation–recombination events occur and that the applied electric field is zero. Take the boundaries of the sample at $x = 0$ and $x = L$. Assume that the excess hole concentration at $x = 0$ can be written as $\delta p(x = 0) = \Delta p_n$. Also assume that the excess carrier concentration at $x = L$ is zero.

2.4 Consider a uniformly doped p-type Si sample with a doping concentration of 10^{14} cm^{-3} at $T = 300$ K. Assume that the sample is uniformly illuminated for $t < 0$ by light such that there are 10^{16} electron–hole pairs (EHPs) generated per cubic centimeter per second. At $t = 0$ the generation rate is halved. Find the excess minority carrier concentration as a function of time in the sample.

2.5 Consider a uniformly doped Si sample that is in steady-state with no applied electric field. Assume that the sample has length L and that at each boundary the excess carrier concentration recombines to equilibrium, implying that the excess carrier concentration is zero at the boundaries. Assume that the sample is uniformly illuminated such that the generation rate is G_L. Determine an expression for the excess minority carrier concentration as a function of position.

2.6 A silicon sample is doped with 10^{16} cm^{-3} boron atoms and a certain number of shallow donors. Note that boron is an acceptor in silicon. The Fermi level is 0.36 eV above the intrinsic level, E_i, at 300 K. What is the donor concentration, N_d? Assume that the free hole concentration is zero.

2.7 In a p-type Si bar excess carriers are generated at the end of the bar at $x = 0$. The doping concentration is $N_a = 5 \times 10^{16}$ cm^{-3}. Neglect the donor concentration. Assume that the excess carriers are only generated at $x = 0$ and that the excess electron concentration there is 10^{15} cm^{-3}. The applied field is zero. Assume that the carrier lifetime is 8.0×10^{-7} s. Determine the excess carrier concentration as a function of x.

3

Junctions

In this chapter we discuss different types of junctions that are of importance in semiconductor devices. Specifically, we examine p–n junctions, Schottky barriers, and ohmic contacts. We will delay discussing metal–insulator–semiconductor, MIS, junctions until Chapter 6. We begin our discussion with p–n homojunctions. Heterojunctions, junctions formed between two dissimilar materials, are discussed briefly in Chapter 11. Devices made using heterojunctions are also presented in Chapter 11.

3.1 p–n homojunction in equilibrium

Before we begin our study of junctions let us make a few definitions. The bulk region is the area far from the junction where the carrier concentrations are equal to their equilibrium values. The metallurgical junction is the physical location of the junction between the n- and p-type regions. The depletion region is the area surrounding the metallurgical junction. It is called the depletion region since the action of the built-in field within the junction sweeps out the free carriers leaving behind immobile space charge.

For simplicity we make the following assumptions:

(i) The junction is one-dimensional, and a one-dimensional analysis can be employed.
(ii) The metallurgical junction is located at $x = 0$.
(iii) The p–n homojunction is a step or abrupt junction with uniformly doped p and n regions.
(iv) There exist perfect ohmic contacts far away from the metallurgical junction.

Let us first consider the p–n junction in equilibrium. As is always the case, in equilibrium no net current flows. For simplicity let us assume that our p–n junction is formed by putting an n-type layer into contact with a p-type layer. Recall that in equilibrium the Fermi level is flat (see Chapter 2). Thus the Fermi level is simply a flat horizontal line in the band diagram. We start with the p and n layers apart as shown in Fig. 3.1(a). As before the Fermi level lies below the intrinsic level for the p-type material and the Fermi level lies above the intrinsic level for the n-type material. Once the two layers are put into contact, there is a net electron transfer from the n-type material to the p-type material in order to equilibrate the Fermi level. The energy band diagram in equilibrium shown in Fig. 3.1(b) can be drawn following three simple rules:

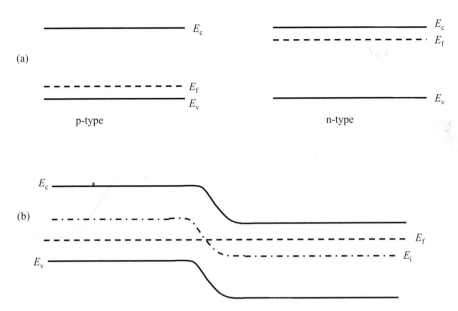

Figure 3.1 p- and n-type semiconductors (a) initially apart and (b) in contact. Notice that the Fermi level is flat in (b). Additionally, far from the metallurgical junction the bulklike properties are recovered.

(i) The Fermi level is flat everywhere in the junction.
(ii) Within the bulk regions far from the metallurgical junction the bulklike properties of the semiconductors are recovered.
(iii) The energy bands are continuous and simply connected across the junction.

We start by first drawing a straight horizontal line for the Fermi level. Next we draw the p- and n-type regions far from the metallurgical junction. Thus the bulklike properties are recovered and the p and n regions far from the junction are like the p and n regions in Fig. 3.1(a). Finally, the energy bands are continuous and simply connected across the junction as shown in Fig. 3.1(b). In this way the correct energy band structure for the homojunction can be drawn.

Notice that the energy bands bend to accommodate the Fermi level. What though is the physical meaning of the band bending? Recall that the bands represent the electron energies. The band bending implies that the electron energies are greater on the p-side than on the n-side. Equivalently, the electrostatic potential is greater on the n-side than on the p-side. How though does this occur? In order that the Fermi level aligns itself throughout the junction, electrons must transfer from the n-side to the p-side and holes must move from the p-side to the n-side. When the electrons move from the n-type material to the p-type material they leave behind ionized donors and positive space charge. Similarly, the transferred holes leave behind ionized acceptors and negative space charge. Therefore, the n-side of the junction has a net positive space charge while the p-side of the junction has a net negative space charge. The resulting system is shown in Fig. 3.2. The presence of a space charge region surrounding the

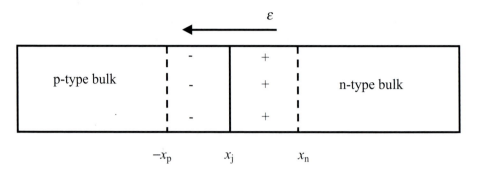

Figure 3.2 A p–n junction showing the bulk and depletion regions: x_j is the metallurgical junction ($x = 0$), x_n is the edge of the depletion region on the n-side of the junction and $-x_p$ is the edge of the depletion region on the p-side of the junction. The space charge region lies between x_n and $-x_p$. The electric field points from the n-side (positive space charge) to the p-side (negative space charge) as shown in the diagram.

metallurgical junction produces an electric field that points from the positive ionized donor atoms on the n-side to the negative ionized acceptors on the p-side as shown in the diagram.

The negative and positive space charge produced from the ionized donors and acceptors (remember that the donors and acceptors are chemically bound to the lattice and cannot freely move) produces a built-in electric field as shown in Fig. 3.2. The built-in field is directed from the n-side to the p-side. The built-in field acts to sweep out the free carriers within the space charge region. As a result the space charge region becomes depleted of free carriers and is called the depletion region. Hence a built-in potential is also created which is higher on the n-side than the p-side as can be seen in Fig. 3.3. In Fig. 3.3 W is the width of the depletion region.

Free electrons and holes are exchanged between the n- and p-sides of the junction until the electrostatic potential difference increases to the point at which further diffusion of net charge across the junction is balanced. The edges of the depletion region are labeled x_n and $-x_p$ on the n-side and p-side respectively. The sum of x_n and $-x_p$ is simply the width of the depletion region, W.

The potential difference between the n- and p-sides of the junction in equilibrium is called the built-in potential, V_{bi}. The built-in potential can be calculated from the diagram in the following way. We start with the band diagrams for the n- and p-sides when they are apart as shown in Fig. 3.4. The built-in potential is equal to the full band bending in equilibrium. The amount by which the bands bend is equal to the difference between the Fermi levels in the n- and p-type material. Setting the zero of potential energy at E_v, then qV_{bi} can be determined from inspection of Fig. 3.4 as

$$q V_{bi} = E_g - E_{fp} - (E_c - E_{fn}) \tag{3.1}$$

We may write (3.1) in a different form using the expressions for the electron and hole concentrations from Chapter 1. The electron concentration is

$$n = N_c e^{-(E_c - E_{fn})/kT} \tag{3.2}$$

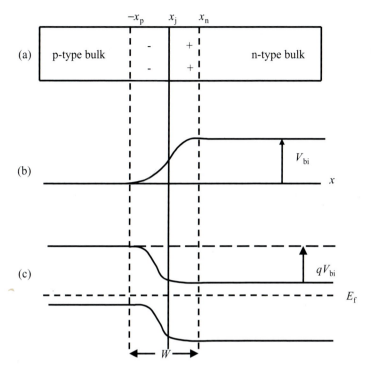

Figure 3.3 p–n homojunction: (a) doping scheme; (b) potential diagram; (c) energy band diagram. W is the width of the depletion region.

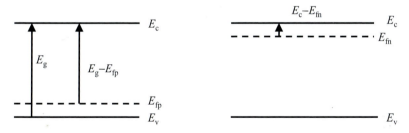

Figure 3.4 p- and n-type material drawn apart prior to contact. The built-in potential is simply equal to the difference between the Fermi levels on either side of the junction. The Fermi levels align when put into contact. Therefore, the amount that the bands have to bend to align the Fermi levels is simply equal to the difference in the Fermi levels on either side of the junction.

Solving (3.2) for $E_c - E_{fn}$ yields

$$E_c - E_{fn} = -kT \ln \left(\frac{n}{N_c} \right) = kT \ln \left(\frac{N_c}{n} \right) \tag{3.3}$$

The hole concentration is given as

$$p = N_v e^{-E_{fp}/kT} \tag{3.4}$$

Solving (3.4) for E_{fp} results in

$$E_{\mathrm{fp}} = -kT \ln \left(\frac{p}{N_{\mathrm{v}}} \right) = kT \ln \left(\frac{N_{\mathrm{v}}}{p} \right) \tag{3.5}$$

Finally, the energy gap E_{g} can be determined from n_{i} the intrinsic concentration as

$$n_{\mathrm{i}}^2 = N_{\mathrm{c}} N_{\mathrm{v}} e^{-E_{\mathrm{g}}/kT} \tag{3.6}$$

Solving (3.6) for E_{g} obtains

$$E_{\mathrm{g}} = -kT \ln \left(\frac{n_{\mathrm{i}}^2}{N_{\mathrm{c}} N_{\mathrm{v}}} \right) = kT \ln \left(\frac{N_{\mathrm{c}} N_{\mathrm{v}}}{n_{\mathrm{i}}^2} \right) \tag{3.7}$$

Using the above results in (3.1) the built-in voltage can be rewritten as

$$q V_{\mathrm{bi}} = E_{\mathrm{g}} - E_{\mathrm{fp}} - (E_{\mathrm{c}} - E_{\mathrm{fn}}) \tag{3.8}$$

$$q V_{\mathrm{bi}} = kT \ln \left(\frac{N_{\mathrm{c}} N_{\mathrm{v}}}{n_{\mathrm{i}}^2} \right) - kT \ln \left(\frac{N_{\mathrm{v}}}{p_{\mathrm{p0}}} \right) - kT \ln \left(\frac{N_{\mathrm{c}}}{n_{\mathrm{n0}}} \right) \tag{3.9}$$

Combining and simplifying (3.9) becomes,

$$q V_{\mathrm{bi}} = kT \ln \left(\frac{p_{\mathrm{p0}} n_{\mathrm{n0}}}{n_{\mathrm{i}}^2} \right) \tag{3.10}$$

where p_{p0} and n_{n0} are the equilibrium hole concentration on the p-side and equilibrium electron concentration on the n-side respectively.

Since the system is in equilibrium the Law of Mass Action holds on each side of the junction. This implies the following relationship between the carrier concentrations

$$n_{\mathrm{n0}} p_{\mathrm{n0}} = n_{\mathrm{i}}^2 = n_{\mathrm{p0}} p_{\mathrm{p0}} \tag{3.11}$$

Using the Law of Mass Action, (3.11) can be rewritten as

$$q V_{\mathrm{bi}} = kT \ln \left(\frac{p_{\mathrm{p0}} n_{\mathrm{n0}}}{n_{\mathrm{n0}} p_{\mathrm{n0}}} \right) \tag{3.12}$$

which simplifies to

$$q V_{\mathrm{bi}} = kT \ln \left(\frac{p_{\mathrm{p0}}}{p_{\mathrm{n0}}} \right) \tag{3.13}$$

Equivalently, the built-in voltage can be written as

$$q V_{\mathrm{bi}} = kT \ln \left(\frac{n_{\mathrm{n0}}}{n_{\mathrm{p0}}} \right) \tag{3.14}$$

If the junction is abrupt with N_{a} and N_{d} acceptors and donors respectively then

$$p_{\mathrm{p0}} = N_{\mathrm{a}} \qquad p_{\mathrm{n0}} = \frac{n_{\mathrm{i}}^2}{n_{\mathrm{n0}}} = \frac{n_{\mathrm{i}}^2}{N_{\mathrm{d}}} \tag{3.15}$$

Finally, the built-in voltage can be written in terms of the doping concentrations as

$$q V_{bi} = kT \ln \left(\frac{N_a N_d}{n_i^2} \right) \tag{3.16}$$

Next we find an expression for the depletion region width, W, and the edges of the depletion region. The electric field within the junction can be determined from Poisson's equation. The one-dimensional Poisson's equation is

$$\frac{d\varepsilon}{dx} = -\frac{d^2 V}{dx^2} = \frac{\rho}{\kappa_s \varepsilon_0} = \frac{q}{\kappa}(p - n + N_d - N_a) \tag{3.17}$$

where κ_s is the dielectric constant (relative permittivity) of the semiconductor, and ε_0 is the permittivity of free space. For simplicity throughout the book we will refer to the product of the dielectric constant and permittivity of vacuum as κ as shown in (3.17). Within the depletion region it is assumed that the free carrier concentrations are zero. This implies that both p and n vanish since no free carriers exist. The assumption that the depletion region has no free carrier concentration is reasonable since it is assumed that the presence of the built-in field will act to sweep out the free carriers. We further make use of the assumption that both the n- and p-sides are uniformly doped. On the n-side of the junction we assume that there are only donors. Thus the Poisson equation becomes

$$-\frac{d^2 V}{dx^2} = \frac{q N_d}{\kappa} = \frac{d\varepsilon}{dx} \qquad (0 \le x \le x_n) \tag{3.18}$$

On the p-side the Poisson equation includes only the ionized acceptors. It is given as

$$-\frac{d^2 V}{dx^2} = -\frac{q N_a}{\kappa} = \frac{d\varepsilon}{dx} \qquad (-x_p \le x \le 0) \tag{3.19}$$

To find the field profile, (3.18) and (3.19) are integrated. For example on the p-side the field is given as

$$\varepsilon(x) = \int_{-x_p}^{x} -\frac{q N_a}{\kappa} dx' = -\frac{q N_a}{\kappa} x' \Big|_{-x_p}^{x} \tag{3.20}$$

Evaluating (3.20) obtains

$$\varepsilon(x) = -\frac{q N_a}{\kappa}(x + x_p) \qquad -x_p \le x \le 0 \tag{3.21}$$

Similarly, the field on the n-side is given as

$$\varepsilon(x) = \frac{q N_d}{\kappa}(x - x_n) \qquad 0 \le x \le x_n \tag{3.22}$$

Notice that the field goes to zero at the edges of the depletion region. The maximum value of the field occurs at $x = 0$. Consider the value of the field at $x = 0$. It is given as

$$\varepsilon(x)\big|_{x=0} = -\frac{q N_a x_p}{\kappa} = -\frac{q N_d x_n}{\kappa} \tag{3.23}$$

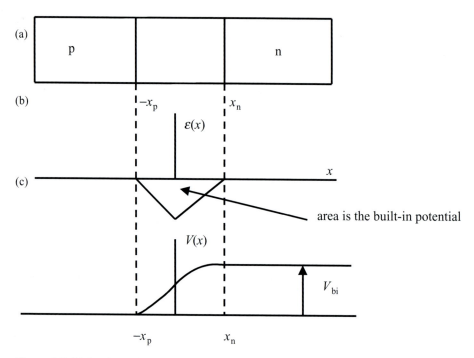

Figure 3.5 (a) Doping scheme within a p–n junction showing the edges of the depletion region; (b) electric field profile within the p–n junction and (c) potential profile within the p–n junction.

which is the maximum value of the electric field within the depletion region. The electric field is plotted in Fig. 3.5.

Equation (3.23) is consistent with Maxwell's equations. This can be shown as follows. Recall that in electromagnetics the normal component of the $\vec{D}$ field is continuous across a boundary provided that there is no surface charge concentration. We have then that

$$\vec{D}_\mathrm{p} = \vec{D}_\mathrm{n} \qquad (3.24)$$

But the $\vec{D}$ field is simply equal to

$$\vec{D} = \kappa \vec{\mathcal{E}} \qquad (3.25)$$

which can be used with (3.22) and (3.23) to determine the magnitude of the $\vec{D}$ field as

$$D_\mathrm{p}(x = 0) = \kappa \left(-\frac{q N_\mathrm{a} x_\mathrm{p}}{\kappa} \right) = -q N_\mathrm{a} x_\mathrm{p} \qquad (3.26)$$

Similarly, on the n-side the magnitude of the $\vec{D}$ field is

$$D_\mathrm{n}(x = 0) = \kappa \left(-\frac{q N_\mathrm{d} x_\mathrm{n}}{\kappa} \right) = -q N_\mathrm{d} x_\mathrm{n} \qquad (3.27)$$

Equating (3.26) and (3.27) yields

$$N_\mathrm{a} x_\mathrm{p} = N_\mathrm{d} x_\mathrm{n} \qquad (3.28)$$

Equation (3.28) implies that the amount of space charge uncovered on the n-side is equal to the amount of space charge uncovered on the p-side. This is necessary in order to maintain charge balance in the semiconductor. Notice that if the doping concentration on one side is far larger than that on the other side most of the depletion region width resides on the lower doped side. For example, in a p^+–n junction, where the p^+ means that the p-side is very heavily doped relative to the n-side, that x_p is much smaller than x_n. Since the depletion region width, W, is equal to the sum of x_p and x_n, most of the depletion region lies within the n-side of the junction and $W \sim x_n$.

The built-in voltage can be obtained in an alternative way from use of the energy band diagram. The built-in voltage is equal to the area under the field vs. position curve. Therefore, through integration of the electric field with respect to x the area can be determined and the built-in voltage evaluated. For uniform doping the electric field vs. x is triangular as shown in Fig. 3.5. Therefore, the area can be determined simply by using the formula for the area of a triangle. The built-in voltage is then

$$V_{bi} = \tfrac{1}{2}\varepsilon_{max} W \tag{3.29}$$

Substituting for ε_{max} the expression given by (3.23), the built-in voltage becomes

$$V_{bi} = \frac{q N_d x_n W}{2\kappa} \tag{3.30}$$

But W the depletion region width is equal to

$$W = x_n + x_p \tag{3.31}$$

and x_n and x_p are related as

$$N_a x_p = N_d x_n \tag{3.32}$$

x_n can now be evaluated using (3.31) and (3.32) as

$$x_n = \frac{N_a}{N_d}(W - x_n) = \frac{N_a W}{(N_a + N_d)} \tag{3.33}$$

Thus the built-in voltage becomes

$$V_{bi} = \frac{q N_a N_d W^2}{2\kappa(N_a + N_d)} \tag{3.34}$$

Equation (3.34) can be solved for the width of the depletion region, W, in equilibrium giving

$$W = \sqrt{\frac{2\kappa V_{bi}}{q} \frac{(N_a + N_d)}{N_a N_d}} \tag{3.35}$$

The edges of the depletion region are

$$x_p = \frac{W N_d}{N_a + N_d} = \sqrt{\frac{2\kappa V_{bi}}{q}\left[\frac{N_d}{N_a(N_a + N_d)}\right]} \tag{3.36}$$

and

$$x_n = \frac{W N_a}{N_a + N_d} = \sqrt{\frac{2\kappa V_{bi}}{q} \left[\frac{N_a}{N_d(N_a + N_d)} \right]} \tag{3.37}$$

Example Problem 3.1 p–n junctions in equilibrium

Consider an abrupt, silicon p–n junction in equilibrium. Assume that the acceptor and donor concentrations are 10^{18} cm^{-3} and 10^{16} cm^{-3} respectively.

(a) Determine the location of E_f in the n- and p-regions.

On the p-side the hole concentration, p_p, is given as

$$p_p = n_i e^{\frac{(E_i - E_{fp})}{kT}}$$

Solving for $E_i - E_{fp}$ obtains

$$E_i - E_{fp} = kT \ln \left(\frac{p_p}{n_i} \right)$$

Substituting in the numbers yields

$$E_i - E_{fp} = 0.0259 \ln \left(\frac{10^{18}}{10^{10}} \right) eV = 0.477 \, eV$$

Similarly, the location of the Fermi level on the n-side can be obtained as

$$E_{fn} - E_i = kT \ln \left(\frac{n_n}{n_i} \right) = 0.0259 \ln \left(\frac{10^{16}}{10^{10}} \right) eV = 0.358 \, eV$$

(b) Determine the built-in voltage from the band diagram.

As discussed in the text the built-in voltage is equal to the full band bending in equilibrium. The full band bending is equal to the difference between the Fermi levels on the n- and p-sides of the junction. The built-in voltage is then

$$V_{bi} = 0.477 + 0.358 = 0.835 \, eV$$

(c) Check the result in part (b) using the formulas for the built-in voltage developed in the text.

Since the doping concentrations are given let us use the formula that involves these quantities. The built-in voltage is then

$$V_{bi} = \frac{kT}{q} \ln \left(\frac{N_a N_d}{n_i^2} \right) = 0.0259 \ln \left(\frac{10^{18} 10^{16}}{(10^{10})^2} \right) eV = 0.835 \, eV$$

which is precisely the same result as found in (b).

Table 3.1 Particle flux and current flow direction for holes
and electrons in a p–n junction.

	Diffusion		Drift	
	Particle flux	Current flow	Particle flux	Current flow
Holes	p → n	p → n	n → p	n → p
Electrons	n → p	p → n	p → n	n → p

3.2 p–n homojunctions under bias

In a p–n junction both drift and diffusion occur. It is best to start our discussion of
the nonequilibrium behavior of a p–n junction by examining in which direction the
currents flow due to diffusion and drift. First note that a diffusion particle flux will flow
from the region of high concentration to the region of low concentration. For holes the
current flow is in the same direction as the particle flux. However, for electrons the
current flow is in the opposite direction of the particle flux. The drift current depends
upon the field. As we determined in Section 3.1 the field points from the n- to the
p-side. Thus holes will move from n to p while electrons move in the opposite direction,
from p to n. The behavior of the carriers and currents is summarized in Table 3.1.

Inspection of Table 3.1 indicates that diffusion currents due to either electrons or
holes flow in the same direction while the drift currents due to each carrier species also
flow in the same direction, but opposite to that of the diffusion currents. In equilibrium
the net current is zero so the diffusion current flowing in one direction is compensated
by the drift current flowing in the opposite direction.

The n-side is at a higher electrostatic potential than the p-side. The drift current is
essentially due to electrons and holes generated in the transition region or depletion
region of the p–n junction or close by the depletion region (within roughly one diffusion
length of the depletion region). For an ideal p–n junction we neglect generation–
recombination within the depletion region itself and attribute all of the drift current to
generation–recombination within a diffusion length of the depletion region on either
side of the junction. In equilibrium of course no current flows. Therefore, the drift and
diffusion currents compensate for one another in equilibrium such that the net current
is zero.

There are two different bias conditions for the p–n junction. These are forward and
reverse bias. Let us consider each biasing separately. Forward bias occurs when a
positive potential is applied to the p-side with respect to the n-side of the junction. We
recall that in equilibrium the electrostatic potential is higher on the n-side than on the
p-side. Therefore, there is a potential barrier separating the p- and n-sides. For a hole to
diffuse across the junction and be collected within the n-region it must have sufficient
kinetic energy in the forward direction to surmount the potential barrier separating the
p- and n-sides of the junction. If the junction is forward biased, then the electrostatic
potential of the p-side is raised. As a result the potential barrier separating the p- and
n-sides is reduced by an amount equal to the forward bias voltage. The lowering of

potential   energy band diagram

(a)
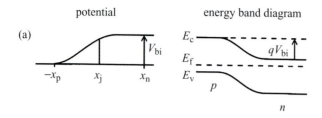

(b)
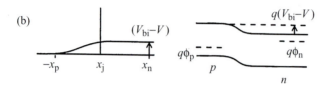

(c)
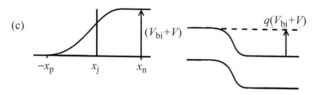

Figure 3.6 Potential and corresponding energy band diagram for a p–n junction diode under three different biasing conditions. These are: (a) equilibrium, (b) forward bias, and (c) reverse bias (Brennan, 1999).

the potential barrier increases the number of holes that can surmount the potential barrier in diffusing from the p-side to the n-side. Thus a greater diffusion current can flow. In addition to lowering the potential barrier between the p- and n-sides a forward bias acts to reduce the width of the depletion region and reduce the magnitude of the electric field between the n- and p-sides. The depletion region width under bias can be expressed as

$$W = \sqrt{\frac{2\kappa}{q}(V_{bi} - V)\frac{(N_a + N_d)}{N_a N_d}} \tag{3.38}$$

Under forward bias the current is dominated by diffusion. The diffusion current increases exponentially with increasing forward bias. In reverse bias the diffusion current is effectively choked off leaving only the drift current, which flows in the opposite direction to the diffusion current. To fully understand the effect that the applied bias has on the diffusion current it is useful to examine the potential and energy band diagrams under equilibrium, forward bias, and reverse bias conditions. Figure 3.6 shows the potential and energy band diagram for a p–n junction under equilibrium, forward, and reverse bias. In equilibrium the potential barrier separating the p- and n-sides is simply equal in magnitude to the built-in potential, V_{bi}. Application of a forward bias acts to reduce this potential barrier resulting in a new barrier of height,

$f(E)$

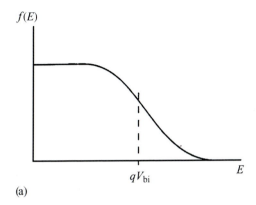

(a)

$f(E)$

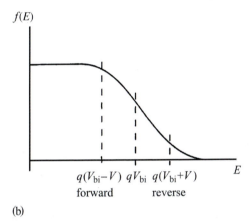

(b)

Figure 3.7 Energy distribution function plotted as a function of energy for (a) equilibrium and (b) forward and reverse bias conditions in a p–n junction diode (Brennan, 1999).

$V_{bi} - V$. Finally, when the junction is reverse biased by applying a positive voltage to the n-side with respect to the p-side the potential difference between the p- and n-sides is increased. As a result, the potential barrier is increased between the two sides of the junction with the n-side lying at the higher potential. Since the potential barrier is increased the diffusion current is decreased and is ultimately choked off.

It is useful to investigate further why increasing the potential barrier chokes off the diffusion current. To understand the effect of the bias on the diffusion current it is useful to examine Fig. 3.7. In Fig. 3.7 the energy distribution function is plotted vs. carrier energy for equilibrium and forward and reverse bias. For simplicity we assume that the distributions plotted represent the kinetic energy in the field direction. The dashed line labeled qV_{bi} in both (a) and (b) represents the energy corresponding to the built-in potential. Carriers with energy less than qV_{bi} have insufficient kinetic energy in the field direction to overcome the potential barrier and thus cannot contribute to the diffusion current. If a bias is applied the bias can act to either reduce the potential barrier height, forward bias, or increase the potential barrier height, reverse bias as shown in Fig. 3.6. Under forward bias a positive voltage is applied to the p-side thus lowering the potential barrier while under reverse bias a positive voltage is applied to the n-side increasing the potential barrier. Inspection of Fig. 3.7(b) shows that there is an exponential increase in the number of carriers that can overcome the potential barrier under forward bias. This leads to an exponential increase in the diffusion current. Conversely, there is an exponential decrease in the number of carriers that can overcome the potential barrier under reverse bias leading to an exponential decrease in the diffusion current. If the reverse bias is sufficiently high, the diffusion current can be effectively choked off leaving only the drift current flowing in the junction.

As mentioned above, the drift current is assumed to be due to the generation–recombination of carriers within a diffusion length of the depletion region. For example, let an electron–hole pair be generated in the p-side within a diffusion length of the depletion region. The electron diffuses to the edge of the depletion region on the p-side of the junction and is then swept by the field to the n-side of the junction. Similarly, an electron–hole pair can be generated on the n-side within a diffusion length of x_n, the edge of the depletion region on the n-side. The generated hole diffuses to x_n after which it is swept by the action of the field across the depletion region and into the p-side of the junction. The magnitude of the drift current depends upon the generation rate rather than on the magnitude of the electric field. This can be understood as follows. If the generation rate decreases, even if the field stays the same, the number of carriers that contribute to the drift current necessarily decreases. As a result, the drift current decreases in magnitude since there are fewer carriers present to contribute to the current. Similarly, if the generation rate increases, there are more carriers that contribute to the drift current leading to an increase in the drift current. The magnitude of the field, which depends upon the applied bias, does not affect the magnitude of the current. It only changes how fast the carriers are swept out of the junction. Thus the magnitude of the drift current is effectively independent of the magnitude of the bias. Consequently, the drift current is basically constant with bias.

Under reverse bias conditions the width of the depletion region increases. This can be understood from the relationship for the depletion region width, W, under bias given by (3.38). For reverse bias the applied voltage is negative and thus the difference between the built-in potential and the applied potential increases (a negative number times a negative number is positive). Thus if we assign V to be $-V$ the depletion region width under reverse bias becomes

$$W = \sqrt{\frac{2\kappa}{q}(V_{bi} + V)\frac{(N_a + N_d)}{N_a N_d}} \tag{3.39}$$

where V is assumed to be a negative reverse bias. Notice that W is larger under this condition than in equilibrium.

Now we can understand the nature of the current–voltage characteristic. Under forward bias the diffusion current increases exponentially while the drift current remains effectively constant. Thus the diffusion current dominates the forward characteristic and the current increases exponentially with bias as shown in Fig. 3.8. Under reverse

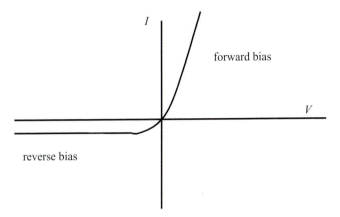

Figure 3.8 Current–voltage characteristic for a p–n junction. Notice that under forward bias the current increases exponentially while under reverse bias the current remains effectively constant.

bias the diffusion current is choked off leaving only the drift current. Since the drift current is effectively independent of bias then under reverse bias a constant small current flows as shown in Fig. 3.8. Notice that the drift current has an opposite sign to that of the diffusion current. This is consistent with Table 3.1 where we found that the drift and diffusion currents flowed in opposite directions.

Let us now consider ideal diode behavior quantitatively. The current flow in an ideal p–n junction can be determined as follows. First we make some simplifying assumptions. These are:

(i) There are no external sources of radiation or generation present.
(ii) The depletion approximation holds.
(iii) The system is in steady-state.
(iv) No generation–recombination processes occur in the depletion region. The only generation–recombination events occur within the bulk regions.
(v) Low level injection conditions hold.
(vi) The electric field is zero within the bulk regions. The only nonzero electric field appears in the depletion region of the diode.
(vii) The bulk regions are uniformly doped.
(viii) The junction is abrupt.

The derivation starts with the one-dimensional continuity equations. These are:

$$\frac{1}{q}\frac{d}{dx}J_n + G - R = \frac{dn}{dt} \tag{3.40}$$

for electrons and

$$-\frac{1}{q}\frac{d}{dx}J_p + G - R = \frac{dp}{dt} \tag{3.41}$$

for holes. The current densities are given as

$$J_n = qn\mu_n\varepsilon(x) + qD_n\frac{dn}{dx}$$

$$J_p = qp\mu_p\varepsilon(x) - qD_p\frac{dp}{dx}$$

(3.42)

First solve the continuity equations in the bulk regions. In the bulk regions the electric field is zero in accordance with assumption (vi) above. The generation rate, G, is also equal to zero as stated in assumption (iv). The current densities become then

$$J_n = qD_n\frac{dn}{dx} \quad \text{and} \quad J_p = -qD_p\frac{dp}{dx}$$

(3.43)

The recombination rates, R_n and R_p, are given as

$$R_n = \frac{\delta n}{\tau_n} \quad \text{and} \quad R_p = \frac{\delta p}{\tau_p}$$

(3.44)

In steady-state the continuity equations become

$$D_n\frac{d^2 n}{dx^2} - \frac{\delta n}{\tau_n} = 0 \quad \text{and} \quad D_p\frac{d^2 p}{dx^2} - \frac{\delta p}{\tau_p} = 0$$

(3.45)

But the total electron and hole carrier concentrations can be written in terms of the equilibrium concentrations and the excess concentration as

$$n = n_0 + \delta n \quad \text{and} \quad p = p_0 + \delta p$$

(3.46)

The derivatives of the equilibrium concentrations with respect to t and x are zero. Thus (3.45) can be rewritten using (3.46) as

$$D_n\frac{d^2\delta n}{dx^2} - \frac{\delta n}{\tau_n} = 0 \quad \text{and} \quad D_p\frac{d^2\delta p}{dx^2} - \frac{\delta p}{\tau_p} = 0$$

(3.47)

Kirchoff's Current Law implies that the current at any point is constant. The total current, J, is

$$J = J_n + J_p$$

(3.48)

Within the depletion region both the generation and recombination rates, G and R, vanish in accordance with assumption (iv). This implies that what goes into the depletion region must come out by the node law. In other words, the electron and hole currents are each constant throughout the depletion region. If the minority carrier current is known at the edges of the depletion region it is known everywhere within the depletion region. The behavior of the electron and hole current densities is shown in Fig. 3.9. The current densities vary within the bulk regions as a function of position due to generation–recombination processes. J_p decreases within the n-region and J_n decreases within the p-regions. In both cases the carriers are the minority carriers and thus recombine with the majority carriers. As such the minority carriers generally fully recombine before they reach the ohmic contacts on either side of the junction. Notice

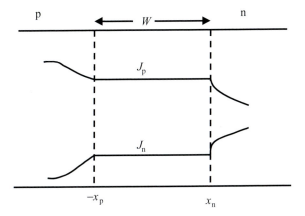

Figure 3.9 Hole and electron current densities plotted vs. position in a p–n junction. The region marked by W is the depletion region. Notice that both current densities are constant within the depletion region but change within the bulk regions. The change in the current densities within the bulk regions is due to generation–recombination processes. Within the depletion region the generation–recombination rates are assumed to be zero.

that the majority carrier current densities increase with position in the bulk regions. The total current remains constant everywhere in the device. Thus within the bulk regions the sum of the current densities is the same as in the depletion region. This of course must be true since the device is assumed to be in steady-state and the node law requires a conservation of current. If the current densities are known at the edge of the depletion region, they are known everywhere within the depletion region. Thus we seek to find the magnitude of the current densities at the edge of the depletion region.

Recall at the edges of the depletion region that the electric field vanishes. It is assumed that all of the voltage drop within the p–n junction occurs across the depletion region and that the bulk regions are field free. Before we begin the solution let us pause and review the notation that will be used. The notation is as follows:

n_{p0} = equilibrium concentration of electrons on the p-side.

p_{n0} = equilibrium concentration of holes on the n-side.

n_p = electron concentration on the p-side of the junction. It is equal to the sum of the equilibrium and excess electron concentrations.

p_n = hole concentration on the n-side of the junction. It is equal to the sum of the equilibrium and excess hole concentrations.

δn_p = excess electron concentration on the p-side of the junction.

δp_n = excess hole concentration on the n-side of the junction.

We next find the boundary conditions that are needed to solve the continuity equations. Consider the boundary conditions at $-x_p$ and x_n. The built-in potential is

$$q V_{bi} = kT \ln \left(\frac{p_{p0}}{p_{n0}} \right) \tag{3.49}$$

Thus the equilibrium hole concentration on the p-side is

$$p_{p0} = p_{n0} e^{\frac{qV_{bi}}{kT}} \tag{3.50}$$

Under bias the potential barrier is changed as $V_{bi} - V$, where V is the applied potential. Therefore, the hole concentration at the edge of the depletion region on the p-side is

$$p(-x_p) = p(x_n) e^{\frac{q(V_{bi}-V)}{kT}} \tag{3.51}$$

Under low level injection, $p_{p0} \sim p(-x_p)$. This assumption is reasonable since the excess hole concentration is small with respect to the equilibrium hole concentration on the p-side, p_{p0}. Dividing (3.51) by (3.50) yields

$$\frac{p(-x_p)}{p_{p0}} = \frac{p(x_n)}{p_{n0}} \frac{e^{\frac{q(V_{bi}-V)}{kT}}}{e^{\frac{qV_{bi}}{kT}}} \tag{3.52}$$

The left hand side of (3.52) is simply one. Simplifying, (3.52) becomes

$$\frac{p(x_n)}{p_{n0}} = e^{\frac{qV}{kT}} \tag{3.53}$$

The excess hole concentration at the edge of the depletion region is defined as Δp_n and can be expressed as

$$\Delta p_n = p(x_n) - p_{n0} = p_{n0}\left(e^{\frac{qV}{kT}} - 1\right) \tag{3.54}$$

Similarly, the excess electron concentration at the edge of the depletion region on the p-side of the junction is

$$\Delta n_p = n(-x_p) - n_{p0} = n_{p0}\left(e^{\frac{qV}{kT}} - 1\right) \tag{3.55}$$

Next consider a long diode, one in which the contacts are far from the edges of the depletion region. The boundary conditions on the excess carrier concentrations are then

$$\delta n_p(x = -\infty) = 0 \quad \text{and} \quad \delta p_n(x = +\infty) = 0 \tag{3.56}$$

The excess minority carrier concentrations are zero far from the junction. Physically, the excess minority carrier concentrations recombine far from the junction.

Now that the boundary conditions for the excess carrier concentrations are determined we next solve the continuity equations and apply the boundary conditions. The continuity equations need to be solved for the minority carriers in both bulk regions. For simplicity we will utilize the coordinate systems shown in Fig. 3.10. Let us first solve the continuity equation for the hole concentration on the n-side of the junction. In terms of the coordinate system x' the hole continuity equation is

$$D_p \frac{d^2}{dx^2}\delta p_n(x') - \frac{\delta p_n(x')}{\tau_p} = 0 \tag{3.57}$$

The diffusion length L_p is defined as

$$L_p = \sqrt{D_p \tau_p} \tag{3.58}$$

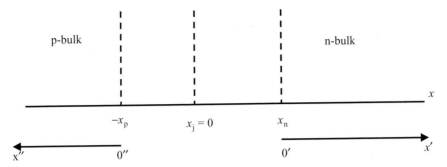

Figure 3.10 p–n junction showing new coordinate systems within each bulk region used for the solution of the continuity equations: x_j is the metallurgical junction, x_n the edge of the depletion region on the n-side of the junction, and $-x_p$ the edge of the depletion region on the p-side of the junction.

With this definition (3.57) becomes

$$\frac{d^2}{dx^2}\delta p_n(x') - \frac{\delta p_n(x')}{L_p^2} = 0 \tag{3.59}$$

The general solution of (3.59) is

$$\delta p_n(x') = A_1 e^{\frac{x'}{L_p}} + A_2 e^{-\frac{x'}{L_p}} \tag{3.60}$$

We then apply the boundary conditions to the excess hole concentration to evaluate the constants A_1 and A_2. These are

(i) At $x' = +\infty$ the excess hole concentration on the n-side is zero. Thus

$$\delta p_n(x' = +\infty) = 0 \tag{3.61}$$

and A_1 must vanish.

(ii) At $x' = 0'$ the excess hole concentration is equal to

$$\delta p_n(x' = 0') = p_{n0}\left(e^{\frac{qV}{kT}} - 1\right) = \Delta p_n \tag{3.62}$$

Therefore, A_2 is given as

$$A_2 = p_{n0}\left(e^{\frac{qV}{kT}} - 1\right) = \Delta p_n \tag{3.63}$$

The excess hole concentration becomes then

$$\delta p_n(x') = \Delta p_n e^{-\frac{x'}{L_p}} = p_{n0}\left(e^{\frac{qV}{kT}} - 1\right)e^{-\frac{x'}{L_p}} \tag{3.64}$$

The corresponding hole current density can be determined as

$$J_p(x' = 0') = J_p(x_n) = -q D_p \frac{d}{dx}\delta p_n(x')\Big|_{x'=0} = \frac{q D_p p_{n0}}{L_p}\left(e^{\frac{qV}{kT}} - 1\right) \tag{3.65}$$

The hole and electron currents in a p–n junction are then

$$J_p(x' = 0') = J_p(x_n) = \frac{q D_p p_{n0}}{L_p}\left(e^{\frac{qV}{kT}} - 1\right) \tag{3.66}$$

and

$$J_n(x'' = 0'') = J_n(-x_p) = \frac{q D_n n_{p0}}{L_n}\left(e^{\frac{qV}{kT}} - 1\right) \tag{3.67}$$

The sum of the current densities at the edges of the depletion region is a constant and equal to the total current density within the junction, J. Thus the total current density is

$$J = q\left(\frac{D_n n_{p0}}{L_n} + \frac{D_p p_{n0}}{L_p}\right)\left(e^{\frac{qV}{kT}} - 1\right) \tag{3.68}$$

The current is obtained by multiplying the current density by the area, A. Defining I_0 as the reverse saturation current, the current can be expressed as

$$I = I_0\left(e^{\frac{qV}{kT}} - 1\right) = q A\left(\frac{D_n n_{p0}}{L_n} + \frac{D_p p_{n0}}{L_p}\right)\left(e^{\frac{qV}{kT}} - 1\right) \tag{3.69}$$

which is called the Shockley equation. Notice that under forward bias, V is positive, so the exponential term is dominant, and therefore the current increases exponentially. Under reverse bias the potential is a negative number and thus the exponential for a sizable V approaches zero. Thus the reverse bias current is effectively constant and is given by I_0.

Example Problem 3.2 Germanium p+–n diode

Consider a germanium p+–n diode. The intrinsic concentration of Ge is 2.5×10^{13} cm^{-3}. Given: $\tau_p = \tau_n = 10 \times 10^{-6}$ s, $N_a = 10^{19}$ cm^{-3}, $N_d = 10^{16}$ cm^{-3}, $T = 300$ K, $kT = 0.0259$ eV, $D_p = 39$ cm^2/s, $D_n = 26$ cm^2/s, $A = 1.25 \times 10^{-4}$ cm^2.

(a) Determine the current I at an applied voltage, V, of 0.2 volts in the forward direction.

Start with the Shockley equation. It is given as

$$I = I_0\left(e^{\frac{qV}{kT}} - 1\right) \qquad I_0 = q A\left(\frac{D_n n_{p0}}{L_n} + \frac{D_p p_{n0}}{L_p}\right)$$

The diffusion lengths are given as

$$L_n = \sqrt{\tau_n D_n} \qquad L_p = \sqrt{\tau_p D_p}$$

and can be evaluated as

$$L_n = \sqrt{1 \times 10^{-5} \times 26} = 0.016 \text{ cm} \qquad L_p = \sqrt{1 \times 10^{-5} \times 39} = 0.0197 \text{ cm}$$

The Law of Mass Action can be used to determine the equilibrium minority carrier concentrations on either side of the junction. This obtains

$$n_{p0} p_{p0} = n_i^2 \qquad p_{p0} = 10^{19} = N_a \qquad n_{n0} = N_d = 10^{16}$$

Thus n_{p0} and p_{n0} can be calculated as

$$n_{p0} = \frac{n_i^2}{p_{p0}} = \frac{(2.5 \times 10^{13})^2}{10^{19}} = 6.25 \times 10^7 \text{ cm}^{-3}$$

$$p_{n0} = \frac{n_i^2}{n_{n0}} = \frac{(2.5 \times 10^{13})^2}{10^{16}} = 6.25 \times 10^{10} \text{ cm}^{-3}$$

The current now can be determined by substituting into the Shockley equation:

$$I = qA \left(\frac{D_n n_{p0}}{L_n} + \frac{D_p p_{n0}}{L_p} \right) \left(e^{\frac{qV}{kT}} - 1 \right)$$

All values within the Shockley equation are known. The current computes to be

$$I = 44.7 \text{ mA}$$

(b) Determine the reverse saturation current at $V = -0.2$ V.

$$I_{rev} \sim I_0 = -1.98 \times 10^{-5} \text{ A} = -19.8 \text{ μA}$$

3.3 **Deviations from ideal diode behavior**

In deriving the Shockley equation we made numerous assumptions. Some of these assumptions are not always valid, leading to significant departures from the Shockley equation. One of the major deviations from ideal behavior is due to reverse bias breakdown. As the diode is further reverse biased a large current can begin to flow if the magnitude of the applied voltage is greater than the magnitude of the breakdown voltage, V_{br}. The breakdown voltage is the reverse bias voltage where the current tends towards negative infinity. Under these conditions the diode becomes unstable. The diode can operate safely up to a point provided that the breakdown is reversible. Once the breakdown becomes irreversible, secondary breakdown is reached, the diode will become unstable.

There are two main mechanisms for reverse breakdown. These are avalanche and zener breakdown. Let us consider each mechanism in turn. Avalanche breakdown is caused by carrier multiplication through impact ionization. At high reverse bias, the field within the depletion region becomes quite large. Therefore, carriers that drift through the depletion region are accelerated to high energies (way up in the band). Some of the carriers reach sufficiently high energy that when they collide with the lattice they can exchange their excess kinetic energy with a valence band electron promoting it to the conduction band. Consequently, two electrons in the conduction band and one hole exist after the event whereas prior to the impact ionization event only one high energy electron was present. Similarly, a high energy hole can initiate the process. The high energy hole makes a collision with the lattice transferring its excess kinetic energy to a valence band electron. Provided the initiating hole has sufficient energy, which must be at least equal to the energy gap, an electron–hole pair can be generated from the collision. The final state is two holes in the valence band, the initiating hole and the

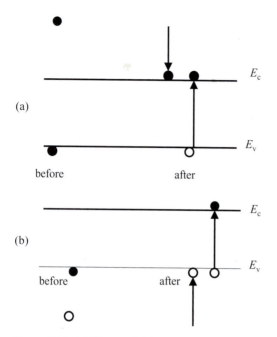

Figure 3.11 (a) Electron initiated impact ionization process. The initial condition (before) consists of a high energy electron within the conduction band and an electron in the valence band. The high energy electron makes a collision with the lattice transferring its excess kinetic energy to the valence band electron. The valence band electron jumps into the conduction band leaving behind a hole in the valence band. Notice that an electron–hole pair is generated. (b) Hole initiated impact ionization process. Same behavior as in (a) except that the hole initiates the event.

secondary hole created by the impact ionization event along with an electron within the conduction band. Both electron initiated and hole initiated impact ionization events are sketched in Fig. 3.11. As can be readily seen from the figure, impact ionization is a threshold process. The initiating carrier must have energy equal to or greater than the energy band gap in order to cause an impact ionization event.

Since avalanche breakdown is a threshold event there is a minimum electric field necessary to heat the initiating carriers to sufficient energy to cause impact ionization. This minimum field is called the critical field and the corresponding voltage is the breakdown voltage of the diode. The maximum electric field in the junction occurs at the metallurgical junction, $x = 0$. The value of the maximum electric field in equilibrium was found in Section 3.1 to be

$$\varepsilon_{\max} = -\frac{q N_a x_p}{\kappa} = -\frac{q N_d x_n}{\kappa} \tag{3.70}$$

But x_n in equilibrium is given as

$$x_n = \sqrt{\frac{2\kappa V_{bi}}{q}\left[\frac{N_a}{N_d(N_a + N_d)}\right]} \tag{3.71}$$

Under bias x_n becomes

$$x_n = \sqrt{\frac{2\kappa(V_{bi} - V)}{q} \left[\frac{N_a}{N_d(N_a + N_d)}\right]} \tag{3.72}$$

The maximum value of the electric field can be found using (3.70) and (3.72) to be

$$\varepsilon_{max} = -\frac{qN_d}{\kappa}\sqrt{\frac{2\kappa(V_{bi} - V)}{q} \left[\frac{N_a}{N_d(N_a + N_d)}\right]} \tag{3.73}$$

which simplifies to

$$\varepsilon_{max} = -\sqrt{\frac{2q(V_{bi} - V)}{\kappa} \left(\frac{N_a N_d}{N_a + N_d}\right)} \tag{3.74}$$

Next let us find a relationship with V_{BR}, the breakdown voltage. The critical field is equal to the maximum field at breakdown. Thus the critical field can be determined from (3.74) as

$$\varepsilon_{crit}^2 = \frac{2q(V_{bi} + V_{BR})}{\kappa} \left(\frac{N_a N_d}{N_a + N_d}\right) \tag{3.75}$$

where it is assumed that the applied voltage (in this case the breakdown voltage, V_{BR}) is negative leading to a positive sign before V. If the junction is p^+–n then (3.75) simplifies to

$$\varepsilon_{crit}^2 = \frac{2q(V_{bi} + V_{BR})N_d}{\kappa} \tag{3.76}$$

Solving (3.76) for $V_{bi} + V_{BR}$ yields

$$V_{bi} + V_{BR} = \frac{\kappa\varepsilon_{crit}^2}{2qN_d} \tag{3.77}$$

Generally, the breakdown voltage is much larger than the built-in voltage. Consequently, the built-in voltage can be neglected with respect to the breakdown voltage. With this assumption the breakdown voltage is given as

$$V_{BR} = \frac{\kappa\varepsilon_{crit}^2}{2qN_d} \tag{3.78}$$

Inspection of (3.78) shows that the breakdown voltage is proportional to $1/N_d$ for a p^+–n junction.

The second breakdown mechanism in a reverse biased p–n junction diode is Zener breakdown. Zener breakdown occurs through carrier tunneling. A carrier, either an electron or hole, exhibits both wavelike and particlelike behaviors that form the basis of quantum mechanics (see Brennan (1999) for a full discussion of quantum effects). These quantum mechanical effects become observable when the dimensions of the system are comparable to the wavelength of the carrier which is called the de Broglie wavelength. Since an electron or hole can have wavelike properties its behavior is quite different from that of classical particles. A classical particle (like a baseball,

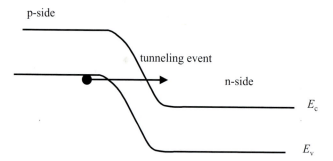

p-side

tunneling event

n-side

E_c

E_v

Figure 3.12 Tunneling event in a reverse biased p–n junction diode.

etc.) cannot penetrate a potential barrier. Instead the classical particle is reflected by a potential barrier provided that the kinetic energy of the particle is less than the barrier height. However, a quantum mechanical object can penetrate a barrier potential. If the barrier is sufficiently thin and not too high, a carrier can not only penetrate the barrier but have a probability of emerging on the other side. This is called quantum mechanical tunneling or just tunneling. Carriers in semiconductors can tunnel through a potential barrier.

Zener tunneling occurs in reverse biased p–n junctions. Carriers tunnel from the filled valence band on the p-side of the junction into the conduction band on the n-side of the junction if the depletion region is small. The situation is shown in Fig. 3.12. As can be seen from the figure, the reverse bias pulls the bands down such that the valence band on the p-side lies above the conduction band on the n-side.

In both, avalanche breakdown and Zener tunneling the reverse current increases dramatically once either multiplication or tunneling begins. The resulting current–voltage characteristic is shown in Fig. 3.13. As can be seen from the figure, the current increases strongly at high bias due to the onset of either multiplication or tunneling.

Example Problem 3.3 Breakdown in a p–n junction diode

Assume that the maximum field in the junction is the critical field for Zener breakdown. Calculate the applied voltage in the junction if the maximum field is 10^5 V/cm. Assume that the junction is abrupt and made from silicon. The intrinsic concentration is 10^{10} cm^{-3}, $N_a = 10^{18}$ cm^{-3}, and $N_d = 10^{15}$ cm^{-3}. Additional information: $q = 1.6 \times 10^{-19}$ C, $\kappa = 11.8$ (8.85 $\times$ 10^{-14} F/cm).

The magnitude of the maximum electric field is

$$|\varepsilon_{max}| = \sqrt{\frac{2q}{\kappa} (V_{bi} - V) \frac{N_a N_d}{(N_a + N_d)}}$$

Substituting in and solving for $V_{bi} - V$ obtains

$$10^5 \text{ V/cm} = \sqrt{\frac{2(1.6 \times 10^{-19} \text{ C})(V_{bi} - V)(10^{18} \text{ cm}^{-3})(10^{15} \text{ cm}^{-3})}{11.8 \ (8.85 \times 10^{-14} \text{ F/cm})(10^{18} \text{ cm}^{-3} + 10^{15} \text{ cm}^{-3})}}$$

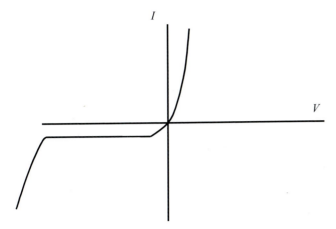

Figure 3.13 Current–voltage characteristic of a p–n junction showing reverse bias breakdown. As can be seen from the figure the reverse current increases dramatically at relatively large reverse bias. There are two principal mechanisms that act to increase the reverse current. These are carrier multiplication through impact ionization and tunneling. In either mechanism the current increases strongly with increasing reverse bias. In the forward direction (first quadrant) the current increases exponentially due to the diffusion current.

$V_{bi} - V$ is given by

$$V_{bi} - V = \frac{10^5 \text{ V}}{1.75 \times 10^4} = 5.7 \text{ V}$$

But V_{BR} is negative since the device is reverse biased. Thus we obtain

$$V_{bi} + V_{BR} = 5.7 \text{ V}$$

and the built-in voltage is

$$V_{bi} = \frac{kT}{q} \ln \left(\frac{N_a N_d}{n_i^2} \right)$$

Substituting in, V_{bi} is equal to

$$V_{bi} = 0.0259 \text{ V} \ln \left(\frac{10^{18} \text{ cm}^{-3} 10^{15} \text{ cm}^{-3}}{(10^{10} \text{ cm}^{-3})^2} \right) = 0.775 \text{ V}$$

The breakdown voltage is then

$$V_{BR} = 5.7 \text{ V} - 0.775 \text{ V} \sim 5.0 \text{ V}$$

3.4 **Carrier injection, extraction, charge control analysis, and capacitance**

Let us next examine some further details of p–n junction behavior. In this section we concentrate on three important attributes of p–n junction operation. These are carrier injection and extraction, charge control analysis, and capacitance. First let us consider

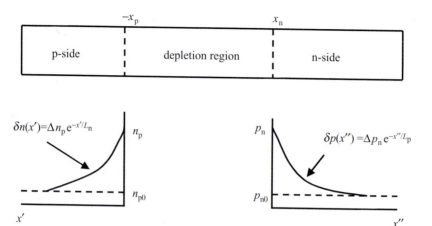

Figure 3.14 Excess electron and hole concentrations on the p- and n-sides respectively. As can be seen from the diagram the excess concentrations decrease exponentially away from the edge of the depletion region on either side. At some distance into the bulk regions the excess concentration vanishes leaving only the equilibrium minority carrier concentration.

charge injection and extraction. Under forward bias minority carriers are injected across the junction. The excess minority carrier profile is sketched in Fig. 3.14. Notice that on the n-side the excess hole concentration decreases exponentially away from x_n, the edge of the depletion region on the n-side of the junction. Similarly, the excess electron concentration on the p-side decreases exponentially away from $-x_p$. In either case reasonably far from the edge of the depletion region the excess minority carrier concentration vanishes and the minority carrier concentration becomes equal to its equilibrium concentration. The excess concentrations are

$$\delta n(x') = \Delta n_p e^{-\frac{x'}{L_p}} \tag{3.79}$$

for electrons on the p-side and

$$\delta p(x'') = \Delta p_n e^{-\frac{x''}{L_p}} \tag{3.80}$$

for holes on the n-side. In (3.79) and (3.80) Δn_p is the excess electron concentration at the edge of the depletion region, $-x_p$, and Δp_n is the excess hole concentration at $x = x_n$. From the definition of the built-in voltage, V_{bi}, the ratio of p_{p0} to p_{n0} is given as

$$\frac{p_{p0}}{p_{n0}} = e^{\frac{qV_{bi}}{kT}} \qquad p_{p0} = p_{n0} e^{\frac{qV_{bi}}{kT}} \tag{3.81}$$

Under bias the potential barrier is changed to $V_{bi} - V$, where V is the applied potential. If V is negative notice that we obtain $V_{bi} + V$. Under low level injection, $p(-x_p) \sim p_{p0}$. Neglecting the equilibrium concentration, we also note that Δp_n is equal to $p(x_n)$. This is reasonable since the minority equilibrium concentration is typically very small with respect to the nonequilibrium concentration. With these assumptions (3.81) becomes

$$p(-x_p) = p(x_n) e^{\frac{q(V_{bi}-V)}{kT}} \tag{3.82}$$

Dividing (3.82) by (3.81) obtains

$$\frac{p(-x_\mathrm{p})}{p_\mathrm{p0}} = \frac{p(x_\mathrm{n})}{p_\mathrm{n0}} \mathrm{e}^{-\frac{qV}{kT}} \tag{3.83}$$

Recall that $p(x_\mathrm{n})$ is equal to

$$\Delta p_\mathrm{n} = p(x_\mathrm{n}) - p_\mathrm{n0} = p_\mathrm{n0}\mathrm{e}^{\frac{qV}{kT}} - p_\mathrm{n0} = p_\mathrm{n0}\!\left(\mathrm{e}^{\frac{qV}{kT}} - 1\right) \tag{3.84}$$

Similarly,

$$\Delta n_\mathrm{p} = n(-x_\mathrm{p}) - n_\mathrm{p0} = n_\mathrm{p0}\!\left(\mathrm{e}^{\frac{qV}{kT}} - 1\right) \tag{3.85}$$

Notice that under forward bias, V is positive and the values of Δp_n and Δn_p are large and essentially equal to

$$\Delta p_\mathrm{n} = p_\mathrm{n0}\mathrm{e}^{\frac{qV}{kT}} \tag{3.86}$$

and

$$\Delta n_\mathrm{p} = n_\mathrm{p0}\mathrm{e}^{\frac{qV}{kT}} \tag{3.87}$$

Under reverse bias V is negative and the exponential term is negligible with respect to the constant term. We have then

$$\Delta p_\mathrm{n} = -p_\mathrm{n0} \tag{3.88}$$

and

$$\Delta n_\mathrm{p} = -n_\mathrm{p0} \tag{3.89}$$

Let us next turn our attention to charge control analysis. The total excess hole charge, Q_p, injected into the n-side of the junction is simply equal to the area under the right hand side of Fig. 3.14. Q_p can be determined then as

$$Q_\mathrm{p} = q A \int_{x_\mathrm{n}}^{\infty} \delta p_\mathrm{n}(x, t)\mathrm{d}x \tag{3.90}$$

where the integral has been extended to infinity. This can be done since the excess hole concentration decays exponentially and as x approaches infinity the excess hole concentration is zero.

Equation (3.90) can be solved by expressing the excess hole concentration using the continuity equation. In the bulk region, from x_n to infinity, the electric field is zero. We start with the one-dimensional continuity equation given as

$$\frac{\partial}{\partial t}\delta p_\mathrm{n} = D_\mathrm{p}\frac{\partial^2}{\partial x^2}\delta p_\mathrm{n} - \frac{\delta p_\mathrm{n}}{\tau_\mathrm{p}} + G_\mathrm{L} \tag{3.91}$$

But the optical generation rate is zero, so G_L in (3.91) vanishes. The current density can be written as

$$J_\mathrm{p} = -q D_\mathrm{p}\frac{\partial \delta p_\mathrm{n}}{\partial x} \tag{3.92}$$

Therefore,

$$D_{p}\frac{\partial^{2}\delta p_{n}}{\partial x^{2}} = -\frac{1}{q}\frac{\partial J_{p}}{\partial x} \tag{3.93}$$

Substituting (3.93) into (3.91) and simplifying yields (where the subscript n has been omitted for notational simplicity)

$$\frac{\partial}{\partial t}(q\delta p) = -\frac{\partial J_{p}}{\partial x} - \frac{q\delta p}{\tau_{p}} \tag{3.94}$$

Multipling (3.94) by Adx and integrating gives

$$\frac{d}{dt}\left(qA\int_{x_{n}}^{\infty}\delta p dx\right) = -A\int_{J_{p}(x_{n})}^{J_{p}(\infty)} dJ_{p} - \frac{qA}{\tau_{p}}\int_{x_{n}}^{\infty}\delta p dx \tag{3.95}$$

but

$$\left(qA\int_{x_{n}}^{\infty}\delta p dx\right) = Q_{p} \tag{3.96}$$

Evaluating (3.95) results in

$$\frac{d}{dt}Q_{p} = -AJ_{p}(\infty) + AJ_{p}(x_{n}) - \frac{Q_{p}}{\tau_{p}} \tag{3.97}$$

But $J_{p}(\infty) = 0$. This is because we have only an electron current on the far end of the n-side. The excess hole concentration has decayed to zero and thus the hole current density is also zero (hole drift and diffusion currents are zero). For a p^{+}–n junction $J_{p}(x_{n}) \gg J_{n}(-x_{p})$. Therefore the total current $i \sim AJ_{p}(x_{n})$. Equation (3.97) thus becomes

$$\frac{d}{dt}Q_{p} = i_{\text{diff}} - \frac{Q_{p}}{\tau_{p}} \tag{3.98}$$

The solution of this differential equation gives the stored charge. An example of how to work with (3.98) is given in Example Problem 3.4.

Finally, let us consider the capacitance of a diode. There are two types of capacitance in a p–n junction. These are the junction capacitance and the charge storage capacitance. The junction capacitance is the capacitance associated with the depletion layer. The charge storage capacitance is the capacitance associated with the stored charge with changing bias.

The junction capacitance can be understood as follows. From Fig. 3.15 we see that there is space charge associated with each side of the depletion region in a p–n junction. The total amount of charge on the n-side is

$$Q_{n} = qN_{d}Ax_{n0} \tag{3.99}$$

and on the p-side is

$$Q_{p} = -qN_{a}x_{p0}A \tag{3.100}$$

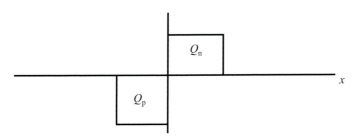

Figure 3.15 p–n junction diode showing the space charge within the depletion region.

The capacitance is defined as

$$C = \left| \frac{dQ}{dV} \right| \tag{3.101}$$

The width of the depletion region under bias is given as

$$W = \sqrt{\left[\frac{2\kappa}{q}(V_{bi} - V) \left(\frac{N_a + N_d}{N_a N_d} \right) \right]} \tag{3.102}$$

V of course can be either forward or reverse biased. If V is reverse biased then it has a negative sign, increasing the value of W. If V is forward biased then it has a positive sign, decreasing the value of W. The net charge on either side of the depletion region is

$$|Q| = q A x_n N_d = q A x_p N_a \tag{3.103}$$

but

$$x_n = \frac{N_a W}{N_a + N_d} \qquad x_p = \frac{N_d W}{N_a + N_d} \tag{3.104}$$

Therefore, substituting into (3.103) the relationships given by (3.104) obtains,

$$|Q| = \frac{q A N_a N_d W}{(N_a + N_d)} = A \sqrt{2q\kappa(V_{bi} - V)\frac{N_d N_a}{(N_d + N_a)}} \tag{3.105}$$

The junction capacitance is given then as

$$C_j = \left| \frac{dQ}{d(V_{bi} - V)} \right| = \frac{A}{2} \sqrt{\frac{2q\kappa}{(V_{bi} - V)} \frac{N_d N_a}{(N_d + N_a)}} \tag{3.106}$$

The junction capacitance is similar to that of a parallel plate capacitor. For a parallel plate capacitor the capacitance is

$$C = \frac{\kappa A}{W} \tag{3.107}$$

Rewriting (3.106) obtains

$$C_j = \kappa A \sqrt{\frac{q}{2\kappa(V_{bi} - V)} \frac{N_d N_a}{(N_d + N_a)}} = \frac{\kappa A}{W} \tag{3.108}$$

The junction capacitance dominates in reverse bias.

Under forward bias the charge storage capacitance dominates. The charge storage capacitance is also often referred to as the diffusion capacitance. Basically the charge storage capacitance arises from the fact that as the current changes due to charge storage effects, the voltage does not change as quickly due to charge storage effects. Recall under forward bias that minority carrier injection occurs. This means that the minority carrier concentration increases above the equilibrium carrier concentration near the edge of the depletion region. When the diode is switched, the current flow changes instantaneously but it takes time before the injected charge or stored charge is removed. The steady-state stored charge is found from Example Problem 3.4 to be

$$Q_p = I \tau_p \tag{3.109}$$

Using (3.80) obtains

$$\delta p(x'') = \Delta p_n e^{-\frac{x''}{L_p}} \tag{3.110}$$

The stored charge at any instant is obtained by integrating (3.110) with respect to x'':

$$Q_p = q A \int_0^\infty \Delta p_n e^{-\frac{x''}{L_p}} dx'' = q A \Delta p_n L_p \tag{3.111}$$

which can be rewritten as

$$Q_p = q A L_p p_n \left(e^{\frac{qV}{kT}} - 1 \right) \tag{3.112}$$

Taking the derivative with respect to V yields the charge storage capacitance as

$$C_s = \frac{dQ_p}{dV} = \frac{q^2}{kT} A L_p p_n e^{\frac{qV}{kT}} \sim \frac{q}{kT} I \tau_p \tag{3.113}$$

where it is assumed that the applied bias is much larger than the thermal voltage.

Example Problem 3.4   Determination of the stored charge in a p⁺–n diode

Determine an expression for the stored charge at the edge of the depletion region on the n-side of a p⁺–n junction as a function of time if the diode is originally in steady-state with a forward current of I and is switched to zero current at $t = 0$.

The starting point is the equation derived in the text for Q_p. This is

$$\frac{d}{dt}Q_p = i_{diff} - \frac{Q_p}{\tau_p}$$

The current switches from I to zero at $t = 0$. With this initial condition solve the differential equation for Q_p.

The homogeneous solution is given as

$$\frac{d}{dt}Q_p + \frac{Q_p}{\tau_p} = 0$$

The solution of this equation is

$$Q_p = Ae^{-\frac{t}{\tau_p}}$$

The general solution is then

$$Q_p = Ae^{-\frac{t}{\tau_p}} + B$$

For $t < 0$ in steady-state,

$$\frac{Q_p}{\tau_p} = I$$

Therefore, at $t = 0$, Q_p equals $I\tau_p$. Applying the boundary conditions to the problem obtains the solution:

at $t = 0$

$$Q_p = I\tau_p = A + B$$

as t approaches infinity,

$$Q_p = 0 = B$$

Therefore, $A = I\tau_p$. The solution for Q_p is then

$$Q_p = I\tau_p e^{-\frac{t}{\tau_p}}$$

Example Problem 3.5 Maximum electric field and capacitance in a p–n junction diode

(a) Determine the n-type doping concentration in a silicon p–n abrupt junction such that the maximum electric field is 3×10^5 V/cm. Assume that the temperature is 300 K and the p-type doping concentration is 10^{18} cm^{-3}. Assume the reverse bias voltage is 25 V, $\kappa_s = 11.7$, and $\kappa T = 0.0259$ eV. Neglect the built-in voltage.

The maximum electric field in the junction is

$$|\varepsilon_{max}| = \sqrt{\frac{2q\,V_{BR}}{\kappa}\frac{N_aN_d}{(N_a + N_d)}}$$

Substituting in for each quantity, plus the electronic charge and ε_0,

$$(3 \times 10^5)^2 = \sqrt{\frac{2(1.6 \times 10^{-19})(25)}{(11.7)(8.85 \times 10^{-14})} \frac{10^{18} N_d}{10^{18} + N_d}}$$

Solving for N_d obtains

$$N_d = 1.174 \times 10^{16} \, \text{cm}^{-3}$$

(b) Determine the capacitance of the junction described in part (a). Neglect V_{bi}.

The capacitance per unit area is

$$\frac{C}{A} = \sqrt{\frac{q\kappa}{2(V_{bi} + V_{BR})} \frac{N_a N_d}{(N_a + N_d)}}$$

where the built-in voltage has been neglected. The capacitance per unit area is calculated then to be

$$\frac{C}{A} = \sqrt{\frac{1.6 \times 10^{-19}(11.7)(8.85 \times 10^{-14})}{2(25)} \left[\frac{(10^{18})(1.17 \times 10^{16})}{(10^{18} + 1.17 \times 10^{16})}\right]}$$

$$= 6.19 \times 10^{-9} \, \text{F/cm}$$

3.5 Schottky barriers

Generally, there are two different types of metal–semiconductor contacts. These are Schottky barriers and ohmic contacts. A Schottky barrier is a rectifying metal–semiconductor contact while an ohmic contact has a linear relationship between the current and voltage. In this section we discuss Schottky barrier contacts in detail along with brief discussing of ohmic contacts.

Let us first consider a Schottky contact in equilibrium. Recall that no current flows in equilibrium and that the gradient of the Fermi level is zero. Thus the Fermi level is flat everywhere in equilibrium. At the metal–semiconductor interface in equilibrium the Fermi levels must align. Consider the behavior of a metal first. As discussed in Chapter 1, the Fermi level lies above the conduction band edge in a metal. The energy needed in order that an electron be ionized, escape from the metal and enter the vacuum level, is called the work function energy or $q\varphi_m$. The metal is characterized by its work function. Most of the conduction electrons within the metal are at an energy reasonably close to the Fermi level. Therefore, the work function is the energy difference between the vacuum level and the Fermi level. In a nondegenerate semiconductor, on the other hand, the Fermi level does not lie above the conduction band edge but is somewhere in the forbidden gap. The semiconductor work function is defined in the same way as for a metal, it is the energy difference between the Fermi level and the vacuum level. However, since most of the electrons in a semiconductor are not at the Fermi level, it is necessary to introduce another quantity that describes the separation of the conduction band from the vacuum level. This is the electron affinity. The electron affinity ($q\varphi_s$) and work function are shown in Fig. 3.16.

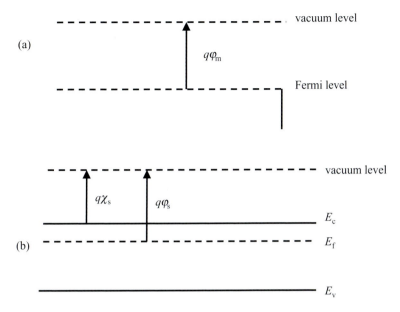

Figure 3.16 Band diagram of (a) a metal in equilibrium and (b) a semiconductor in equilibrium, showing the work functions and electron affinity within the semiconductor.

Let us now consider an n-type semiconductor with a work function less than that of the metal in equilibrium. The band diagrams for the metal and semiconductor while apart are shown in Fig. 3.17(a) and in contact are shown in Fig. 3.17(b). As we have mentioned when the two materials are brought into contact in equilibrium the Fermi level is flat as shown in the figure. In order to equilibrate the Fermi level, electrons must transfer from the semiconductor to the metal. This can be clearly understood from the fact that the Fermi level is higher in the semiconductor than the metal. Therefore, to equilibrate the Fermi level everywhere there must be a net electron transfer from the semiconductor to the metal. A simple analogy with water tanks is useful. Consider two tanks of water separated by a partition. Let the water level be higher in one tank than the other. Once the partition is removed between the two tanks the water level of course equalizes with water moving from the tank with the higher level to the tank with the lower level. A similar situation occurs for the metal and semiconductor. The major difference is that we are dealing with charged particles in the metal–semiconductor junction as opposed to the water tanks. After electrons are transferred to the metal, a positive space charge due to ionized donors is produced within the semiconductor. Conversely, a net negative charge is deposited on the metal. If the metal is a perfect conductor it cannot sustain a tangential field, and thus the negative charge lies on its surface. The band diagram shown in Fig. 3.17(b) can now be understood. The easiest way to determine how the bands bend is to recall that electrons roll downhill in potential energy diagrams. The presence of a net negative charge on the surface of the metal acts to repel the electrons from the surface. As a result the bands must bend such that the electrons roll away from the interface as shown in Fig. 3.17(b).

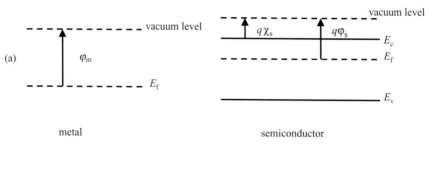

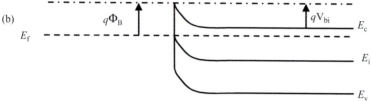

Figure 3.17 (a) Band diagrams of a metal and semiconductor while apart. (b) Band diagram of the resulting metal–semiconductor junction in equilibrium. This type of junction is called a Schottky barrier.

The net transfer of electrons from the semiconductor to the metal forms a depletion region of ionized donors within the semiconductor near the surface. A built-in field and built-in potential result from the formation of the depletion region. The built-in voltage of the junction is equal to the full band bending in equilibrium. The total amount by which the bands bend in the semiconductor is equal to the built-in potential as shown in the figure. The barrier potential, shown as Φ_B in the diagram, is called the Schottky barrier height. The amount of potential energy blocking the transfer of electrons from the metal to the semiconductor is $q\Phi_B$. The magnitude of the built-in potential and Schottky barrier can be obtained from inspection of Fig. 3.17. Notice that the full band bending can be determined from the difference between the Fermi levels on either side of the junction. Thus the built-in potential is given as

$$q V_{bi} = q\varphi_m - q\varphi_s \tag{3.114}$$

The built-in potential can be determined in another way using Fig. 3.17. Inspection of the figure shows that the difference of the Schottky barrier height and the separation between E_c and E_f in the semiconductor yields the built-in potential. The separation between E_c and E_f can be found as follows. The electron concentration is given as

$$n = N_c e^{-\frac{(E_c - E_f)}{kT}} \tag{3.115}$$

Solving (3.115) for $E_c - E_f$ yields

$$E_c - E_f = kT \ln\left(\frac{N_c}{n}\right) \tag{3.116}$$

The built-in potential is given then as

$$q V_{bi} = q \Phi_B - kT \ln \left(\frac{N_c}{n} \right) \tag{3.117}$$

The Schottky barrier height is equal to the difference between the metal work function and the electron affinity:

$$q \Phi_B = q \varphi_m - q \chi_s \tag{3.118}$$

So we see that for electrons within the semiconductor there exists a potential barrier, the built-in potential, between the semiconductor and the metal. For electrons within the metal there exists a potential barrier, the Schottky barrier, between the metal and the semiconductor.

In order for the carriers to move from one material to the other under bias they must have sufficient kinetic energy to overcome the barrier height. The semiconductor is at a higher potential than the metal. This is because the semiconductor has a positive space charge depletion region producing a positive potential. If we apply a negative bias to the metal with respect to the semiconductor the potential difference between the metal and semiconductor is increased. This corresponds to a reverse bias on the Schottky barrier. Hence the energy bands of the semiconductor bend down more producing an even higher barrier for electron emission into the metal. Conversely, if a positive bias is applied to the metal with respect to the semiconductor the junction becomes forward biased since the potential difference is reduced between the metal and semiconductor; the bands bend down less on the semiconductor side. It is important to note that only the potential barrier due to the band bending changes under bias. The Schottky barrier height is relatively unchanged with bias. The reason why the Schottky barrier height does not change with bias is because all of the voltage drop appears in the semiconductor. The metal is assumed to be a perfect conductor so it cannot support a tangential field. All of the charge deposited on the metal from the applied bias lies on the surface of the metal. Though we assume that the Schottky barrier does not change with bias, there is a small bias dependent effect that lowers the Schottky barrier height. The details of this mechanism are presented in Brennan (1999) and will not be repeated here. Suffice it to say that to a good approximation the Schottky barrier height is voltage independent. The resulting band diagrams for forward and reverse biased Schottky barriers are shown in Fig. 3.18.

In summary, when a negative voltage is applied to the metal with respect to the semiconductor the potential barrier in the semiconductor is increased since the potential difference between the metal and semiconductor is increased. This corresponds to a reverse bias. The Schottky barrier is relatively unchanged. The new potential barrier on the semiconductor side is given as $q(V_{bi} + V)$, where V is the applied reverse bias. Notice that the reverse biased potential barrier is greater than the equilibrium potential barrier on the semiconductor side of the junction. The device becomes forward biased when a positive potential is applied to the metal with respect to the semiconductor. The potential barrier in the semiconductor is decreased since the potential difference between the metal and semiconductor is decreased. Again the Schottky barrier is

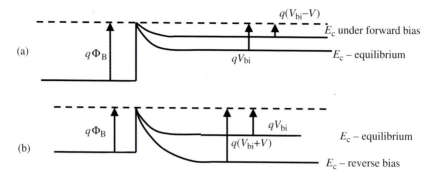

Figure 3.18 Metal–semiconductor Schottky barrier diode band diagram under (a) forward bias and (b) reverse bias. The position of the conduction band edge is shown in both diagrams for equilibrium as a basis of comparison for the biased conditions. Notice that the Schottky barrier itself, $q\Phi_B$, does not change with bias. Only the band bending on the semiconductor side is a function of the applied bias.

essentially unchanged. The new potential barrier on the semiconductor side is given as $q(V_{bi} + V)$, where V is the applied forward bias. In this case the forward biased potential barrier is less than that in equilibrium.

The next question is how is the current flow affected by the biasing condition? There are two different currents in the Schottky barrier. These are due to a particle flux from the metal into the semiconductor and a particle flux from the semiconductor into the metal. A Schottky barrier is a unipolar device in that essentially only electrons contribute to the current. The current flows in the opposite direction from the particle flux. The current that arises from particle flux from the semiconductor to the metal is basically a diffusion current. As in a p–n junction the diffusion current is a strong function of the barrier height separating the semiconductor from the metal. If the barrier height is lowered by the action of a forward bias, there is an exponential increase in the number of carriers that have sufficient kinetic energy in the field direction to surmount the potential barrier and contribute to the diffusion current. If the potential barrier is raised by reverse biasing the junction the diffusion current is choked off. It is important to recognize that the particle flux from the semiconductor to the metal is highly dependent upon the bias conditions. The current that arises from the particle flux from the metal into the semiconductor is called a thermionic emission current. The full details of thermionic emission are presented in Brennan (1999) and will not be repeated here. In equilibrium the particle fluxes from either side of the junction are equal resulting in no net current flow.

Since the current flows in the opposite direction from the particle flux it is useful to define which currents dominate the device under different biasing conditions. The metal–semiconductor current density, J_{m-s}, is due to a particle flux from the semiconductor into the metal. The semiconductor–metal current density, J_{s-m}, is due to a particle flux from the metal into the semiconductor. Under forward bias, J_{m-s} dominates since it represents the particle flux from the semiconductor to the metal

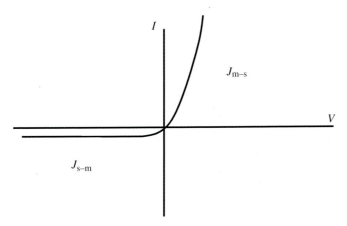

Figure 3.19 The current–voltage characteristic for a Schottky barrier diode showing the dominant currents in each quadrant. The metal–semiconductor current dominates the forward bias characteristic while the semiconductor–metal current dominates the reverse bias characteristic.

that arises from diffusion. This current is highly bias dependent. Under reverse bias, J_{s-m} dominates since it arises from a particle flux from the metal into the semiconductor. The metal–semiconductor current is choked off in reverse bias leaving only the relatively constant semiconductor–metal current. The reverse current is essentially constant since the Schottky barrier height is bias independent. Thus as the reverse bias increases the Schottky barrier height remains unchanged and the carrier flux also remains unchanged. Therefore, under reverse bias the current is essentially constant as shown in Fig. 3.19. Under forward bias the current increases exponentially as shown in the figure. The dominant currents for different biasing conditions are also shown in Fig. 3.19.

The depletion region width in a Schottky barrier can be determined as follows. For a p–n junction diode we found that the depletion region width is given as

$$W = \sqrt{\frac{2\kappa}{q}(V_{bi} - V)\frac{(N_a + N_d)}{N_a N_d}} \tag{3.119}$$

In the limit a Schottky barrier is a highly doped asymmetric junction. The metal takes the role of the highly doped side of the junction. All of the depletion region extends within the semiconductor layer and none is within the metal. Assuming that the semiconductor is doped n-type (3.119) becomes for a highly doped asymmetric junction,

$$W = \sqrt{\frac{2\kappa}{q}(V_{bi} - V)\frac{1}{N_d}} \tag{3.120}$$

We conclude this section with a brief discussion of ohmic contacts. An ohmic contact can be formed between a metal and a semiconductor if no potential barrier exists between the two materials. This can be achieved if the metal work function

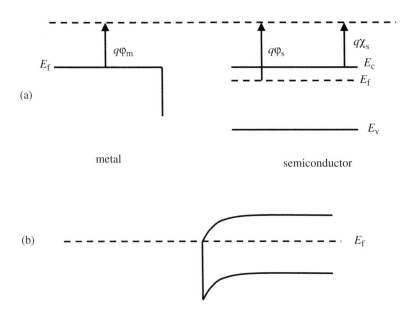

Figure 3.20 Equilibrium band diagrams showing ohmic contact formation: (a) metal and semiconductor apart and (b) in contact.

is less than the semiconductor work function. Consider for example a metal–n-type-semiconductor contact. The band diagrams for when the materials are apart and in contact are shown in Fig. 3.20. When in contact electrons must transfer from the metal into the semiconductor to equilibrate the Fermi level. As a result the semiconductor has a net negative charge while the metal has a net positive charge. Recall that electrons roll downhill in potential energy diagrams. Thus the bands must bend such that the electrons are attracted to the positive potential on the metal as shown in Fig. 3.20. Further inspection of Fig. 3.20 shows that the barrier to electron flow is small and is easily overcome by the application of a small bias. As such the contact is ohmic, the current and voltage vary linearly.

Example Problem 3.6 Schottky barrier diode

Consider an Al–Si Schottky barrier diode. The doping concentration within the silicon is 10^{15} cm^{-3}. The effective density of states N_c is given as 3.22×10^{19} cm^{-3} and the Schottky barrier height, Φ_B is 0.72 eV.

(a) Determine the built-in voltage, V_{bi}.

$$V_{bi} = \varphi_m - \varphi_s$$

But the built-in voltage can be found in an alternative way using (3.117) which is

$$q V_{bi} = q \Phi_B - kT \ln \left(\frac{N_c}{n} \right)$$

Assuming that all of the donors are ionized then $n = 10^{15}$ cm^{-3}. Substituting in, the built-in potential is found as

$$V_{bi} = 0.72 \text{ V} - 0.0259 \text{ V} \ln \left(\frac{3.22 \times 10^{19}}{10^{15}} \right) = 0.451 \text{ V}$$

(b) Determine the depletion region width, W, in equilibrium.

$$W = \sqrt{\frac{2\kappa}{qN_d}(V_{bi} - V)}$$

In equilibrium $V = 0$. Substituting in obtains,

$$W = \sqrt{\frac{2(11.8)(8.85 \times 10^{-14})}{(1.6 \times 10^{-19})(10^{15})} 0.451} = 7.67 \times 10^{-5} \text{ cm}$$

Problems

3.1 Find an expression for the electron current in the n-type material of a forward biased p–n junction.

3.2 A p$^+$–n silicon junction is doped with $N_d = 1 \times 10^{16}$ cm^{-3} on the n-side, where $D_p = 10$ cm^2/s and $\tau_p = 0.1$ μs. The junction area is 10^{-4} cm^2. Calculate the reverse saturation current and the forward current for $V = 0.5$ volts.

3.3 An abrupt Si p–n junction has $N_a = 1 \times 10^{15}$ cm^{-3} and $N_d = 1 \times 10^{16}$ cm^{-3}. Use $n_i = 10^{10}$ cm^{-3}.
(a) Calculate the Fermi level positions at 300 K in both the p- and n-regions.
(b) Determine the built in potential.

3.4 A p$^+$–n abrupt junction diode is made from silicon with $N_d = 1 \times 10^{15}$ cm^{-3}. In equilibrium, the depletion region has a total length of 1.013 μm, and area of 0.002 cm^2. Determine the electron concentration on the p-side in equilibrium, n_{p0}. Use $n_i = 10^{10}$ cm^{-3}.

3.5 The same diode described in Problem 3.4 is forward biased such that a current of 1.0×10^{-2} A flows through the device. If the diode is ideal, find the voltage V across the junction.
Assume: $D_p = 5.2$ cm^2/s; $D_n = 13$ cm^2/s; $\tau_p = 10^{-8}$ s, $n_i = 10^{10}$ cm^{-3}.

3.6 Consider a simple p–n Si homojunction with a p-side doping concentration of $N_a = 10^{19}$ cm^{-3}, and an n-side doping concentration of $N_d = 10^{17}$ cm^{-3}. Assume $n_i = 10^{10}$ cm^{-3}.
(a) Determine the built-in voltage for the junction.
(b) Determine the width of the depletion region if a reverse bias of –2 volts is applied.
(c) If the breakdown field strength is 5.0×10^5 V/cm, determine the maximum allowable reverse bias. Assume that the breakdown field is the maximum electric field in the junction. Do not neglect V_{bi}.

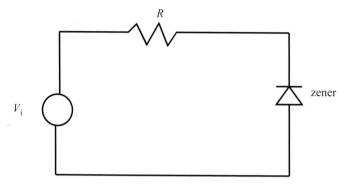

Figure 3.21 The diode is a reverse biased Zener diode. The resistance is equal to R and the voltage is V_i.

3.7 A Schottky barrier diode is made on n-type Si with a doping concentration of $N_d = 1.5 \times 10^{15}$ cm^{-3}. The metal work function is 5.1 V. The electron affinity of silicon is 4.15 V. Given: n_i for Si is 10^{10} cm^{-3}, $kT = 0.0259$ eV, $q = 1.6 \times 10^{-19}$ C, E_i lies at midgap, (1.1 eV)/2, $\varepsilon_0 = 8.85 \times 10^{-14}$ F/cm, $\kappa_{Si} = 11.8$. Determine:
  (a) the barrier height;
  (b) the built-in voltage;
  (c) the depletion layer width at zero bias.

3.8 Consider a Si p$^+$–n diode with $N_a = 10^{19}$ cm^{-3}, $N_d = 10^{15}$ cm^{-3}, $A = 1$ cm^2. Useful information: $n_i = 10^{10}$ cm^{-3}, $kT = 0.0259$ eV, $q = 1.6 \times 10^{-19}$ C, $\varepsilon_0 = 8.85 \times 10^{-14}$ F/cm, $\kappa_{Si} = 11.8$,
  (a) Determine the depletion layer thickness in equilibrium.
  (b) If the ideal reverse saturation current in the diode is calculated to be 0.13 nA, determine the magnitude of the carrier concentration within the depletion region necessary to support this current. Assume that the carriers have a constant, saturation velocity of 10^7 cm/s.

3.9 An n-type silicon Schottky barrier junction has $N_d = 5 \times 10^{15}$ cm^{-3} and $\varphi_m = 4.5$ V. Given an electron affinity for silicon of 4.15 V, $N_c = 2.8 \times 10^{19}$ cm^{-3}, determine ($T = 300$ K):
  (a) the Schottky barrier height;
  (b) the built-in voltage.

3.10 In a p–n homojunction the electron concentration on the n-side is 10^{17} cm^{-3}, the electron concentration on the p-side is 10^7 cm^{-3}, and n_i is 10^{10} cm^{-3}. Determine the built-in potential for the diode. $kT = 0.0259$ eV.

3.11 Find the edge of the depletion region on the n-side of a p–n homojunction if it is doped with donors to 5×10^{17} cm^{-3} and doped with acceptors to 1×10^{18} cm^{-3}. Assume that the edge of the depletion region on the p-side is at 0.25 μm relative to the metallurgical junction.

3.12 Consider a germanium p$^+$–n diode. The intrinsic concentration of Ge is 2.5×10^{13} cm^{-3}. Given: $\tau_p = \tau_n = 10 \times 10^{-6}$ s, $N_a = 10^{19}$ cm^{-3}, $N_d = 10^{16}$ cm^{-3},

$T = 300$ K, $kT = 0.0259$ eV, $D_p = 39$ cm^2/s, $D_n = 26$ cm^2/s, $A = 1.25 \times 10^{-4}$ cm^2, $q = 1.6 \times 10^{-19}$ C.

Determine the current I at an applied voltage, V, of 0.2 V in the forward direction.

3.13 An abrupt p$^+$–n junction is grown using InP with $N_d = 10^{15}$ cm^{-3}. If the avalanche breakdown voltage of the device is -80 V, determine the minimum thickness of the n-region such that the device exhibits avalanche breakdown prior to punchthrough. Neglect the built-in voltage of the device. Given: $n_i = 1.2 \times 10^8$ cm^{-3}, $E_g = 1.34$ eV, $\kappa_s = 12.56$, $q = 1.6 \times 10^{-19}$ C, $\varepsilon_0 = 8.85 \times 10^{-14}$ F/cm.

3.14 Consider a metal–p-type silicon contact with the following information: $n_i = 10^{10}$ cm^{-3}, $N_a = 10^{17}$ cm^{-3}, $\varphi_m = 3$ V, $\chi_s = 2.5$ V, $E_g = 1.1$ eV, $q = 1.6 \times 10^{-19}$ C, $\kappa_i = 11.9$, $\varepsilon_0 = 8.85 \times 10^{-14}$ F/cm, $kT = 0.0259$ eV.
  (a) Determine the built-in voltage of the junction.
  (b) Determine the Schottky barrier height.

3.15 Determine the current, I_d, in the p$^+$–n Zener diode circuit shown in Fig. 3.21 given the following information. Assume that breakdown occurs when the field within the junction reaches its critical value and that at this point no further voltage drop occurs across the diode. Given: $R = 1$ kΩ, $V_i = 5$ V, $F_{crit} = 10^5$ V/cm, $N_d = 10^{16}$ cm^{-3}, $n_i = 10^{10}$ cm^{-3}, $q = 1.6 \times 10^{-19}$ C, $\kappa_i = 11.9$, $\varepsilon_0 = 8.85 \times 10^{-14}$ F/cm, $kT = 0.0259$ eV. Neglect the built-in voltage of the diode.

3.16 A p$^+$–n junction diode is switched from zero bias ($I = 0$) to a forward current I at $t = 0$. Find an expression for the excess hole charge $Q_p(t)$ during the turn-on transient. Assume there is no built-up charge at the edge of the depletion region at $t = 0$.

3.17 Find the electric field profile within a p–n homojunction if the charge density within the depletion region is given as

$$\rho(x) = qax \quad \text{for} \quad -\frac{W}{2} \le x \le \frac{W}{2}$$

and is zero elsewhere. Assume that the depletion layer extends an equal amount into both the n- and p-sides of the junction.

Bipolar junction transistors

A bipolar junction transistor (BJT) consists of two p–n junctions with three terminals. The device comprises three different doped layers. These are alternating p, n, and p layers or n, p, and n layers. The first type of device is called a pnp BJT, while the second device is an npn BJT. The first layer is called the emitter. The second layer is called the base and the third layer is called the collector. Thus for a pnp BJT, the emitter and collector regions are p-type while the base is n-type. The two p–n junctions are then formed between the emitter and base and collector and base. The vast majority of BJT applications are in analog electronics, but BJTs can also be employed in digital circuitry.

4.1 BJT operation

A BJT can be used in several circuit configurations. In most applications the input signal is across two of the BJT leads while the output signal is extracted from a second pair of leads. Since there are only three leads for a BJT, one of the leads must be common to both the input and the output circuitry. Hence we call the different circuit configurations common emitter, common base, and common collector to identify the lead common to both the input and the output. These configurations are shown in Fig. 4.1. The most commonly employed configuration is the common emitter configuration.

A BJT can be biased into one of four possible modes. These are saturation, active, inverted, and cutoff. In the saturation mode both the emitter–base and collector–base junctions are forward biased. The active mode consists of a forward biased emitter–base junction (V_{EB}) and a reverse biased collector–base junction (V_{CB}). The inverted mode has the emitter–base junction reverse biased while the collector–base junction is forward biased. Finally, in the cutoff mode both the emitter–base and collector–base junctions are reverse biased. The biasing conditions and corresponding modes of operation are shown in Fig. 4.2. In addition Table 4.1 lists the polarity as a function of the biasing mode.

The behavior of the device under each of the four biasing modes can be understood as follows. The active mode is the most common mode for transistor amplification. In the saturation mode both junctions are forward biased which corresponds to a high current flow with a concomitant low voltage drop. The saturation mode corresponds to the "on" state of the transistor in digital applications. Since the voltage of the device operated in the saturation mode is low (current is high) the saturation mode corresponds to "zero" in digital binary logic. Conversely, in the cutoff mode both junctions are reverse biased. As a result very little current flows and there is a high voltage drop. The cutoff mode

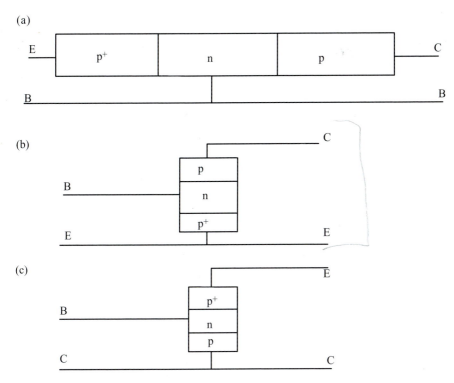

Figure 4.1 Sketch of the three configuration of a BJT: (a) common base configuration; (b) common emitter configuration; (c) common collector configuration.

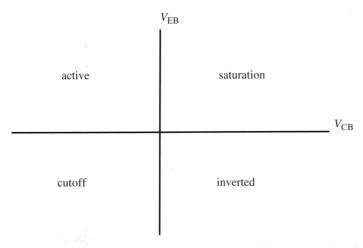

Figure 4.2 Different biasing modes for a pnp BJT. The active mode is the one used for amplification.

Table 4.1 The four polarity combinations and their resulting biasing mode

Biasing mode	Polarity emitter–base junction	Polarity collector–base junction
Saturation	Forward	Forward
Active	Forward	Reverse
Inverted	Reverse	Forward
Cutoff	Reverse	Reverse

corresponds to the off state of the device. Since the voltage drop is high in the cutoff mode (current is low) cutoff represents a "1" in digital binary logic. When the device switches from "1" to "0" it switches from cutoff to saturation. Conversely, the device switches from saturation to cutoff when switching from "0" to "1."

The active mode is used for amplification. Under active mode biasing the output collector current, I_C, is approximately equal to the input emitter current, I_E. However, a small portion of the emitter current is not collected at the output due to the presence of the base current. The base current, I_B, consists of three components. These arise from the following:

1. Recombination of injected holes from the emitter with electrons stored in the base. The electrons lost through recombination must be restored due to space-charge neutrality requirements. I_B restores the electrons lost in the base from recombination through the base contact.
2. Electron injection into the emitter from the base in the forward biased p–n emitter–base junction. I_B must supply the electrons lost.
3. The reverse biased base–collector junction sweeps electrons into the base. These electrons reduce the magnitude of the base current, I_B.

In the common emitter configuration, the input base current is amplified by a factor β which is the ratio of the collector current to the base current. The forward biased emitter–base junction of a pnp BJT produces hole injection into the base from the emitter. The base must remain space-charge neutral. Therefore, the injected excess holes from the emitter that enter the base require compensating excess electrons to enter the base. These electrons enter the base mainly from the base contact, though a small portion enters from the reverse biased collector–base junction. The injected holes spend less time in the base than the electrons since they are injected at higher velocity and energy. The holes spend on average a time called the base transit time, τ_B, in the base. The base transit time is essentially the time it takes the holes to transit the neutral base region and enter the collector–base depletion region. Typically, the base width, W_B, is small compared with the hole diffusion length, L_p. Hence the base transit time is much less than the hole recombination lifetime in the base, τ_p. Recall that the hole recombination lifetime is the average time before a hole recombines with an electron within the base. Since the base transit time is much less than the hole recombination lifetime, there are many holes that pass through the base for each recombination event and hence electron that enters the base. Thus τ_p/τ_B holes pass through the base for each electron which enters the base. The device has a gain then of

$$\frac{i_C}{i_B} = \frac{\tau_p}{\tau_B} = \beta_{dc} \tag{4.1}$$

Equation (4.1) can be derived in an alternative way under steady-state conditions as follows. The dc current gain in a BJT can be determined using the charge control model. The excess minority carrier charge in the base region of the BJT is called Q_B and this can be determined by integrating the excess minority carrier concentration, $\delta p(x)$ over the base volume. Q_B becomes

$$Q_B = \int_0^{W_B} q A \delta p(x) \mathrm{d}x \tag{4.2}$$

where A is the area and W_B the base width. The collector current, I_C, arises from the flow of this injected charge. Using (3.98) Q_B and I_C can be related as

$$\frac{\mathrm{d}Q_B}{\mathrm{d}t} = I_C - \frac{Q_B}{\tau_B} \tag{4.3}$$

Assuming steady-state conditions, the collector current can be expressed in terms of the mean base transit time, τ_B, as

$$I_C = \frac{Q_B}{\tau_B} \tag{4.4}$$

The steady-state base current, I_B, is in turn equal to the amount of recombination current in the base. This is given as

$$I_B = \frac{Q_B}{\tau_p} \tag{4.5}$$

(Throughout this chapter, our discussion and analysis assumes a pnp configuration. If, in fact, the physical device is in an npn configuration, our arguments have to be altered (i.e. electrons become holes, and vice versa).) Therefore, the dc current gain, I_C/I_B, becomes

$$\beta = \frac{I_C}{I_B} = \frac{Q_B/\tau_p}{Q_B/\tau_B} = \frac{\tau_B}{\tau_p} \tag{4.6}$$

From (4.6), it is clear that the dc gain of a BJT is proportional to the ratio of the transit time to the lifetime.

The currents in a BJT can be determined from the solution of the continuity equation. The simplest formulations of the currents in a BJT are made using the following assumptions:

(1) Drift is negligible in the base region. Holes diffuse from the emitter to the collector.
(2) The emitter current comprises entirely holes. This implies that the emitter injection efficiency is unity.

(3) The collector saturation current is negligible.

(4) Current flow in the base can be treated as one-dimensional from the emitter to the collector.

(5) All currents and voltages are in steady-state.

For simplicity we will continue to work with a pnp transistor since, in this case, the hole current flows in the same direction as the hole flux. We further assume that the BJT is biased in the active mode with the emitter–base junction strongly forward biased, and the collector–base junction strongly reverse biased. Generally, the excess hole carrier concentrations on the n-side of the emitter–base junction and on the n-side of the collector–base junction are determined from the boundary conditions for the excess carrier concentration in a diode derived in Chapter 3. The excess hole concentration at the edge of the depletion region on the n-side of the junction is given as

$$\Delta p_E = p_{B0}\left(e^{\frac{qV_{EB}}{kT}} - 1\right)$$
$$\Delta p_C = p_{B0}\left(e^{\frac{qV_{CB}}{kT}} - 1\right) \tag{4.7}$$

where p_{B0} is the equilibrium hole concentration within the n-type base region. The value of p_{B0} can be found in terms of the base donor doping concentration, N_{dB}, as $p_{B0} = n_i^2/N_{dB}$. If the junctions are strongly forward and reverse biased, Eqs. (4.7) become

$$\Delta p_E \sim p_{B0}e^{\frac{qV_{EB}}{kT}}$$
$$\Delta p_C \sim -p_{B0} \tag{4.8}$$

The one-dimensional continuity equation assuming no drift component is

$$\frac{d}{dt}\delta p = D_B \frac{\partial^2}{\partial x^2}\delta p - \frac{\delta p}{\tau_p} + G_L \tag{4.9}$$

where D_B is the hole diffusion constant within the n-type base, δp is the excess hole concentration, τ_p is the hole lifetime, and G_L is the generation rate. In our notation, we will use for the diffusion coefficients and the diffusion lengths D and L respectively subscripted by E, B, or C to denote the emitter, base, or collector. It is further understood that D and L correspond to the minority carrier within each of these regions. For example, D_B is the diffusion coefficient within the base of the minority carrier. For a pnp device the minority carriers within the base are holes. Thus D_B for the pnp structure corresponds to the diffusion coefficient for holes within the base. Under steady-state conditions, with no illumination (4.9) becomes

$$\frac{d^2\delta p}{dx^2} = \frac{\delta p}{L_B^2} \tag{4.10}$$

where L_B is the hole diffusion length within the base. The general solution of (4.10) is

$$\delta p = C_1 e^{\frac{x}{L_B}} + C_2 e^{-\frac{x}{L_B}} \tag{4.11}$$

The boundary conditions are applied at the edge of the depletion region of the emitter–base junction within the base, called $x = 0$, and at the end of the quasi-neutral

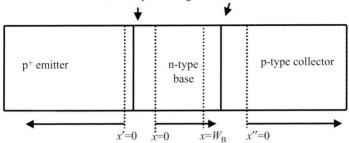

emitter–base depletion region   collector–base depletion region

p$^+$ emitter   n-type base   p-type collector

$x'=0$ $x=0$ $x=W_B$ $x''=0$

Figure 4.3 BJT device structure showing the coordinate systems and the depletion regions used in calculating the device currents.

base region, W_B, as shown in Fig. 4.3. The end of the quasi-neutral base region occurs at the edge of the collector–base depletion region within the base. The boundary conditions are then:

$$\delta p(x = 0) = \Delta p_E = C_1 + C_2$$
$$\delta p(x = W_B) = \Delta p_C = C_1 e^{\frac{W_B}{L_B}} + C_2 e^{-\frac{W_B}{L_B}} \tag{4.12}$$

Solving for the excess hole concentration as a function of x yields

$$C_1 = \frac{\Delta p_C - \Delta p_E e^{-\frac{W_B}{L_B}}}{e^{\frac{W_B}{L_B}} - e^{-\frac{W_B}{L_B}}}$$
$$C_2 = \frac{\Delta p_E e^{\frac{W_B}{L_B}} - \Delta p_C}{e^{\frac{W_B}{L_B}} - e^{-\frac{W_B}{L_B}}} \tag{4.13}$$

The emitter hole current can now be determined from the relation

$$I_p = -q D_B A \frac{d}{dx} \delta p\Big|_{x=0} \tag{4.14}$$

using the expression for δp given by (4.11) and the expressions for the coefficients given by (4.13). The emitter hole current is calculated at $x = 0$ and is found to be

$$I_{Ep} = \frac{q A D_B}{L_B} \left(\Delta p_E \coth \frac{W_B}{L_B} - \Delta p_C \operatorname{cosech} \frac{W_B}{L_B} \right) \tag{4.15}$$

The collector hole current can be found as follows. If the current component corresponding to electron injection from the collector back into the base can be neglected, the total collector current is given by the hole current as

$$I_{Cp} = I_p(x = W_B) = -q A D_B \frac{d}{dx} \delta p(x)\Big|_{x=W_B} \tag{4.16}$$

Substituting into (4.16) for δp yields

$$I_{Cp} = -q A D_B \frac{d}{dx} \left(C_1 e^{\frac{x}{L_B}} + C_2 e^{-\frac{x}{L_B}} \right)\Big|_{x=W_B} \tag{4.17}$$

which reduces to

$$I_{Cp} = \frac{qAD_B}{L_B}\left(C_2 e^{-\frac{W_B}{L_B}} - C_1 e^{\frac{W_B}{L_B}}\right) \tag{4.18}$$

Substituting in for C_2 and C_1 the expressions given by (4.13) and simplifying the result (4.18) becomes

$$I_{Cp} = \frac{qAD_B}{L_B}\left[\Delta p_E \operatorname{cosech}\frac{W_B}{L_B} - \Delta p_C \coth\frac{W_B}{L_B}\right] \tag{4.19}$$

The base current can be readily found from the node law. If the electron currents are neglected, and the total emitter and collector currents are assumed to be those corresponding to the hole currents, the base current is given as

$$I_B = I_{Ep} - I_{Cp} \tag{4.20}$$

Substituting (4.15) and (4.19) into (4.20) gives

$$I_B = \frac{qAD_B}{L_B}\left[(\Delta p_E + \Delta p_C)\left(\coth\frac{W_B}{L_B} - \operatorname{cosech}\frac{W_B}{L_B}\right)\right] \tag{4.21}$$

Equation (4.21) can be simplified using the relationship between the hyperbolic functions

$$\coth u - \operatorname{cosech} u = \frac{\cosh u - 1}{\sinh u} = \tanh\frac{u}{2} \tag{4.22}$$

to be

$$I_B = \frac{qAD_B}{L_B}\left[(\Delta p_E + \Delta p_C)\tanh\frac{W_B}{2L_B}\right] \tag{4.23}$$

It is useful to determine the total currents within the device including the electron component. The total emitter and collector currents are found simply by adding the electron components to the hole components:

$$I_E = I_{En} + I_{Ep} \qquad I_C = I_{Cn} + I_{Cp} \tag{4.24}$$

Let us first consider the electron component to the emitter current. Generally, the excess electron concentration in the emitter (recall this is a pnp device) quasi-neutral region, $\delta n_E(x')$ is

$$\delta n_E(x') = A_1 e^{-\frac{x'}{L_E}} \tag{4.25}$$

where x' is the coordinate within the emitter as shown in Fig. 4.3 ($x' = 0$ defines the edge of the emitter–base depletion region on the emitter side of the junction) and L_E is the electron diffusion length within the p-type emitter. Notice that only the exponentially decaying solution exists since it is assumed that the emitter region width is sufficiently large that the excess electron concentration decays to zero. At $x' = 0$ the excess electron concentration is given by

$$\Delta n_E = n_{E0}\left(e^{qV_{EB}/kT} - 1\right) \tag{4.26}$$

where n_{E0} is the equilibrium electron concentration within the emitter. Recall for this example that the structure is a pnp transistor. Therefore, n_{E0} can be related to the emitter acceptor doping concentration, N_{aE}, as $n_{E0} = n_i^2/N_{aE}$, where n_i is the intrinsic carrier concentration within the emitter. The excess electron concentration in the emitter as a function of position x' is then

$$\delta n_E = \Delta n_E e^{-\frac{x'}{L_E}} = n_{E0}\left(e^{\frac{qV_{EB}}{kT}} - 1\right)e^{-\frac{x'}{L_E}} \tag{4.27}$$

The electron component of the emitter current is

$$I_{En} = -qAD_E\frac{d}{dx'}\delta n_E\big|_{x'=0} \tag{4.28}$$

where D_E is the electron diffusion coefficient within the emitter. Substituting into (4.28) the expression for δn_E and performing the derivative obtains

$$I_{En} = \frac{qAD_E}{L_E}n_{E0}\left(e^{\frac{qV_{EB}}{kT}} - 1\right) \tag{4.29}$$

A similar analysis can be performed to find the electron component of the collector current. In the collector region, we utilize the x'' coordinate axis as shown in Fig. 4.3. The origin, $x'' = 0$, occurs at the edge of the collector–base depletion region on the collector side of the junction. The excess electron concentration within the collector at $x'' = 0$, Δn_C, is

$$\Delta n_C = n_{C0}\left(e^{\frac{qV_{CB}}{kT}} - 1\right) \tag{4.30}$$

where n_{C0} is the equilibrium electron concentration within the p-type collector. The electron collector current, I_{Cn} is then

$$I_{Cn} = -\frac{qAD_C}{L_C}n_{C0}\left(e^{\frac{qV_{CB}}{kT}} - 1\right) \tag{4.31}$$

where L_C and D_C are the electron diffusion length and electron diffusion coefficient within the collector respectively. The total emitter and collector currents are determined by adding the electron components to the hole components. The total emitter current is then

$$I_E = qA\left[\left(\frac{D_E}{L_E}n_{E0} + \frac{D_B}{L_B}p_{B0}\coth\frac{W_B}{L_B}\right)\left(e^{\frac{qV_{EB}}{kT}} - 1\right)\right.$$
$$\left. - \left(\frac{D_B}{L_B}p_{B0}\operatorname{cosech}\frac{W_B}{L_B}\right)\left(e^{\frac{qV_{CB}}{kT}} - 1\right)\right] \tag{4.32}$$

The total collector current is

$$I_C = qA\left[\left(\frac{D_B}{L_B}p_{B0}\operatorname{cosech}\frac{W_B}{L_B}\right)\left(e^{\frac{qV_{EB}}{kT}} - 1\right)\right.$$
$$\left. - \left(\frac{D_C}{L_C}n_{C0} + \frac{D_B}{L_B}p_{B0}\coth\frac{W_B}{L_B}\right)\left(e^{\frac{qV_{CB}}{kT}} - 1\right)\right] \tag{4.33}$$

From knowledge of the currents, the performance parameters that characterize the device operation can be determined. There are four useful quantities that are typically used to describe the static characteristics of a BJT. These are for a pnp BJT:

(1) the emitter injection efficiency, γ, defined as the ratio of the emitter hole current to the total emitter current:

$$\gamma = \frac{I_{Ep}}{I_{Ep} + I_{En}} \tag{4.34}$$

(2) the base transport factor, α_T, defined as the ratio of the collector hole current to the injected emitter hole current:

$$\alpha_T = \frac{I_{Cp}}{I_{Ep}} \tag{4.35}$$

(3) the common base current gain, α_{dc}:

$$\alpha_{dc} = \gamma \alpha_T \tag{4.36}$$

(4) the common emitter current gain, β_{dc}:

$$\beta_{dc} = \frac{\alpha_{dc}}{1 - \alpha_{dc}} \tag{4.37}$$

Equation (4.37) is equivalent to (4.1). Therefore, the common emitter current gain can be determined from α_{dc} or from the ratio of the collector and base currents.

Let us consider each of the performance parameters in turn. Typically, these parameters are defined for the device while operating in the active mode. Under these conditions it is acceptable to approximate the currents by assuming that the emitter–base voltage is sufficiently large that $e^{q V_{EB}/kT} \gg 1$. Additionally, the collector–base junction is assumed to be sufficiently reverse biased such that the second term in (4.15) and (4.19) can be neglected. With these assumptions I_{Ep}, I_{Cp}, and I_{En} are

$$I_{Ep} \sim \frac{q A D_B}{L_B} p_{B0} e^{\frac{q V_{EB}}{kT}} \coth \frac{W_B}{L_B}$$

$$I_{Cp} = \frac{q A D_B}{L_B} p_{B0} e^{\frac{q V_{EB}}{kT}} \operatorname{cosech} \frac{W_B}{L_B} \tag{4.38}$$

$$I_{En} = \frac{q A D_E}{L_E} n_{E0} e^{\frac{q V_{EB}}{kT}}$$

The emitter injection efficiency is then

$$\gamma = \frac{I_{Ep}}{I_{Ep} + I_{En}} = \frac{\dfrac{p_{B0} D_B}{L_B} \coth \dfrac{W_B}{L_B}}{\dfrac{p_{B0} D_B}{L_B} \coth \dfrac{W_B}{L_B} + \dfrac{n_{E0} D_E}{L_E}} = \frac{1}{1 + \dfrac{D_E n_{E0} L_B}{L_E p_{B0} D_B} \tanh \dfrac{W_B}{L_B}} \tag{4.39}$$

The base transport factor is

$$\alpha_T = \frac{1}{\cosh \dfrac{W_B}{L_B}} \tag{4.40}$$

Inspection of (4.40) indicates that the base transport factor approaches 1 with decreasing base width. This is as expected. As the base width decreases most of the injected holes survive their flight through the base prior to recombining. Similarly, if the hole diffusion length within the base, L_B, increases, the holes diffuse further on average before recombining. Provided the base width is small with respect to L_B, then most of the injected holes are collected at the collector, implying a high base transport factor. The common base current gain is

$$\alpha_{dc} = \gamma \alpha_T = \cfrac{1}{\cosh \cfrac{W_B}{L_B} + \cfrac{D_E L_B n_{E0}}{D_B L_E p_{B0}} \sinh \cfrac{W_B}{L_B}} \tag{4.41}$$

Finally, the common emitter current gain is found using (4.41) as

$$\beta_{dc} = \cfrac{1}{\cosh \cfrac{W_B}{L_B} + \cfrac{D_E L_B n_{E0}}{D_B L_E p_{B0}} \sinh \cfrac{W_B}{L_B} - 1} \tag{4.42}$$

The common emitter current gain can be further simplified if it is assumed that W_B is much less than L_B. Under this assumption, the cosh and sinh functions can be approximated as

$$\cosh \frac{W_B}{L_B} \sim 1 + \frac{1}{2}\left(\frac{W_B}{L_B}\right)^2$$
$$\sinh \frac{W_B}{L_B} \sim \frac{W_B}{L_B} \tag{4.43}$$

Substituting the relations given by (4.43) into (4.42) obtains

$$\beta_{dc} = \cfrac{1}{\cfrac{D_E n_{E0} W_B}{D_B p_{B0} L_E} + \cfrac{1}{2}\left(\cfrac{W_B}{L_B}\right)^2} \tag{4.44}$$

Notice that the common emitter dc current gain depends strongly on the base width, W_B. As the base width decreases, the common emitter gain increases. The common emitter dc current gain can be rewritten in terms of the emitter and base doping concentrations, N_{aE} and N_{dB}, as follows. Using the argument on p. 85, we can express the equilibrium electron concentration within the p-type emitter, n_{E0}, and similarly the equilibrium hole concentration within the n-type base, p_{B0}, as

$$n_{E0} = \frac{n_{iE}^2}{N_{aE}} \qquad p_{B0} = \frac{n_{iB}^2}{N_{dB}} \tag{4.45}$$

where n_{iE} and n_{iB} are the intrinsic concentrations within the emitter and base regions respectively. In an ordinary BJT, $n_{iE} = n_{iB}$, since the emitter and base are made of the same semiconductor material. Using (4.45), β_{dc} becomes

$$\beta_{dc} = \cfrac{1}{\cfrac{D_E N_{dB} W_B}{D_B N_{aE} L_E} + \cfrac{1}{2}\left(\cfrac{W_B}{L_B}\right)^2} \tag{4.46}$$

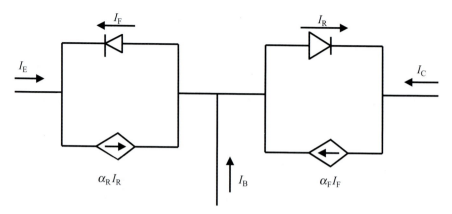

Figure 4.4 Ebers–Moll model equivalent circuit for an npn device.

If W_B is further assumed to be small with respect to L_B, then the last term in the denominator of (4.46) can be neglected. Under this assumption, (4.46) becomes

$$\beta_{dc} \sim \frac{1}{\dfrac{D_E N_{dB} W_B}{D_B N_{aE} L_E}} \tag{4.47}$$

Inspection of (4.47) indicates that β_{dc} is inversely proportional to the product of the base doping concentration and the base width. Thus β_{dc} is inversely proportional to the total number of donors within the base region which is often called the Gummel number.

One of the most useful models of a BJT is the Ebers–Moll model. Essentially, this model treats a BJT as two coupled diodes. The basic model is shown in Fig. 4.4. As can be seen from Fig. 4.4, the basic model consists of two diodes connected back to back with two current sources. The total emitter current is given by (4.32) as

$$I_E = q A \left[\left(\frac{D_E}{L_E} n_{E0} + \frac{D_B}{L_B} p_{B0} \coth \frac{W_B}{L_B} \right) \left(e^{\frac{q V_{EB}}{kT}} - 1 \right) \right.$$
$$\left. - \left(\frac{D_B}{L_B} p_{B0} \operatorname{cosech} \frac{W_B}{L_B} \right) \left(e^{\frac{q V_{CB}}{kT}} - 1 \right) \right] \tag{4.48}$$

and the total collector current is given by (4.33) as

$$I_C = q A \left[\left(\frac{D_B}{L_B} p_{B0} \operatorname{cosech} \frac{W_B}{L_B} \right) \left(e^{\frac{q V_{EB}}{kT}} - 1 \right) \right.$$
$$\left. - \left(\frac{D_C}{L_C} n_{C0} + \frac{D_B}{L_B} p_{B0} \coth \frac{W_B}{L_B} \right) \left(e^{\frac{q V_{CB}}{kT}} - 1 \right) \right] \tag{4.49}$$

Setting the collector–base voltage, V_{CB}, equal to zero in the expression for I_E, the emitter current becomes

$$I_E = q A \left[\left(\frac{D_E}{L_E} n_{E0} + \frac{D_B}{L_B} p_{B0} \coth \frac{W_B}{L_B} \right) \left(e^{\frac{q V_{EB}}{kT}} - 1 \right) \right] \tag{4.50}$$

which resembles an ideal diode equation for I_E. This is represented by the ideal diode in the emitter model of Fig. 4.4. We make the following definition:

$$I_{F0} = q A \left[\left(\frac{D_E}{L_E} n_{E0} + \frac{D_B}{L_B} p_{B0} \coth \frac{W_B}{L_B} \right) \right] \tag{4.51}$$

The ideal diode current component of the emitter current is called I_F and has a magnitude then of

$$I_F = I_{F0} \left(e^{\frac{q V_{EB}}{kt}} - 1 \right) \tag{4.52}$$

Similarly, if $V_{EB} = 0$ in (4.49), an ideal diode equation is obtained for the collector current, I_C. This is represented as an ideal diode in the collector model of Fig. 4.4. The ideal diode current component of the collector current is called I_R. With this assumption, the ideal diode current in the collector region becomes

$$I_R = q A \left[\left(\frac{D_C}{L_C} n_{C0} + \frac{D_B}{L_B} p_{B0} \coth \frac{W_B}{L_B} \right) \left(e^{\frac{q V_{CB}}{kT}} - 1 \right) \right] \tag{4.53}$$

We make the following definition:

$$I_{R0} = q A \left[\left(\frac{D_C}{L_C} n_{C0} + \frac{D_B}{L_B} p_{B0} \coth \frac{W_B}{L_B} \right) \right] \tag{4.54}$$

The corresponding ideal diode current, I_R, is then

$$I_R = I_{R0} \left(e^{\frac{q V_{CB}}{kT}} - 1 \right) \tag{4.55}$$

Using Fig. 4.4, the emitter and collector currents can be written from the node law as

$$I_E = \alpha_R I_R - I_F \tag{4.56}$$

and

$$I_C = \alpha_F I_F - I_R \tag{4.57}$$

But using (4.48) and (4.49), the emitter and collector currents can be written as

$$I_E = a_{11} \left(e^{\frac{q V_{EB}}{kT}} - 1 \right) + a_{12} \left(e^{\frac{q V_{CB}}{kT}} - 1 \right) \tag{4.58}$$

and

$$I_C = a_{21} \left(e^{\frac{q V_{EB}}{kT}} - 1 \right) + a_{22} \left(e^{\frac{q V_{CB}}{kT}} - 1 \right) \tag{4.59}$$

Comparing (4.56) and (4.57) with (4.58) and (4.59) and using the definitions of I_F and I_R, we can make the assignments,

$$\begin{aligned} a_{11} &= -I_{F0} & a_{12} &= \alpha_R I_{R0} \\ a_{21} &= \alpha_F I_{F0} & a_{22} &= -I_{R0} \end{aligned} \tag{4.60}$$

Consequently, we arrive at the Ebers–Moll equations which are

$$\begin{aligned} I_E &= -I_{F0} \left(e^{\frac{q V_{EB}}{kT}} - 1 \right) + \alpha_R I_{R0} \left(e^{\frac{q V_{CB}}{kT}} - 1 \right) \\ I_C &= \alpha_F I_{F0} \left(e^{\frac{q V_{EB}}{kT}} - 1 \right) - I_{R0} \left(e^{\frac{q V_{CB}}{kT}} - 1 \right) \end{aligned} \tag{4.61}$$

We further notice that the reciprocity of the device implies that $a_{12} = a_{21}$. Thus we have

$$\alpha_R I_{R0} = \alpha_F I_{F0} \tag{4.62}$$

The parameter α_F is defined as

$$\alpha_F = \frac{\dfrac{q A D_B}{L_B} \dfrac{p_{B0}}{\sinh(W_B/L_B)}}{q A \left[\dfrac{D_E}{L_E} n_{E0} + \dfrac{D_B}{L_B} p_{B0} \dfrac{\cosh(W_B/L_B)}{\sinh(W_B/L_B)} \right]} \tag{4.63}$$

Similarly, α_R is defined as,

$$\alpha_R = \frac{\dfrac{q A D_B}{L_B} \dfrac{p_{B0}}{\sinh(W_B/L_B)}}{q A \left[\dfrac{D_C}{L_C} n_{C0} + \dfrac{D_B}{L_B} p_{B0} \dfrac{\cosh(W_B/L_B)}{\sinh(W_B/L_B)} \right]} \tag{4.64}$$

With these definitions the total emitter and collector currents are recovered.

Example Problem 4.1 Use of Ebers–Moll equations

The collector current can be written as

$$I_C = \alpha_{dc} I_E + I_{CB0}$$

Use the Ebers–Moll equations to obtain the same result for active mode biasing. Under active biasing conditions $V_{EB} > 0$ and $V_{CB} < 0$. Thus

$$e^{\frac{q V_{CB}}{kT}} \ll 1$$

The Ebers–Moll equations are

$$I_E = I_{F0} \left(e^{\frac{q V_{EB}}{kT}} - 1 \right) - \alpha_R I_{R0} \left(e^{\frac{q V_{CB}}{kT}} - 1 \right)$$

and

$$I_C = \alpha_F I_{F0} \left(e^{\frac{q V_{EB}}{kT}} - 1 \right) - I_{R0} \left(e^{\frac{q V_{CB}}{kT}} - 1 \right)$$

Under active biasing the Ebers–Moll equations simplify to

$$I_E = I_{F0} \left(e^{\frac{q V_{EB}}{kT}} - 1 \right) + \alpha_R I_{R0}$$

$$I_C = \alpha_F I_{F0} \left(e^{\frac{q V_{EB}}{kT}} - 1 \right) + I_{R0}$$

Solve the first of the two simplified Ebers–Moll equations for the V_{EB} factor to give

$$\frac{I_E - \alpha_R I_{R0}}{I_{F0}} = \left(e^{\frac{q V_{EB}}{kT}} - 1 \right)$$

Substitute the above relationship into the second of the simplified Ebers–Moll equations to obtain,

$$I_C = \alpha_F I_{F0} \left(\frac{I_E - \alpha_R I_{R0}}{I_{F0}} \right) + I_{R0}$$

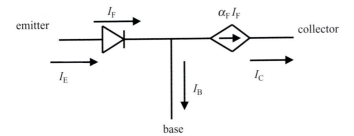

Figure 4.5 Ebers–Moll model for a pnp BJT when $V_{CB} = 0$. Notice that for a pnp BJT that the emitter current flows into the device, the collector and base currents flow out of the device.

which simplifies to

$$I_C = \alpha_F I_E + (1 - \alpha_F \alpha_R) I_{R0}$$

In the text we showed that $\alpha_F = \alpha_{dc}$. Thus I_C can be written as

$$I_C = \alpha_{dc} I_E + (1 - \alpha_{dc} \alpha_R) I_{R0}$$

Defining

$$I_{CB0} = (1 - \alpha_{dc} \alpha_R) I_{R0}$$

then the collector current is given as

$$I_C = \alpha_{dc} I_E + I_{CB0}$$

which is the desired result.

Example Problem 4.2 Determination of the emitter–collector voltage using the Ebers–Moll model

By using the Ebers–Moll model of a BJT, determine the emitter–collector voltage V_{EC}. Consider the operation of a BJT when the collector–base voltage is zero, $V_{CB} = 0$.

When $V_{CB} = 0$, $I_R = 0$. This implies physically that the collector–base junction is in equilibrium and that the device is operated at the border between the active and saturation modes. Clearly, if $V_{CB} = 0$, then $\alpha_R I_R = 0$. The resulting Ebers–Moll model for the BJT is shown in Fig. 4.5. The base current is readily given from the model as

$$I_B = I_E - I_C$$

but from inspection of Fig. 4.5 $I_E = I_F$ and $I_C = \alpha_F I_F$. Thus the base current can be written as

$$I_B = (1 - \alpha_F) I_F = (1 - \alpha_F) I_{F0} \left(e^{\frac{q V_{EB}}{kT}} - 1 \right)$$

V_{EC} can be expressed as

$$V_{EC} = V_{EB} - V_{CB}$$

But $V_{CB} = 0$. Therefore

$$V_{EC} = V_{EB}$$

The base current is then

$$I_B = (1 - \alpha_F)I_{F0}\left(e^{\frac{qV_{EC}}{kT}} - 1\right)$$

Solving for V_{EC} obtains

$$V_{EC} = \frac{kT}{q} \ln\left[1 + \frac{I_B}{(1 - \alpha_F)I_{F0}}\right]$$

4.2 Secondary effects in BJTs

We next examine some of the secondary effects within BJTs. The secondary effects that we consider are drift aided diffusion within the base, base narrowing, and avalanche breakdown. In Chapter 11, we consider the effect of a heterostructure on the performance of a BJT. Such devices are called heterostructure bipolar junction transistors, or HBTs.

4.2.1 Drift in the base region

The first effect that we consider is drift in the base region. There are two general methods by which field aided diffusion can be built into the base of a BJT. These are compositional grading and use of a nonuniform doping concentration. Compositional grading can be achieved using a ternary or quaternary alloy and slowly varying its composition to produce a built-in field. The details of this approach are treated in the book by Brennan and Brown (2002). Here we focus on the use of a nonuniform doping concentration to produce field aided diffusion within the base. Uniform doping within the base region is not always achievable. A typical doping profile for a pnp BJT that arises from successive dopant diffusions is shown in Fig. 4.6. As can be seen from Fig. 4.6 the doping concentration varies with distance from the surface within the emitter and base regions. The field aided diffusion arising from a nonuniform doping concentration is derived in Brennan and Brown (2002) for an npn BJT. Here we will consider only a pnp BJT.

The doping concentration is typically larger near the emitter side of the base than at the collector side of the base. A consequence of the difference in doping concentration across the base is that a built-in electric field is produced at the base. Let $N_B(x)$ be the base donor concentration. In equilibrium no current flows. Assuming that $n(x)$ is equal to $N_B(x)$, the current can be written as

$$I_n(x) = qA\mu_n N_B(x)\varepsilon(x) + qAD_n\frac{dN_B(x)}{dx} = 0 \tag{4.65}$$

Solving for the electric field and using the Einstein relation yields

$$\varepsilon(x) = -\frac{kT}{q}\frac{1}{N_B(x)}\frac{dN_B(x)}{dx} \tag{4.66}$$

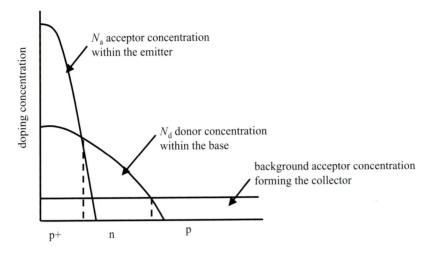

N_a acceptor concentration
within the emitter

N_d donor concentration
within the base

background acceptor concentration
forming the collector

p+   n   p

Position relative to the surface

Figure 4.6 Typical doping profile formed within a BJT showing the emitter, base, and collector regions.

The base doping concentration can be assumed to vary exponentially as

$$N_B(x) = N_B(x=0)e^{-\frac{ax}{W_B}} \tag{4.67}$$

where $x = 0$ is the edge of the base on the emitter side. W_B is the base width and a is defined as

$$a \equiv \ln\left[\frac{N_B(0)}{N_B(W_B)}\right] \tag{4.68}$$

The electric field given by (4.66) can be expressed as

$$\varepsilon(x) = -\frac{kT}{q}\frac{1}{N_B(x)}\frac{d}{dx}\left[N_B(0)e^{-\frac{ax}{W_B}}\right] \tag{4.69}$$

Taking the derivative and simplifying obtains

$$\varepsilon(x) = \frac{kT}{q}\frac{a}{W_B} \tag{4.70}$$

Thus the nonuniform doping concentration produces an electric field within the base quasi-neutral region. The field is positive and points from $x = 0$ to $x = W_B$. For a pnp device, the field acts to push the injected holes from the emitter through the base to the collector depletion region. As a result the built-in field due to the doping inhomogeniety aids hole transport through the base thus reducing the base transit time. The reduction in the base transit time acts to increase the cutoff frequency of the device. Thus for high frequency operation, it is desirable to use a nonuniform base doping concentration to produce field aided diffusion.

4.2.2 **Base narrowing or the Early Effect**

The base width of a BJT can be narrowed by the action of the reverse biased collector–base junction. When the reverse bias on the collector–base junction is sufficiently high and the base doping is sufficiently low, the width of the base region can be reduced by the encroachment of the collector–base depletion region. The variation of the base width with applied bias is called the Early Effect, base width modulation, or base narrowing. The effects of base narrowing can be understood as follows.

Consider the emitter current, I_E, given by (4.48) as

$$
I_E = q A \left[\left(\frac{D_E}{L_E} n_{E0} + \frac{D_B}{L_B} p_{B0} \coth \frac{W_B}{L_B} \right) \left(e^{\frac{q V_{EB}}{kT}} - 1 \right) \right.
$$
$$
\left. - \left(\frac{D_B}{L_B} p_{B0} \operatorname{cosech} \frac{W_B}{L_B} \right) \left(e^{\frac{q V_{CB}}{kT}} - 1 \right) \right] \tag{4.71}
$$

If the collector–base voltage is negative (reverse biased) and significantly larger than kT/q, then I_E becomes

$$
I_E = q A \left[\left(\frac{D_E}{L_E} n_{E0} + \frac{D_B}{L_B} p_{B0} \coth \frac{W_B}{L_B} \right) \left(e^{\frac{q V_{EB}}{kT}} - 1 \right) \right.
$$
$$
\left. + \left(\frac{D_B}{L_B} p_{B0} \operatorname{cosech} \frac{W_B}{L_B} \right) \right] \tag{4.72}
$$

Equation (4.72) can be further simplified if the electron contribution to the emitter current is neglected, implying that the first term in (4.72) is negligible. With this assumption, the assumption that the base width is much less than the base diffusion length, and the fact that under active biasing the emitter–base voltage is much greater than kT/q, (4.72) becomes

$$
I_E \sim q A \frac{D_B}{W_B} p_{B0} e^{\frac{q V_{EB}}{kT}} \tag{4.73}
$$

As the magnitude of the collector–base voltage increases, the base width decreases. Therefore at fixed V_{EB}, the emitter current, I_E increases with increasing magnitude of V_{CB}.

The base narrowing also affects the value of β_{dc}. The value of β_{dc} is given by (4.46) to be

$$
\beta_{dc} = \frac{1}{\dfrac{D_E n_{E0} W_B}{D_B p_{B0} L_E} + \dfrac{1}{2} \left(\dfrac{W_B}{L_B} \right)^2} \tag{4.74}
$$

Inspection of (4.74) shows that as W_B decreases β_{dc} increases. Thus the common emitter current gain increases with increasing collector–base bias.

The limiting situation of the base narrowing effect is called punchthrough. In punchthrough the reverse bias of the collector–base junction is increased to the point that the depletion region extends entirely through the base region. In other words, the base region becomes completely depleted and there no longer exists a base quasi-neutral region. Often the device breaks down through carrier multiplication via impact ionization before punchthrough is reached. In the next section we discuss avalanche breakdown initiated by impact ionization.

4.2.3 Avalanche breakdown

As in a diode, a high reverse bias of the collector–base junction can lead to avalanche breakdown. Avalanche breakdown is triggered by impact ionization events that occur when the electrons or holes reach energies within the conduction or valence bands respectively that are greater than or equal to the energy band gap. A single impact ionization event produces one electron–hole pair at a time. If the initiating carrier is an electron, following the impact ionization event there are two electrons, the primary and secondary electrons, and one hole, a secondary hole. These carriers continue their flights through the depletion region, holes moving towards the p-side and electrons towards the n-side. If the electric field within the depletion region is very high, then during the course of their flights through the depletion region the primary and secondary carriers can be heated to high energy and additional impact ionization events can occur. In this way, the number of carriers increases and the current is multiplied. Ultimately, if the multiplication rate is high the device will break down and it is said to be in avalanche breakdown.

In the common base configuration the maximum collector–base voltage with the emitter open circuited is referred to as BV_{CBO}. It is important to recognize that in the common base configuration the behavior of the collector–base junction is essentially like that of a simple p–n junction diode. Therefore, as the collector–base voltage increases to breakdown the collector current increases rapidly. Provided avalanche breakdown occurs prior to punchthrough, the maximum voltage at the output of the device is just BV_{CBO}. The multiplication factor, M, can be expressed using an empirically based formula in terms of BV_{CBO} and the collector–base voltage, V_{CB}, as

$$M = \frac{1}{1 - \left(\dfrac{V_{CB}}{BV_{CBO}}\right)^n} \tag{4.75}$$

where n is a constant and varies with the junction. The common base current gain, α_{dc}, is defined from

$$I_C = \alpha_{dc} I_E + I_{CBO} \tag{4.76}$$

where I_{CBO} is the collector current that flows when $I_E = 0$. In the presence of multiplication, the emitter current is multiplied by M. Therefore, the common base current gain can be rewritten as

$$\alpha'_{dc} = M \alpha_{dc} \tag{4.77}$$

In (4.77), α'_{dc} is the common base current gain in the presence of multiplication.

The effect of multiplication on the common emitter configuration can be understood from the definition of β_{dc} and (4.77). The common emitter current gain is given by (4.37) as

$$\beta_{dc} = \frac{\alpha_{dc}}{1 - \alpha_{dc}} \tag{4.78}$$

Substituting for α_{dc} its value under multiplication, α'_{dc}, β_{dc} becomes

$$\beta'_{dc} = \frac{M\alpha_{dc}}{1 - M\alpha_{dc}} \tag{4.79}$$

Notice that the common emitter current gain goes to infinity when the product $M\alpha_{dc} = 1$ which is the breakdown condition.

The common emitter breakdown voltage is always less than the common base breakdown voltage, i.e., $BV_{CEO} < BV_{CBO}$. This can be understood as follows. For simplicity let us consider a pnp structure. Generally, the emitter–collector voltage, V_{EC}, can be related to the emitter–base voltage, V_{EB}, and the collector–base voltage, V_{CB}, as

$$V_{EC} = V_{EB} - V_{CB} \tag{4.80}$$

Under active mode biasing the emitter–base junction is forward biased. The emitter–base voltage is much smaller than the reverse biased collector–base voltage. Therefore, $V_{EC} \sim -V_{CB}$. One would then expect that $BV_{CEO} \sim BV_{CBO}$. However, this is not the case. BV_{CEO} is always less than BV_{CBO}. As the V_{CB} becomes appreciable, but still less than BV_{CBO}, some of the holes that traverse the collector–base depletion region gain sufficient energy to impact ionize. Secondary electrons and holes are produced following these impact ionization events. The secondary holes along with the primary holes continue their flight through the depletion region and are collected. The secondary electrons, however, move in the opposite direction and enter the base. In the common emitter configuration, the base current is held constant. Therefore, the additional electrons that enter the base from the base–collector depletion region produced from impact ionization events cannot exit the base through the base contact. Instead, they must be back injected into the emitter. This leads to an increase in the emitter hole current, I_{Ep}. As discussed in Section 4.1, $\tau_p/\tau_B = \beta_{dc}$ holes pass through the base for each electron that enters the base resulting in a large increase in the collector current. Therefore, the hole collector current increases from the transistor gain caused by feedback from the impact ionization process. As a result, the device breaks down at a lower voltage than in the common base configuration.

Quantitatively, the common emitter and common base breakdown voltages can be related as follows (see Brennan and Brown (2002), Section 4.4),

$$BV_{CEO} = BV_{CBO}\beta_{dc}^{-\frac{1}{n}} \tag{4.81}$$

In (4.81), n is a parameter that is larger than 1 and generally has a value between 3 and 6 depending upon the semiconductor material used in the junction. Inspection of (4.81) shows that BV_{CEO} is generally significantly less than BV_{CBO}.

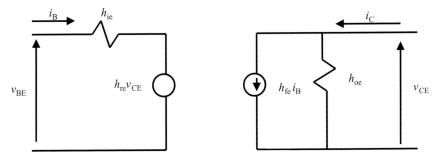

Figure 4.7 Low frequency equivalent circuit for a BJT based on the h parameters discussed in the text.

4.3 **High frequency operation of a BJT**

The common emitter configuration is the most generally employed bipolar transistor connection used in radio frequency (RF) amplifiers. For low frequency operation a BJT can be represented as a simple two-port network characterized by four network parameters. These parameters are called hybrid or h parameters since they are not all dimensionally the same. The h parameters are used to relate the independent variables i_B, the small signal base current, and v_{CE}, the small signal collector–emitter voltage, to the dependent variables v_{BE} and i_C as (see Brennan and Brown (2002), Section 4.5):

$$
\begin{aligned}
v_{BE} &= h_{ie}i_B + h_{re}v_{CE} \\
i_C &= h_{fe}i_B + h_{oe}v_{CE}
\end{aligned}
\tag{4.82}
$$

Using the relationships given in (4.82) a low frequency equivalent circuit for the BJT can be constructed as shown in Fig. 4.7. Using the loop law, the first of equations (4.82) can be used to construct the equivalent circuit for v_{BE} shown in Fig. 4.7. Similarly, from the node law, the second of equations (4.82) can be used to construct the equivalent circuit for i_C. The quantity h_{fe} is equal to the common emitter current gain β_{dc} at low frequencies.

At higher frequency operation, the simple equivalent circuit shown in Fig. 4.7 is no longer useful since it neglects junction capacitances that become more important with increasing frequency. The most commonly employed circuit model for the common emitter configuration at higher frequencies is the hybrid-pi model shown in Fig. 4.8. The quantities used in the hybrid-pi model are defined as follows. The transconductance, g_m, is

$$
g_m \equiv \frac{dI_C}{dV_{EB}}
\tag{4.83}
$$

The transconductance can be shown to be given by (see Brennan and Brown (2002))

$$
g_m = \frac{qI_C}{kT}
\tag{4.84}
$$

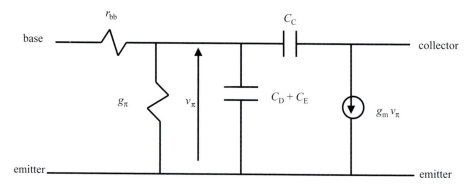

Figure 4.8 Hybrid-pi model equivalent circuit for a BJT.

The small signal input conductance, g_π, is defined as

$$g_\pi \equiv \frac{d I_B}{d V_{EB}} \tag{4.85}$$

which becomes

$$g_\pi = \frac{q I_B}{kT} = \frac{g_m}{h_{fe}} \tag{4.86}$$

The quantity r_{bb} is the base resistance. It specifies the voltage drop between the base contact and the active base region. Finally, the capacitances in the model represent the collector capacitance, C_C, the emitter capacitance, C_E, and the diffusion capacitance, C_D. Using these parameters and the hybrid-pi equivalent circuit model, the frequency behavior of the BJT can be analyzed.

The two most important figures of merit that describe the high frequency behavior of a BJT are the cutoff frequency f_t and the maximum frequency of oscillation f_{max}. The cutoff frequency is defined as the frequency at which the common emitter short-circuit current gain is equal to 1. Similarly f_{max} is defined as the frequency at which the unilateral power gain of the transistor approaches unity. The derivations of both f_t and f_{max} are beyond the level of this book. The interested reader is referred to the book by Brennan and Brown (2002) for the derivation of these quantities. The cutoff frequency for a BJT is

$$f_t = \frac{1}{2\pi} \frac{g_m}{C} \tag{4.87}$$

where C is the sum of C_C, C_E, and C_D. Alternatively, the cutoff frequency can be written in terms of the base transit time, τ_B, (see Brennan and Brown (2002)) as

$$f_t = \frac{1}{2\pi \tau_B} \tag{4.88}$$

Equation (4.88) provides only an estimate of the cutoff frequency since it is based on the approximate hybrid-pi model. A more accurate expression for the cutoff frequency is obtained using the emitter–collector delay time, τ_{EC}, as

$$f_t = \frac{1}{2\pi \tau_{EC}} \tag{4.89}$$

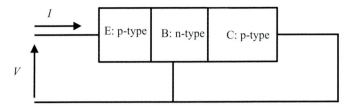

Figure 4.9 Figure for Problem 4.4. The BJT structure is a pnp silicon device.

The emitter–collector delay time consists of four quantities. These are: the emitter junction charging time, τ_E, the base transit time τ_B, the collector–base junction depletion layer transit time, τ_{dC}, and the collector junction charging time τ_C. Though (4.89) is more accurate than (4.88), the base transit time is typically the dominant parameter and thus the use of (4.88) is justified.

The maximum frequency of operation of the BJT is

$$
f_{max} = \sqrt{\frac{f_t}{8\pi r_{bb} C_C}} \tag{4.90}
$$

where r_{bb} is the base resistance. Inspection of (4.90) shows that f_{max} increases with increasing cutoff frequency and reduction in the base resistance. The base resistance in turn depends upon the base doping concentration. To maximize f_{max}, it is important to use a very high base doping concentration and a concomitant low base resistance.

Problems

4.1 Consider a p^+np BJT. Assume that the base is open-circuited and that a current, I, flows into the emitter. If the voltage across the emitter–collector junction is V, determine the emitter–base voltage, V_{EB}. Assume that the collector–base junction is reverse biased, $V_{CB} \ll 0$, and that the magnitude of the voltage, V, is much greater than kT/q.

4.2 Consider a silicon pnp BJT in which the base width is 0.5 μm, the base doping is 5×10^{16} cm^{-3}, the collector doping concentration is 10^{15} cm^{-3} and the collector base bias is -10 V. Determine whether it is in punchthrough conditions.

4.3 Determine the reverse bias voltage required for the following BJT to be in punchthrough conditions. The BJT is a pnp silicon structure with a base doping of 10^{15} cm^{-3}, collector doping of 10^{17} cm^{-3}, and a base width of 0.5 μm. Comment on whether the device will break down prior to punchthrough or not.

4.4 Determine an expression for the base current in the BJT shown in the circuit configuration given in Fig. 4.9.

4.5 If the breakdown electric field in the collector–base junction is 400 kV/cm in the common base configuration, determine the breakdown voltage of the BJT, BV_{CBO}. The device is a pnp silicon BJT with a base doping concentration of 10^{16} cm^{-3}, and a collector doping concentration of 10^{17} cm^{-3}.

4.6. A BJT is configured in the common base mode. Determine the breakdown voltage of the BJT, BV_{CBO} given the following information: $\varepsilon_{\text{crit}} = 400 \text{ kV/cm}$, $N_{\text{aE}} = 10^{19} \text{ cm}^{-3}$, $N_{\text{dB}} = 10^{17} \text{ cm}^{-3}$, $N_{\text{aC}} = 10^{17} \text{ cm}^{-3}$, $\kappa(\text{Si}) = 11.8$, $D_{\text{E}} = 38.8 \text{ cm}^2/\text{s}$, $D_{\text{B}} = 11.6 \text{ cm}^2/\text{s}$, $L_{\text{E}} = 0.2 \text{ μm}$, and $W_{\text{B}} = 0.1 \text{ μm}$. Assume that the value of n in (4.81) is 4. Neglect the built-in voltage of the collector–base junction.

4.7. If the built-in field in the base of a BJT is made via a doping concentration gradient, determine the doping concentration profile needed to produce a field of 100 kV/cm in a silicon pnp device. Assume that the doping concentration within the base region has an exponential dependence and that the base width is 0.1 μm.

4.8. Consider a silicon pnp BJT. Neglect I_{Cn} with respect to I_{Cp} and calculate the cutoff frequency if the total capacitance of the structure is 1.0×10^{-8} F. Given: $A = 1 \text{ mm}^2$, $D_{\text{B}} = 11.6 \text{ cm}^2/\text{s}$, $L_{\text{B}} = 0.15 \text{ μm}$, $W_{\text{B}} = 0.1 \text{ μm}$, $N_{\text{dB}} = 10^{17} \text{ cm}^{-3}$, $V_{\text{EB}} = 0.7 \text{ V}$, $V_{\text{CB}} = -10.0 \text{ V}$ (reverse biased), and $kT = 0.0259 \text{ eV}$.

4.9. Consider a silicon pnp BJT. Determine the dc current gain, β_{dc}, of the device given the following information: $N_{\text{aE}} = 10^{18} \text{ cm}^{-3}$, $N_{\text{dB}} = 10^{16} \text{ cm}^{-3}$, $\tau_{\text{B}} = \tau_{\text{E}} = 10^{-6} \text{ s}$, $\mu_{\text{n}} = 1500 \text{ cm}^2/(\text{V s})$, $\mu_{\text{h}} = 450 \text{ cm}^2/(\text{V s})$, $W_{\text{B}} = 0.5 \text{ μm}$, and $kT = 0.0259 \text{ eV}$.

4.10. Determine the base transport factor for a silicon pnp BJT given that the hole lifetime in the base is 10^{-6} s, the hole mobility is $450 \text{ cm}^2/(\text{V s})$, and $kT = 0.0259 \text{ eV}$. Assume that the base width is
   (a) $W_{\text{B}} = 0.1 \text{ μm}$
   (b) $W_{\text{B}} = 5.0 \text{ μm}$

5

JFETs and MESFETs

In this chapter we discuss the two field effect transistors (FETs) that utilize either p–n junctions or Schottky barriers to provide gating action. These devices are junction field effect transistors, JFETs, and metal semiconductor field effect transistors, MESFETs. Both types of device are based on the field effect, except that they use different mechanisms to provide gating action. In the field effect, the conductivity of the underlying semiconductor can be altered by the presence of an electric field, in this case, produced by the application of a gate bias voltage. The gate typically lies on the top of the device and thus the conductivity can be controlled from the top of the device, making the contacting of the device relatively straightforward. For this reason high levels of integration have been achieved using field effect devices. The most important of these devices is the metal oxide semiconductor field effect transistor, MOSFET, which will be discussed in detail in Chapter 6.

5.1 JFET operation

A JFET uses two p–n junctions on the top and bottom of the device as gates as shown in Fig. 5.1. The top and bottom gates are biased in the same manner. Notice that the gate p–n junctions are p$^+$–n junctions. Thus most of the depletion region width forms within the more lightly doped n-region rather than the p$^+$-region. The depletion regions from the top and bottom gates encroach on the conducting channel formed in the n-type material between the two gates. If the gates are reverse biased, then the depletion region will increase in width below each gate. The depletion region will again lie almost entirely within the n-type part of the gate p–n junctions. Thus with increasing reverse bias the gate depletion regions enlarge further constricting the channel. In order to produce a current, the drain to source voltage is positive so that electrons are accelerated from the source to the drain by the source-drain field. Let us consider two modes of operation. These are with and without an applied gate bias.

Case 1: No applied gate bias, $V_G = 0$. The gates are grounded and the source to drain voltage is increased starting from $V_D = 0$. At $V_D = 0$ the device is in equilibrium and the p$^+$–n junctions at the gate region (since $V_G = 0$) are at zero bias. Therefore only the zero bias depletion region widths extend into the n-region of the device. As the drain voltage is increased, electrons begin to flow from the source to the drain with a concomitant current flow from the drain to the source. The current flows through the undepleted n-type region between the two gate depletion regions. This region is called the channel. For small drain bias the channel behaves as a simple resistor with a relatively small resistance. As the drain voltage increases the depletion region increases

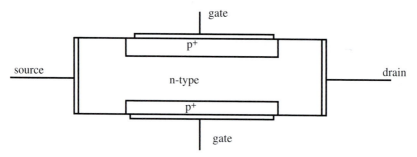

Figure 5.1 JFET, showing the two gates formed on top and bottom of the device. The n-type channel lies between the depletion regions formed by the two p$^+$–n junction gates. The JFET is a unipolar device in that only electrons contribute to the current flow in the structure.

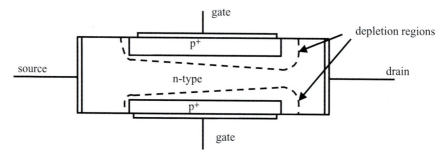

Figure 5.2 JFET with grounded gates and source but with a positive potential applied to the drain. Notice that the depletion regions are wider near the drain end of the device than near the source end. Consequently, the n-type conducting channel is more constricted near the drain producing a higher channel resistance at the drain end of the device. For either gate the depletion region forms mainly within the n-type layer since the doping is far lighter in the n-layer than in the p$^+$-layers.

near the drain end of the device as shown in Fig. 5.2. The channel at the drain end becomes further constricted raising its resistance. As a result the slope of the current–voltage characteristic becomes smaller. The effect of the drain voltage on the device can be understood as follows. Near the drain end of the device the n-region of the channel is at a higher potential than the p$^+$-regions of the two gates. This is due to the fact that the drain has a positive voltage applied to it. So near the drain the n-type region has a positive voltage applied to it relative to the p$^+$-regions of the gates resulting in reverse biasing the gate junctions. As discussed in Chapter 3, the reverse bias acts to increase the depletion region widths of the two gate p$^+$–n junctions. Since the gate junctions are asymmetrically doped the depletion region extends mainly within the n-type layer. Consequently, near the drain the depletion region corresponding to each gate widens and the channel becomes constricted by the encroaching gate depletion regions from both the top and bottom gates. Since the source is grounded like the gates there is little potential difference near the source region of the device. Thus the

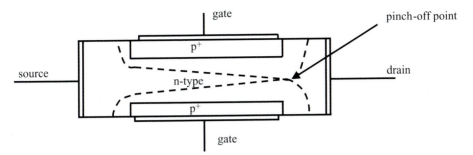

Figure 5.3 JFET biased into pinch off. In this case the gate bias is zero for both the top and bottom gates while a positive voltage is applied to the drain. Ultimately, the drain bias is sufficiently large that the two depletion regions formed by the top and bottom gates touch. When this occurs the device is said to be in pinch off and the conducting channel near the drain end of the device is pinched off. The depletion regions are asymmetric since the source and gates are grounded while the drain is positively biased.

p^+–n junctions near the source are less reverse biased. The effective channel width is significantly wider near the source than near the drain as shown in Fig. 5.2.

As the drain voltage is further increased the channel region near the drain becomes even more constricted since the reverse bias is greater. Finally, the drain voltage becomes sufficiently large that both gates are highly reverse biased and the two gate depletion regions near the drain extend out into the channel and ultimately touch. When this happens the channel is said to be pinched off by the two touching gate depletion regions at the drain end. This condition is called pinch off and is shown in Fig. 5.3.

The significance of pinch off is that the resistance of the device increases greatly since the conducting channel no longer extends completely from the source to the drain end of the device. Consequently, the slope of the current–voltage characteristic becomes progressively smaller until the characteristic ultimately flattens out as shown in Fig. 5.4. Once the device reaches pinch off, the drain current saturates, i.e., it no longer increases with increasing gate voltage. The saturation of the drain current can be physically understood as follows. After pinch off further increase in the drain voltage simply extends the width of the pinched-off region of the channel. If the channel length is much larger than the pinch-off region then to a good approximation the device is essentially unchanged. In essence the distance from the source to the pinch-off point has changed only negligibly and thus essentially the same boundary conditions exist. In other words, the source to pinch-off point separation is essentially the same length and exactly the same voltages are present throughout the device as at the start of saturation. Hence the drain current should be the same since the shape of the conducting region and the potential applied across it do not change. As a result the same current flows leading to saturation. It is important to recognize that upon pinching off the channel a constant current continues to flow. At first one might expect that the current is shut off. This is incorrect. The current saturates since during pinch-off operation in order to induce and maintain the voltages present in the device a current must flow. The voltage

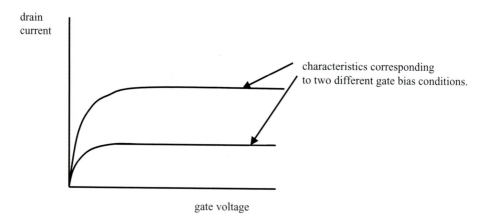

Figure 5.4 Current–voltage characteristic for a JFET showing two different gate biases. The current is said to saturate once it no longer changes with increasing voltage. Saturation occurs once the channel is pinched off.

upon saturation applied to the drain is called the saturation voltage and is written as $V_{D,sat}$.

Case 2: Gate biasing. When the gate bias is less than zero, $V_G < 0$, the p^+–n regions forming the gates are of course reverse biased. The gate bias acts to increase the depletion region widths resulting in a net decrease in the channel width. Therefore the channel resistance is larger for a given drain voltage when the gates are reverse biased, $V_G < 0$, than when $V_G = 0$. The channel will become pinched off at a lower source-drain voltage since it is narrower to begin with.

The gate voltage at which pinch-off occurs at zero drain voltage is called V_p. The JFET is shown under both zero gate bias and reverse gate bias at different drain voltages in Figs. 5.5 and 5.6. As can be seen from the figures initially the channel width is large and a linear relationship exists between the current and the voltage. As the drain voltage increases the channel constricts near the drain end of the device. The action of a reverse bias on the gates is simply to constrict the channel at lower drain voltage as shown in Fig. 5.6. Since the channel is more constricted the resistance increases and the slope of the current–voltage characteristic becomes smaller (lower conductance, higher resistance). Finally, at sufficiently high drain voltage the channel is pinched off near the drain end leading to current saturation. In the next section we discuss the qualitative behavior of MESFETs.

5.2 MESFET and MODFET operation

One of the primary disadvantages of JFETs is that they require top and bottom p^+–n junction gates. This complicates their fabrication since they cannot have a strictly two-dimensional planar structure. An alternative device that functions very much like a JFET is a MESFET. This uses a reverse biased Schottky barrier as a gate instead of two reverse biased p^+–n junctions. MESFETs are typically used for compound

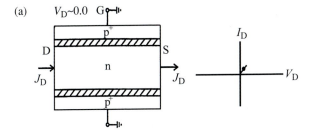

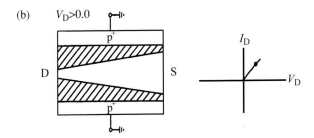

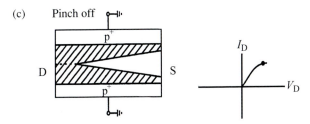

Figure 5.5 JFET device and corresponding current–voltage characteristic under three different conditions, (a) low drain voltage, (b) significant drain voltage and (c) pinch off. In all three cases the gates are grounded. The dashed region represents the depletion region of the device (Brennan, 1999).

semiconductors such as GaAs and InP. These materials have higher mobilities and hence a higher frequency of operation than Si. Additionally, unlike Si, GaAs and InP do not have native oxides so metal–insulator–semiconductor field effect transistors, MISFETs, cannot be readily made as they are for Si. We will discuss the most popular MISFET, the silicon based MOSFET in Chapter 6.

A typical MESFET structure is shown in Fig. 5.7. Inspection of Fig. 5.7 shows that the MESFET device is a two-dimensional planar structure; all of the contacts are on the top surface. The structure consists of a thin epitaxial layer of conducting semiconductor material grown on top of a semi-insulating substrate. The conduction all occurs within the top epitaxial layer. Therefore, the gate depletion region formed under the Schottky contact and the semi-insulating layer constrict the conducting channel.

Schottky barrier formation was discussed at length in Chapter 3. As we found in Chapter 3 a Schottky barrier will form between a metal and semiconductor if the

(a)

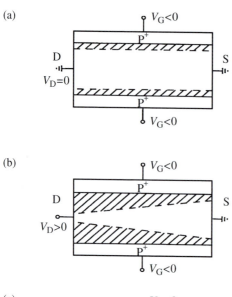

(b)

(c)

Figure 5.6 JFET device with a nonzero gate bias applied under three different conditions. These are: (a) zero drain bias, (b) applied drain bias such that the device is below pinch-off and (c) after pinch off. The dashed region represents the depletion region (Brennan, 1999).

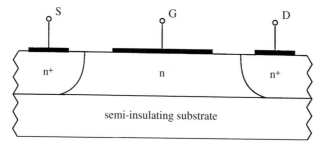

Figure 5.7 Simple MESFET device structure showing the source, gate, and drain contacts labeled as S, G, and D respectively. The active region is n-type and is grown on top of the semi-insulating substrate. The current flow in the device occurs in the active n-type layer between the source and drain (Brennan, 1999).

metal work function is greater than the semiconductor work function. Then, in order to equilibrate the Fermi levels when the metal and semiconductor are brought into contact, electrons must transfer from the semiconductor into the metal, leaving behind positive space charge within the semiconductor and producing a net negative charge on the surface of the metal. The net negative charge on the metal acts to repel electrons from the surface causing the bands to bend away from the interface. As we have mentioned electrons roll downhill in potential energy diagrams so the bands bend within the semiconductor such that the electrons will roll away from the interface. The potential barrier blocking electron flow from the metal into the semiconductor is called the Schottky barrier and is essentially independent of the applied bias.

On the semiconductor side of the junction the potential barrier is highly bias dependent. In equilibrium it is equal to the built-in potential of the junction. Under forward bias, application of a positive voltage to the metal with respect to the semiconductor, the potential barrier on the semiconductor side is lowered. Since a flux due to diffusion flows from the semiconductor into the metal, the diffusion current increases exponentially with the lowered potential barrier since many more carriers have sufficient kinetic energy in the forward direction to overcome the potential barrier. Reverse bias, application of a negative bias on the metal with respect to the semiconductor, increases the potential barrier on the semiconductor side thus choking off the diffusion current leaving only the voltage independent reverse bias thermionic emission current. As in a p^+–n junction under reverse bias the depletion region widens extending mainly within the n-type region of the device. Increased reverse bias of the Schottky barrier further increases the depletion region width. The conducting layer is made relatively thin and there is basically no conduction within the highly resistive semi-insulating substrate. Therefore, the Schottky barrier depletion region and the semi-insulating layer bound the conducting channel on the top and bottom respectively. Application of a reverse bias on the Schottky barrier gate acts to increase the depletion region constricting the channel formed between the top depletion region and the bottom semi-insulating substrate material. As in the JFET device, sufficient reverse bias of the Schottky barrier gate can ultimately lead to pinching off the conducting channel. As a result the current saturates once pinch off is achieved.

The behavior of a MESFET under different gate bias conditions is shown in Figs. 5.8 and 5.9. Figure 5.8 shows the behavior of the device under zero gate bias and a positive drain voltage. As is the case for a JFET, the application of a positive drain voltage (needed to produce a current flow in the device) acts to reverse bias the gate junction near the drain. Consequently, the width of the depletion region increases near the drain end of the device constricting the conducting channel formed between the depletion region and the semi-insulating substrate. As the channel width decreases the source-drain resistance increases causing the slope of the current–voltage characteristic to decrease as shown in Fig. 5.8. Since the source is grounded it is at the same potential as the Schottky barrier gate. Therefore, the depletion region near the source is essentially unchanged from its equilibrium value.

Further rise in the drain current increases the Schottky barrier depletion region width until it finally touches the semi-insulating substrate as shown in Fig. 5.9. At this point

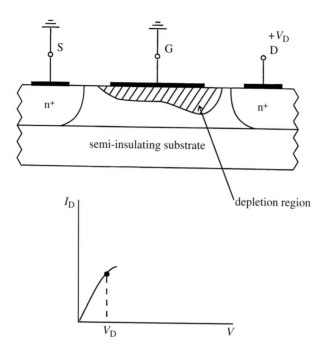

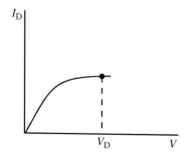

Figure 5.8 Sketch of a MESFET under a drain bias of V_D below pinch off and its corresponding current–voltage characteristic. Notice that the current has not yet saturated (Brennan, 1999).

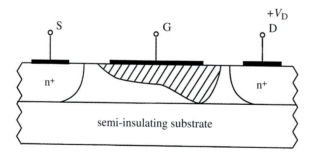

Figure 5.9 Sketch of a MESFET biased at pinch off along with its corresponding current–voltage characteristic. In this case, the current has saturated (Brennan, 1999).

the conducting channel is pinched off and the current saturates; the current no longer increases with increasing drain voltage. As is the case of the JFET once pinch off is reached, further increase in the drain voltage results in an increased pinch-off region. If the device has a long channel length, then the increase in the pinch-off region is small with respect to the channel length. Thus the device essentially has the same dimensions and voltages and current saturation occurs.

Though MESFETs are especially important for compound semiconductors, most compound semiconductor FETs are somewhat different from the MESFET structure outlined above. An improved version of a MESFET is currently the most ubiquitous FET made from compound semiconductors. This structure is called a MODFET, for modulation doped field effect transistor. We will discuss different MODFET structures in detail in Chapter 11. Nevertheless, we introduce the topic here and discuss the basic operation of a MODFET device. Let us first consider the processes of modulation doping and then we will present a short discussion of MODFETs.

Modulation doping is achieved by doping a wide band gap semiconductor layer grown adjacent to a narrow gap semiconductor. The two dissimilar semiconductor materials when grown on top of one another are called a heterostructure. A more detailed discussion of heterostructures and their related devices is presented in Chapter 11. Here we will give only a very brief discussion. An example of a material system used for modulation doping is the AlGaAs/GaAs system. AlGaAs has an energy gap larger than that of GaAs. The mobility of intrinsic GaAs is very high and thus GaAs is suitable for high frequency device applications. However, if the GaAs is doped, then the dopants will create scattering centers, thereby reducing the mobility. An additional advantage of the GaAs/AlGaAs materials system is that these two materials are lattice matched. This means that the lattice constants of the two materials are essentially identical. As a result when the two materials are grown one on top of the other the two lattices match quite closely and few dangling bonds are present. Alternatively, if the materials have different lattice constants then if the top layer is thin enough it will adjust to the lattice constant of the bottom layer. In so doing however, the top layer becomes strained (see Brennan and Brown (2002) for a detailed discussion of strain in semiconductors). If the top layer is thicker, beyond a critical thickness, the top layer relaxes and there are dislocations formed at the heterointerface if the lattice constants of the two semiconductors are different.

The processes of modulation doping can be understood as follows. The wide band gap semiconductor (in this case AlGaAs) is doped n-type while the narrow gap material (in this case GaAs) is undoped as shown in Fig. 5.10(a). When the two semiconductors are placed into contact in equilibrium the Fermi level must be flat everywhere as shown in Fig. 5.10(b). In addition, far from the metallurgical junction the bulklike properties of the materials are recovered. Finally, at the junction itself the bands must connect but there is a discontinuity in the conduction and valence bands due to the difference in the energy gaps between the two constituent semiconductors. In order for the Fermi levels to align in equilibrium, electrons must flow from the n-type doped wide band gap semiconductor into the intrinsic narrow gap semiconductor. This leaves behind positively charged ionized donors in the AlGaAs and adds negative charge to the GaAs.

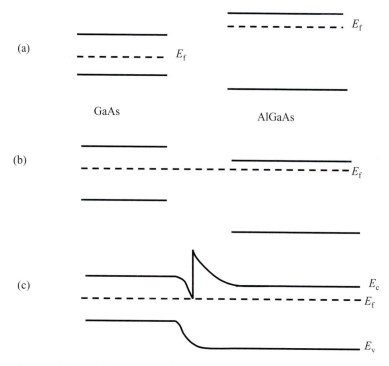

Figure 5.10 Band diagrams showing modulation doping in a heterostructure: (a) two materials apart (notice that the wide band gap material is doped n-type); (b) application of the first two rules for junction formation in equilibrium; and (c) resulting band diagram for the system.

At the heterointerface electrons within the GaAs layer are attracted to the positive charge within the AlGaAs layer and thus the bands bend down towards the interface as shown in Fig. 5.10(c). In other words the bands bend such that electrons within the GaAs will roll downhill towards the interface from the attraction due to the positive charge in the AlGaAs.

The specific advantage of modulation doping is that the ionized donors and free electrons are spatially separated from one another. Notice that the free electrons are transferred into the intrinsic GaAs leaving behind ionized donors within the AlGaAs. The electron transfer occurs as mentioned above to equilibrate the Fermi level throughout the junction. As a result the free electron carriers within the GaAs are spatially separated from the ionized donors. The importance of spatially separating the carriers and donor ions can be understood as follows. As discussed in Chapter 2, the mobility is a function of the scattering rate. The higher the scattering rate, the lower the mobility. A high mobility is desired for high frequency device operation. The scattering rate is increased in the presence of donor ions since they provide ionized impurity scattering centers. These centers act to scatter the electron carriers. The ionized impurity scattering rate depends upon the spatial separation of the carriers and ions. The smaller that separation the higher the ionized impurity scattering rate is (essentially ionized impurity scattering is Coulombic in nature and thus varies inversely with the separation

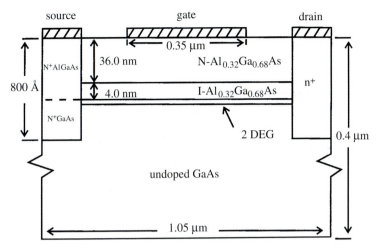

Figure 5.11 MODFET or high electron mobility transistor (HEMT) device structure formed using GaAs and AlGaAs. The substrate consists of an undoped GaAs layer. The two-dimensional electron gas, 2 DEG, is formed at the GaAs/AlGaAs heterointerface as shown in the diagram. The AlGaAs layers comprise an intrinsic spacer layer followed by an n-type doped AlGaAs layer.

distance, see Brennan (1999), Chapter 9). By spatially separating the carriers and the ionized donors the ionized impurity scattering rate is greatly reduced thus maintaining a high mobility within the GaAs. Thus modulation doping provides a means of increasing the free carrier concentration without compromising the mobility.

A MODFET utilizes modulation doping in a FET structure. The basic structure is sketched in Fig. 5.11. As can be seen from the figure, a MODFET is similar to a MESFET in that again there are a source, a drain, and a gate. The conduction pathway is different though between the two devices. In a MESFET the channel is in the bulk epitaxial layer, sandwiched between the Schottky gate depletion layer and the semi-insulating substrate. The channel is constricted by the application of a reverse gate bias. In a MODFET the conducting channel is at the AlGaAs/GaAs interface. At the interface free electron carriers accumulate from the modulation doping and it is these carriers that conduct the current in the device. As mentioned above modulation doping provides a means of increasing the free carrier concentration without increasing the ionized impurity scattering rate and thereby sacrificing the mobility. Further inspection of Fig. 5.11 shows that there is an undoped 4.0 nm thick AlGaAs layer separating the GaAs channel region and the doped AlGaAs layer. The undoped layer is called the spacer layer and acts to further separate the conducting channel from the doped layer. As a result the ionized impurity scattering rate is further lowered and the mobility becomes close to that of intrinsic GaAs making the device very useful for high frequency operation. As we will see in Chapter 11, due to their high frequency operation, MODFETs are extensively used in wireless telecommunications systems.

The channel region of the device is a two-dimensional electron gas, 2 DEG as marked in Fig. 5.12. The full details of a two-dimensional electron gas are discussed

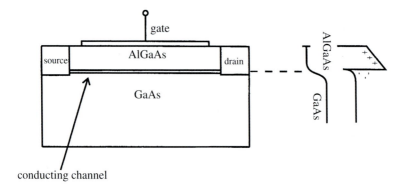

conducting channel

Figure 5.12 MODFET or HEMT device structure and the corresponding energy band structure. Notice that the bands within the GaAs bend down at the heterointerface due to the presence of the ionized donor atoms in the AlGaAs layer. The conducting channel is formed near the interface and depending upon the amount of band bending a two-dimensional electron gas can form (Brennan, 1999).

in Brennan and Brown (2002) and will not be repeated here. Suffice it to say that due to the strong band bending in the GaAs channel at the interface, spatial quantization effects occur (see Brennan (1999)). Thus instead of a three-dimensional bulk system the band bending results in quantum confinement in the direction perpendicular to the interface such that only two spatial degrees of freedom of motion exist. This is the origin of the name two-dimensional electron gas. Figure 5.12 clearly shows the device and the band bending at the AlGaAs/GaAs interface. As can be seen from Fig. 5.12 the electrons are confined between the conduction band edge discontinuity on one side and the band bending on the other producing a "notch." Since the notch has a size comparable to the electron wavelength the electron behaves like a wave and exhibits quantum mechanical effects. One result is that the allowed energy levels of the electron within the notch are quantized. In other words, the allowed energies no longer form a continuum with all values of kinetic energy allowed, but instead the energies are discrete, quantized levels.

In Chapter 11 we will return to our discussion of MODFETs when we address power transistors and amplifiers. In that chapter we will examine the figures of merit that make a transistor suitable for high power, high frequency operation as is needed in power amplifiers. In addition we will examine various MODFET device designs and compare their performance.

5.3 **Quantitative description of JFETs and MESFETs**

In this section we develop a quantitative description of the processes of a JFET and extend it to include a MESFET. These two types of devices operate in essentially the same manner and thus can be quantitatively described in a similar way. The geometry of a JFET showing the different variables of importance is sketched in Fig. 5.13. The

(a)

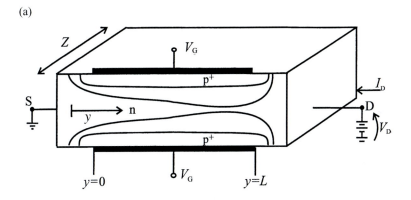

(b)

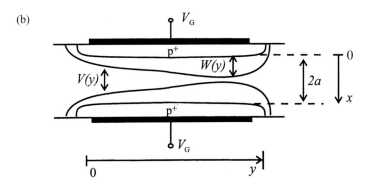

Figure 5.13 JFET device design used in determining the quantitative performance of the device. Part (a) shows the three-dimensional structure while part (b) shows the cross-sectional layout and corresponding parameters of interest in the device (Brennan, 1999).

direction along the channel is the y-direction in the figure, while the z-direction is along the width and the x-direction points from one depletion region to the next. In deriving the expression for the drain current we make several simplifying assumptions. These are:

(i) The p^+–n junctions are assumed to be abrupt and are uniformly doped.
(ii) The device is symmetric about the plane $x = a$.
(iii) The current flow occurs within only the undepleted portion of the channel.
(iv) Breakdown conditions are avoided within each junction.
(v) The voltage drop between the contacts and between the source region and $y = 0$ and the drain region and $y = L$ is negligible.

We further assume that both the gate and drain voltages are below pinch off. This condition implies

$$0 \leq V_D \leq V_{D,\text{sat}} \qquad 0 \geq V_G \geq V_p \tag{5.1}$$

We start the derivation with the expression for the electron current density:

$$\vec{J}_n = q\mu_n n\vec{\varepsilon} + qD_n\vec{\nabla}n \tag{5.2}$$

Within the conducting channel of the JFET the electron concentration is given as N_d, where it is assumed that all of the donors are ionized. The diffusion current is neglected with respect to the drift current. Though the diffusion current can be important, its inclusion necessitates a numerical solution which is beyond the level of this book. Inclusion of only the drift current results in a closed form analytical result for the drain current. Neglecting the diffusion current, (5.2) becomes

$$|\vec{J}_n| = J_{ny} = q\mu_n N_d\varepsilon_y = -q\mu_n N_d\frac{dV}{dy} \tag{5.3}$$

where V is the voltage drop along the y-direction from the source end to the drain end of the channel. Any gate leakage current is neglected. Thus there are no sources or sinks of current in the device. What flows in the source must flow out the drain. The current through any cross-sectional plane within the channel must be equal to I_D, the drain current. I_D is

$$I_D = -\iint J_{ny}\,dx\,dz \tag{5.4}$$

The negative sign in (5.4) arises since I_D flows in the $-y$-direction. Recall that the electron flow is from the source to the drain in the positive y-direction but the current flow is in the opposite direction, from the drain to the source. Therefore, the current flow is in the $-y$-direction. The cross-sectional area in which the current flows is in the x–z plane where Z is the width of the device and $2a$ is the distance between the top and bottom metallurgical junctions forming the gates. The only real difference between the expressions for the drain current for a JFET and a MESFET is that the distance between the Schottky barrier gate and the semi-insulating substrate is a instead of $2a$. The current I_D can be written as

$$I_D = -Z\int_{W(y)}^{2a-W(y)} J_{ny}\,dx \tag{5.5}$$

The bounds in the integral reflect the boundaries of the conducting channel in the device as shown in Fig. 5.13(b). The integral over x is only over the width of the channel in accordance with assumption (iii) listed above, i.e., the current flow only occurs within the undepleted portion of the channel until pinch off is achieved. The integral in (5.5) can be simplified by recognizing that by symmetry the integral is equal to one bounded from $W(y)$ to a multiplied by 2. Equation (5.5) is then

$$I_D = 2Z\int_{W(y)}^{a} q\mu_n N_d\frac{dV}{dy}\,dx \tag{5.6}$$

All of the quantities in (5.6) are independent of x so the integration can be easily performed to yield

$$I_D = 2qZ\mu_n N_d\frac{dV}{dy}x\Big|_{W(y)}^{a} \tag{5.7}$$

which becomes

$$I_D = 2qZ\mu_n N_d \frac{dV}{dy}[a - W(y)] \tag{5.8}$$

or

$$I_D = 2qZ\mu_n N_d \frac{dV}{dy}a\left[1 - \frac{W(y)}{a}\right] \tag{5.9}$$

But $W(y)$ is a function of V as well as y so it can be written as $W(V)$. Integrating both sides of (5.9) with respect to y obtains

$$\int I_D dy = 2qZ\mu_n N_d a \int \left[1 - \frac{W(V)}{a}\right]dV \tag{5.10}$$

The bounds on the integrals in (5.10) are found as follows. The channel extends from $y = 0$ to $y = L$, which are the bounds on the integral on the left hand side of (5.10). The corresponding bounds on the integral on the right hand side are the potential at $y = 0$, which is the source voltage. In most situations the source is grounded which is assumed here. Therefore, at $y = 0$ the potential, V, is zero. At $y = L$ at the drain end of the channel the voltage is the drain voltage, V_D. Recall that we assume that there is no voltage drop from the drain contact to the point $y = L$.

In order to integrate (5.10), it is necessary to obtain an expression for $W(V)$, the depletion region width. We make a further assumption called the gradual channel approximation. The gradual channel approximation assumes that the rate of change of the potential V is small in the y-direction with respect to the x-direction. When the channel length L is very much larger than a, $L \gg a$, it is reasonable to assume that the rates of change of V and ε are small in the y-direction as compared to the x-direction. Notice that when $V_G = 0$, it is clear that the same potential is dropped from the center of the channel region to the p^+ contact as is dropped from 0 to L in the y-direction along the channel. Based on these observations the depletion region width can be approximated to be one-dimensional. Thus $W(V)$ is given as

$$W(V) = \sqrt{\frac{2\kappa}{qN_d}(V_{bi} - V)} \tag{5.11}$$

where $N_a \gg N_d$, which holds for a p^+–n junction. V is the applied potential drop across the p^+–n junction.

The potential drop across the gate junction can be determined with help from Fig. 5.14. Inspection of Fig. 5.14 shows that the voltage drop across the junction, V, is simply equal to $V_G - V(y)$. Therefore, (5.11) for $W(V)$ can be written as

$$W(V) = \sqrt{\frac{2\kappa}{qN_d}[V_{bi} + V(y) - V_G]} \tag{5.12}$$

The channel becomes pinched off when the width of the depletion region, $W(V)$, is equal to a. Pinch off can occur under different conditions. One condition is when the drain voltage is zero, $V_D = 0$, and the gate voltage, V_G, is equal to the pinch-off voltage,

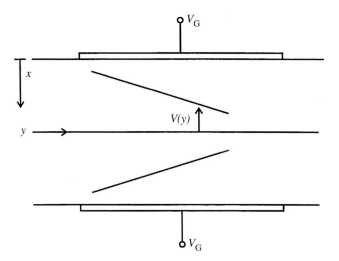

Figure 5.14 Sketch of the channel region of a JFET device. The slanted lines represent the edge of the depletion region beneath each gate (Brennan, 1999).

V_p. The pinch-off voltage V_p is defined as the gate voltage necessary to pinch off the channel when $V_D = 0$. Under these biasing conditions, $W(V(y) = 0)$ and $V_G = V_p$, a is given as

$$a = \sqrt{\frac{2\kappa}{qN_d}(V_{bi} - V_p)} \tag{5.13}$$

Dividing (5.12) by (5.13) obtains

$$\frac{W(V)}{a} = \sqrt{\frac{V_{bi} + V(y) - V_G}{V_{bi} - V_p}} \tag{5.14}$$

The drain current can now be written as

$$I_D = \frac{2q\mu_n Z N_d a}{L} \int_0^{V_D} \left[1 - \sqrt{\frac{V_{bi} + V(y) - V_G}{V_{bi} - V_p}} \right] dV \tag{5.15}$$

To perform the integration in (5.15) we observe that it is of the form:

$$\int 1 - \sqrt{c_1 + c_2 u} \, du \tag{5.16}$$

where c_1 and c_2 are

$$c_1 = \frac{V_{bi} - V_G}{V_{bi} - V_p} \qquad c_2 = \frac{1}{V_{bi} - V_p} \tag{5.17}$$

Note that the second term in (5.16) integrates out to be

$$\int \sqrt{c_1 + c_2 u} \, du = \frac{2}{3c_2} \sqrt{(c_1 + c_2 u)^3} \tag{5.18}$$

The integral of the square root term within the bracketed integrand in (5.15) becomes

$$\frac{2}{3}(V_{bi} - V_p)\left[\left(\frac{V_{bi} - V_G}{V_{bi} - V_p} + \frac{V_D}{V_{bi} - V_p}\right)^{\frac{3}{2}} - \left(\frac{V_{bi} - V_G}{V_{bi} - V_p}\right)^{\frac{3}{2}}\right] \tag{5.19}$$

Finally, the drain current becomes

$$I_D = \frac{2q\mu_n Z N_d a}{L}\left\{V_D - \frac{2}{3}(V_{bi} - V_p)\left[\left(\frac{V_{bi} - V_G + V_D}{V_{bi} - V_p}\right)^{\frac{3}{2}} - \left(\frac{V_{bi} - V_G}{V_{bi} - V_p}\right)^{\frac{3}{2}}\right]\right\} \tag{5.20}$$

which holds below pinch off. To approximate the pinch-off voltage, V_p, we can neglect the built-in voltage in (5.13) and solve to give

$$V_p = \frac{q N_d a^2}{2\kappa} \tag{5.21}$$

Next we consider the operation of the JFET after pinch off in the saturation region. The drain voltage, V_D, is greater than the drain saturation voltage, $V_{D,sat}$ in saturation. The saturation drain current is referred to as $I_{D,sat}$. To obtain the saturation drain current, we set V_D equal to $V_{D,sat}$, giving

$$I_{D,sat} = \frac{2q\mu_n Z N_d a}{L}\left\{V_{D,sat} - \frac{2}{3}(V_{bi} - V_p)\left[\left(\frac{V_{bi} - V_G + V_{D,sat}}{V_{bi} - V_p}\right)^{\frac{3}{2}}\right.\right.$$
$$\left.\left. - \left(\frac{V_{bi} - V_G}{V_{bi} - V_p}\right)^{\frac{3}{2}}\right]\right\} \tag{5.22}$$

At the drain end of the channel, W approaches a when $V(L) = V_{D,sat}$.

The device can become pinched off in several different ways. Let us consider two different cases. The first case is when the drain voltage is zero and pinch-off occurs strictly from the gate voltage. Thus $V_D = 0$ and $V_G = V_p$. The width of the depletion region at pinch off is of course simply equal to a. Thus

$$W(V) = a = \sqrt{\frac{2\kappa}{q N_d}(V_{bi} - V_p)} \tag{5.23}$$

The second case is when both drain and gate voltages are applied to the device. Again $W(V) = a$ but the relationship for $W(V)$ is given by (5.12). Denoting $V(y = L)$ as $V_{D,sat}$ as done above, $W(V)$ becomes

$$W(V) = a = \sqrt{\frac{2\kappa}{q N_d}(V_{bi} + V_{D,sat} - V_G)} \tag{5.24}$$

Equations (5.23) and (5.24) must of course be equal. Thus

$$\sqrt{\frac{2\kappa}{q N_d}(V_{bi} + V_{D,sat} - V_G)} = \sqrt{\frac{2\kappa}{q N_d}(V_{bi} - V_p)} \tag{5.25}$$

Simplifying, (5.25) becomes

$$V_{D,sat} = |V_G - V_p| \tag{5.26}$$

Using (5.26) the saturation drain current can now be determined as

$$I_{D,sat} = \frac{2q\mu_n Z N_d a}{L} \left\{ (V_G - V_p) - \frac{2}{3}(V_{bi} - V_p) \left[\left(\frac{V_{bi} - V_G + V_G - V_p}{V_{bi} - V_p} \right)^{\frac{3}{2}} - \left(\frac{V_{bi} - V_G}{V_{bi} - V_p} \right)^{\frac{3}{2}} \right] \right\} \tag{5.27}$$

which simplifies to

$$I_{D,sat} = \frac{2q\mu_n Z N_d a}{L} \left\{ (V_G - V_p) - \frac{2}{3}(V_{bi} - V_p) \left[1 - \left(\frac{V_{bi} - V_G}{V_{bi} - V_p} \right)^{\frac{3}{2}} \right] \right\} \tag{5.28}$$

Often (5.28) is approximated by an empirical relationship called the square law. The square law gives for the saturated drain current

$$I_{D,sat} = I_{D0} \left(1 - \frac{V_G}{V_p} \right)^2 \tag{5.29}$$

where I_{D0} is equal to the drain saturation current at zero gate bias.

The above results hold as well for MESFETs as for JFETs. The only difference is that the distance in the x-direction between the metallurgical junction and the semi-insulating layer is a rather than $2a$ as assumed for a JFET. If this definition is utilized the results are simply a factor of 2 smaller for the MESFET than the JFET. Thus I_D and $I_{D,sat}$ are given by (5.20) and (5.28) except divided by 2. Most modern MESFETs have a short-channel length. In Chapter 7 we will discuss challenges that short-channel devices encounter. Much of the analysis of a short-channel device requires a two-dimensional potential profile which is typically determined numerically. The interested reader is referred to the book by Brennan (1999) for a discussion of numerical models of a MESFET.

The resulting current vs. voltage characteristic for a JFET is shown in Fig. 5.15. Different values of the gate voltage correspond to different lines in the characteristic. At low drain voltages the current–voltage characteristic is linear. As the drain voltage increases the current saturates in each case.

Example Problem 5.1 Determination of the pinch-off voltage in a MESFET

Derive an expression for the pinch-off voltage in a MESFET with an active layer thickness of t.

In a MESFET the depletion region is produced by a reverse biased Schottky barrier that is used to form the gate of the device. The conducting channel is formed between the edge of the depletion region on one side and the edge of the active layer thickness/insulator layer interface on the other side. As the reverse bias is increased,

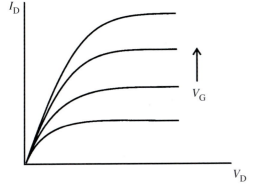

Figure 5.15 Current–voltage characteristic for a JFET device. Notice that there is a family curves, each corresponding to a different gate bias (Brennan, 1999).

the depletion region expands choking off the conducting channel. Ultimately the conducting channel is choked off at a single point as shown in Fig. 5.9. The applied bias to the Schottky gate contact that produces pinch-off is called the pinch-off voltage. Therefore, we start with the expression for the depletion region width of a Schottky barrier. This is given by (5.11) as

$$W = \sqrt{\frac{2\kappa(V_{bi} - V)}{qN_d}}$$

where V_{bi} is the built-in voltage. At pinch-off the depletion region width is equal to the width of the active region of the MESFET, t. Thus the expression for the depletion region can be rewritten as

$$t = \sqrt{\frac{2\kappa(V_{bi} - V)}{qN_d}}$$

Solving for $(V_{bi} - V)$ obtains

$$(V_{bi} - V) = \frac{qN_d t^2}{2\kappa}$$

Neglecting the built-in voltage the magnitude of the pinch-off voltage is

$$V = \frac{qN_d t^2}{2\kappa}$$

which is essentially the same expression as that for a JFET except that we use t, the thickness of the active layer of the MESFET, instead of a, half the separation between the two p–n junction gates.

Example Problem 5.2 Calculation of the saturation current of silicon JFET

Determine the magnitude of the saturation current in the following silicon JFET assuming the empirical square law. Given: $N_d = 10^{17}$ cm^{-3}, $I_{D0} = 0.50$ mA, $a = 0.1$ μm, $\kappa = 11.9$ (8.85×10^{-14} F/cm). Consider operation at 2 V for V_G.

The square law for the current saturation is given by (5.29) as,

$$I_{D,sat} = I_{D0} \left(1 - \frac{V_G}{V_p} \right)^2$$

Everything is known except V_p. The magnitude of the pinch-off voltage can be obtained from

$$V_p = \frac{q N_d a^2}{2\kappa}$$

Substituting in the parameters of the problem gives

$$V_p = \frac{1.6 \times 10^{-19}\,\text{C}(10^{17}\,\text{cm}^{-3})(0.1 \times 10^{-4}\,\text{cm})^2}{2(11.8)(8.85 \times 10^{-14}\,\text{F/cm})}$$
$$= 0.766\,\text{V}$$

The saturated drain current can now be found as follows

$$I_{D,sat} = 0.50\,\text{mA} \left(1 - \frac{2\,\text{V}}{0.766\,\text{V}} \right)^2$$
$$= 1.30\,\text{mA}$$

Example Problem 5.3 JFET circuit design

Consider the simple JFET circuit made with an n-channel device shown in Fig. 5.16. The value of V_{DD} is 10 V, $V_p = -4$ V, and the transistor is in saturation. Use the simple square law formula given by (5.29) with a value of 2 mA for I_{D0}. Determine the values of the drain and source resistances if the circuit has a drain current of 1 mA and a drain voltage $V_D = 5.0$ V.

Given that the transistor is biased into saturation, the drain current is

$$I_{D,sat} = I_{D0} \left(1 - \frac{V_G}{V_p} \right)^2$$

Substituting in for each term obtains

$$1\,\text{mA} = 2\,\text{mA} \left(1 - \frac{V_G}{-4\,\text{V}} \right)^2$$

which gives

$$V_G = -1.17\,\text{V}$$

From inspection of the figure, the current through the source resistor

$$I_D = -\frac{V_G}{R_S}$$

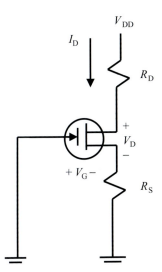

Figure 5.16 Simple JFET circuit used in Example Problem 5.3. The object of the problem is to determine the values of the drain and source resistances given that the device is biased into saturation.

The value of the source resistance is then

$$R_S = -\frac{-1.17\,\text{V}}{1\,\text{mA}} = 1.17\,\text{k}\Omega$$

Using the loop law we have

$$V_{DD} = I_D R_D + V_D + I_D R_S$$

Solving for R_D obtains

$$R_D = \frac{V_{DD} - V_D - I_D R_S}{I_D}$$

Substituting for each value yields

$$R_D = \frac{10.0\,\text{V} - 5.0\,\text{V} - (1\,\text{mA})(1.17\,\text{k}\Omega)}{1\,\text{mA}} = 3.83\,\text{k}\Omega$$

5.4 **Small signal model for a JFET**

The small signal behavior of a JFET can be modeled as a two-port network as shown in Fig. 5.17. The inputs to the device are the gate and source while the outputs are the drain and source as shown in the figure. The gate source is a reverse biased p^+–n junction diode. At low frequencies a reverse biased diode behaves much like an open circuit since there is very little current flow. Therefore, the input to the JFET is an open circuit. At the output the dc drain current, I_D, is a function of V_D and V_G. The dc drain current can be written then as

$$I_D = I_D(V_D, V_G) \tag{5.30}$$

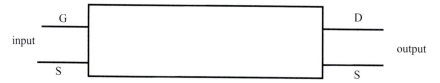

Figure 5.17 Two-port representation of the small signal model of a JFET. The inputs are the gate and source while the outputs are the drain and source as shown in the diagram.

The total drain current is equal to the sum of the ac and dc parts. Calling i_d the ac drain current the total current is

$$I_D(V_D + v_d, V_G + v_g) \tag{5.31}$$

and thus

$$I_D(V_D + v_d, V_G + v_g) = I_D(V_D, V_G) + i_d \tag{5.32}$$

or equivalently

$$i_d = I_D(V_D + v_d, V_G + v_g) - I_D(V_D, V_G) \tag{5.33}$$

Using a first order Taylor series expansion for $I_D(V_D + v_d, V_G + v_g)$ obtains

$$I_D(V_D + v_d, V_G + v_g) = I_D(V_D, V_G) + \left.\frac{\partial I_D}{\partial V_D}\right|_{V_G} v_d + \left.\frac{\partial I_D}{\partial V_G}\right|_{V_D} v_g \tag{5.34}$$

Substituting (5.34) into (5.33) yields

$$i_d = \left.\frac{\partial I_D}{\partial V_D}\right|_{V_G} v_d + \left.\frac{\partial I_D}{\partial V_G}\right|_{V_D} v_g \tag{5.35}$$

The transconductance and drain or channel conductance are defined as

$$g_m = \left.\frac{\partial I_D}{\partial V_G}\right|_{V_D} \qquad \text{(transconductance)} \tag{5.36}$$

and

$$g_d = \left.\frac{\partial I_D}{\partial V_D}\right|_{V_G} \qquad \text{(drain conductance)} \tag{5.37}$$

The small signal drain current can now be written as

$$i_d = g_d v_d + g_m v_g \tag{5.38}$$

The corresponding equivalent circuit can be constructed such that the input is an open circuit and the drain current is given by (5.38). The simplest equivalent circuit that satisfies these criteria is shown in Fig. 5.18. The equivalent circuit of Fig. 5.18 holds for low frequency operation where the capacitances can be ignored. As can be seen from the figure, the input is an open circuit while the output is such that the ac drain current is equal to $g_d v_d + g_m v_g$.

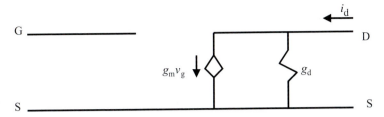

Figure 5.18 Low frequency model of a JFET. Notice that the input is an open circuit corresponding to the reverse biased source to gate junctions. The output part of the model is consistent with the expression for the ac drain current derived in the text.

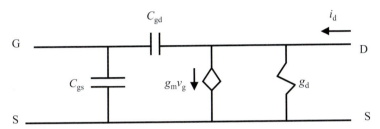

Figure 5.19 High frequency equivalent circuit for a JFET. Two capacitances are added to the circuit, C_{gd} and C_{gs}, for the gate drain and gate source respectively.

A more complete small signal model is shown in Fig. 5.19. This model holds for high frequency operation. Two capacitances have been added to the equivalent circuit. These are the gate drain and gate source capacitances, C_{gd} and C_{gs}, respectively.

One of the more important figures of merit that characterizes the high frequency performance of a FET is the cutoff frequency. The cutoff frequency is defined as the frequency at which the magnitude of the forward current gain is unity with the output short circuited. We will have more to say about the cutoff frequency in later chapters, particularly Chapters 6 and 11. The cutoff frequency can be described as (see Problem 5.9),

$$f_t = \frac{g_m}{2\pi C_G} \tag{5.39}$$

where C_G is the gate capacitance and g_m is the transconductance. Inspection of (5.39) shows that high frequency behavior occurs at a high transconductance and low gate capacitance.

Example Problem 5.4 Calculation of the transconductance of a JFET

Determine the transconductance of a JFET above pinch off.

This can be done as follows. The saturation drain current of a simple JFET is given by (5.28) as

$$I_{D,sat} = \frac{2qZ\mu N_d a}{L} \left\{ V_G - V_p - \frac{2}{3}(V_{bi} - V_p)\left[1 - \left(\frac{V_{bi} - V_G}{V_{bi} - V_p}\right)^{\frac{3}{2}}\right]\right\}$$

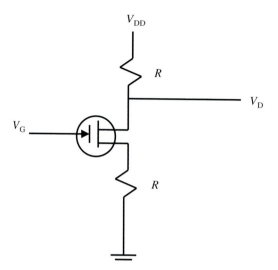

Figure 5.20 JFET circuit configuration for Problem 5.1. The resistances, R, have identical value. V_{DD} is given as 5 V.

For simplicity in writing, let us define G_0 as

$$G_0 = \frac{2q Z \mu N_d a}{L}$$

V_p is the pinch-off voltage of the JFET, L the channel length, a the half-width of the channel in equilibrium, W the width of the device, N_d the doping concentration, V_{bi} the built-in voltage of the p$^+$–n junctions forming the gates, and μ the electron mobility. The transconductance is obtained by taking the derivative of $I_{D,sat}$ with respect to V_G as

$$g_m = G_0 \left[1 - \left(\frac{V_{bi} - V_G}{V_{bi} - V_p} \right)^{\frac{1}{2}} \right]$$

Clearly, from the above expression, g_m is certainly less than or equal to G_0. Therefore, we can approximate g_m by G_0 to give an upper bound on the transconductance. Therefore

$$g_m \leq G_0 = \frac{2q W \mu N_d a}{L}$$

Problems

5.1 Given the following information about a JFET determine whether it is in saturation or not: $N_d = 5 \times 10^{17}$ cm^{-3}, $a = 0.1 \times 10^{-4}$ cm, Si based with $\kappa = 11.8$ (8.85 $\times$ 10^{-14} F/cm). The JFET is connected to the circuit shown in Fig. 5.20. Assume that the resistances are the same, that V_{DD} is equal to 5 V, $V_G = -2$ V and that there is no voltage drop in the JFET itself.

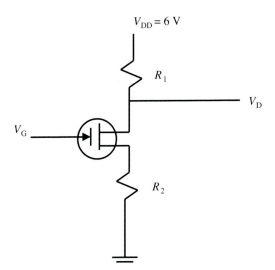

Figure 5.21 JFET circuit configuration for Problem 5.4. The two resistances shown in the figure are different. The value of V_{DD} is 6 V.

5.2 An ideal n-channel JFET has $N_a = 10^{18}$ cm^{-3}, $N_d = 10^{16}$ cm^{-3}, a channel length of 25 μm, $Z = 250$ μm, $a = 1$ μm, an electron mobility, $\mu_n = 1300$ cm^2/V s. Given that $kT = 0.0259$ eV, $q = 1.6 \times 10^{-19}$C, $\varepsilon_0 = 8.85 \times 10^{-14}$ F/cm, $\kappa_s = 11.8$, and $n_i = 10^{10}$ cm^{-3}, calculate the following:
(a) the pinch-off voltage, V_p. Neglect the built-in voltage in determining V_p;
(b) the built-in voltage;
(c) the channel conductance, g_d, defined as

$$g_d = \frac{\partial I_D}{\partial V_D} \text{ at constant } V_G$$

when the device is biased above the pinch-off condition.

5.3 An n-channel Si JFET has $N_a = 10^{19}$ cm^{-3}, $N_d = 10^{16}$ cm^{-3}, a (half the channel width) $= 1$ μm, $L = 25$ μm, $Z = 1$ mm. Assume $kT = 0.0259$ eV, $n_i = 1.0 \times 10^{10}$ cm^{-3}, the relative dielectric constant $= 11.8$, the free space dielectric constant $= 8.85 \times 10^{-14}$ F/cm, $q = 1.6 \times 10^{-19}$ C, and $\mu_n = 1300$ cm^2/(V s). Determine the following:
(a) the built-in voltage, V_{bi};
(b) the magnitude of the pinch-off voltage, V_p. Neglect the built-in voltage;
(c) the breakdown voltage of one of the junctions if the critical electric field is 300 kV/cm. Neglect the built-in voltage in this calculation.

5.4 Find the magnitude of the gate voltage on the JFET within the circuit shown in Fig. 5.21 such that the device is just biased into saturation. Assume that $V_{DD} = 6$ V, $R_1 = 5$ kΩ and $R_2 = 10$ kΩ. The JFET has the following material and fabrication parameters: $N_d = 7.5 \times 10^{17}$ cm^{-3}, $a = 10^{-5}$ cm, $\kappa = 11.8$ (8.85×10^{-14} F/cm). Assume that there is no voltage drop in the JFET itself.

5.5 A JFET has $N_a = 10^{19}$ cm^{-3}, $N_d = 10^{16}$ cm^{-3}, $a = 0.1$ μm, $L = 10$ μm and $Z = 1$ mm. Assume the device is Si with an intrinsic concentration of 10^{10} cm^{-3} and a relative dielectric constant of 11.8. Assume also that in saturation the square law can be used and that the value of I_{D0} is 5.0 μA. Determine the following:
(a) the built-in voltage;
(b) the pinch-off voltage;
(c) the drain saturation voltage if the gate voltage is -2 V;
(d) the drain current if $V_D = 1$ V and the gate bias is -2 V. Is the device in saturation?

5.6 An n-channel Si JFET has $Z = 24$ μm, $L = 4$ μm, $a = 1.2$ μm, and $N_a = 10^{19}$ cm^{-3}, $N_d = 5 \times 10^{15}$ cm^{-3}, and $\mu_n = 1200$ cm^2/(V s). Assume the gate voltage is -3 V. Determine:
(a) the drain current at saturation;
(b) the drain voltage at saturation.

5.7 A Si JFET is to be constructed that has an n-type doping of 10^{16} cm^{-3}. Determine the maximum width of the device as measured from the metallurgical junctions of the top and bottom gate electrodes. This quantity is simply $2a$. Neglect the built-in voltage in the junction and note that the breakdown voltage for a p$^+$–n junction doped 10^{16} cm^{-3} in the n-region is 55 V.

5.8 Consider the simple ac model of a JFET shown in Fig. 5.18. If the device is biased into saturation, consider its behavior if parasitic resistances are added at the drain and source regions of the device. What is the corresponding value of the transconductance?

5.9 The cutoff frequency is defined as the frequency at which the magnitude of the forward current gain is unity with the output short circuited. Using Fig. 5.19 determine an expression for the cutoff frequency in terms of the transconductance.

5.10 Using the result of Problem 5.9 determine the cutoff frequency of a JFET in terms of the gate capacitance. Simplify your result for g_m by considering its maximum value.

5.11 Determine the breakdown voltage of an n-channel Si JFET if the critical field of silicon is 300 kV/cm. Assume that the pinch-off voltage of the device is -7.0 V and that the value of a is 1.0 μm. Assume for Si that the intrinsic concentration is 10^{10} cm^{-3}, the relative dielectric constant is 11.8, and the electronic charge, q, is 1.6×10^{-19} C. Assume that the gates are p$^+$–n junctions. Neglect V_{bi} in the expression for the critical field.

5.12 Determine the gate capacitance of a JFET device if it has a cutoff frequency of 50 GHz. Assume that the device is a Si n-channel JFET, with a carrier mobility of 1450 cm^2/(V s), width of 0.1 mm, channel length $L = 1.0$ μm, $a = 0.5$ μm, and a pinch-off voltage of -6.0 V.

6

Metal–insulator–semiconductor structures and MOSFETs

In this chapter we discuss metal–insulator–semiconductor, MIS, structures. The most important MIS structure is the metal-oxide-semiconductor, or MOS, structure. The MOS structure is used to provide gating action in MOSFETs. Here we will discuss both ideal and realistic systems and examine their behavior both in equilibrium and under bias. The chapter concludes with a discussion of the workings of MOSFETs.

6.1 MIS systems in equilibrium

The basic device structure of interest is sketched in Fig. 6.1. Inspection of Fig. 6.1 shows that the MIS structure consists of three different layers: metal, insulator, and semiconductor layers. In this section we examine the operation of MIS systems in equilibrium. We start with a discussion of ideal MIS systems that have the following properties:

 (i) The metal–semiconductor work function difference, φ_{ms}, is zero at zero applied bias.
 (ii) The insulator is perfect; it has zero conductivity, $\sigma = 0$.
(iii) No interface states located at the semiconductor–oxide interface are assumed to exist.
(iv) The semiconductor is uniformly doped.
 (v) There is a field free region between the semiconductor and the back contact. In other words, there is no voltage drop within the bulk semiconductor.
(vi) The structure is essentially one-dimensional.
(vii) The metal gate can be treated as an equipotential surface.

To draw the band diagram of the MIS structure in equilibrium we first recognize that the Fermi level must be flat everywhere in the structure as shown in Fig. 6.2. The Fermi level lies the same distance from the vacuum level in the semiconductor and the metal since we assume that the work function difference is zero. When the layers of the MIS structure are put into contact the Fermi level naturally aligns without any charge transfer between the metal and semiconductor. As a result, there is no band bending in the diagram to accommodate the alignment of the Fermi level. The third assumption made above implies that the interface between the semiconductor and oxide is perfect: no dangling bonds, impurities, etc. exist at the interface. Additionally, the second assumption implies that the oxide is perfect. It also does not contain any defects or charged impurities. As a result there are no defects in the system. Coupled with the fact that there is no work function difference these assumptions imply that

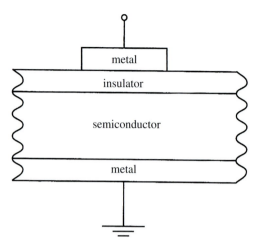

Figure 6.1 MIS structure. The top metal contact is often referred to as the gate. The most common MIS structure is that with silicon and silicon dioxide forming the semiconductor and insulator layers respectively.

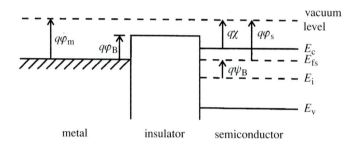

Figure 6.2 Energy band diagram for an ideal MIS structure in equilibrium. Notice that the metal and semiconductor work functions are equal thus ensuring that the Fermi level lies flat in equilibrium with no band bending (Brennan, 1999).

the system exhibits no band bending. The system is said to be in flat band since all of the bands are flat and no band bending occurs.

Though the ideal MIS system is quite simple it is somewhat oversimplified. Practical systems typically do not have a zero work function difference, contain interface states, and the oxide often has defects within it. All of these effects act to change the operation of a MIS structure. Let us consider in some detail how the structure changes upon the relaxation of some of the assumptions made above.

The first assumption that comes into question is that of equal work functions leading to a zero work function difference between the semiconductor and the metal. As an example let us consider the original MIS system, the MOS system with Al as the metal, silicon dioxide (SiO_2) as the insulating oxide and Si as the semiconductor. Al was used as the gate metal until the late 1970s, when it was replaced by highly doped polysilicon. Polysilicon can be readily deposited over the SiO_2 layer immediately after the oxide is grown. Thus its use serves to reduce contamination of the gate oxide. In addition, polysilicon can withstand high temperature processing and can be readily

(a)

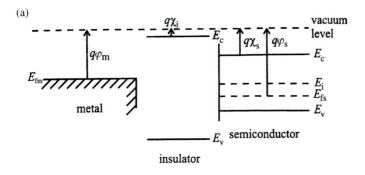

$\vec{F}$ →

(b)

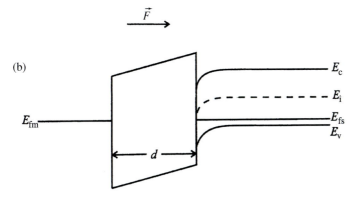

Figure 6.3 MIS system for a nonideal system where $\varphi_m < \varphi_s$: (a) band diagram of the system before contact and (b) band diagram of the system after contact. Notice that the bands necessarily bend in this case to ensure that the Fermi levels align (Brennan, 1999).

defined ensuring sharp feature lines. The polysilicon is highly doped either n- or p-type to give it a high conductivity. Though the Al–SiO$_2$–Si MOS system is not used extensively in modern MOS devices, it serves as an excellent example to illustrate the workings of a MOS system. The main advantage of illustrating the workings of the MIS junction using Al for the gate metal is that the work function of the metal gate remains fixed. For this reason, we will use it in our discussion within this chapter. In the case of polysilicon, the work function depends upon the doping type. The work function varies significantly for n- and p-type doping thus changing the behavior of the MOS system. For either n-type or p-type Si the work function difference, $\varphi_m - \varphi_s$ is always negative in an Al–SiO$_2$–Si system implying that the work function of the Si is greater than that of the Al. The work function difference is largest for highly doped p-type Si and is smallest for highly doped n-type Si. This is quite obvious from the fact that for highly doped n-type Si the Fermi level moves closer to the vacuum level thus reducing the Si work function. Conversely, for highly doped p-type Si the Fermi level moves closer to the valence band thus increasing the separation of the Fermi level and the vacuum level.

The band diagram for a nonideal MOS system under equilibrium is shown in Fig. 6.3. Part (a) of the figure shows the materials when kept apart. Notice that the work function difference is clearly nonzero and that the work function of the metal is smaller than that of the semiconductor. For simplicity we have selected p-type Si as the semiconductor material. When the materials are all placed in contact the Fermi level is flat everywhere. In order to equilibrate the Fermi level there must be a net transfer of electrons out of the metal and into the semiconductor. This cannot occur through the insulator since it does not conduct any current. Instead charge is transferred in and out from the metal contacts on either end of the MOS structure or in the case of the semiconductor free carriers are pushed away from the interface. The transfer of electrons results in the deposition of negative charge on the semiconductor and positive charge on the metal. Since the metal is assumed to be a perfect conductor it cannot sustain any tangential electric fields. Thus the positive charge on the metal resides entirely on its surface at the metal–oxide interface. Depending upon the magnitude of the work function difference, the compensating negative charge in the semiconductor can arise from ionized acceptors only or from both the inversion layer charge produced by the minority carrier electrons and ionized acceptors. In any event, the negative charge induced within the semiconductor balances that produced on the metal thus maintaining space charge neutrality within the structure.

Figure 6.3(b) shows that there is a band bending within the semiconductor and the insulator. Again it is useful to recall that electrons roll downhill in energy band diagrams. The positive charge on the metal surface attracts electrons within the semi-conductor. Consequently, electrons roll towards the oxide–semiconductor interface due to the attraction of the positive charge on the metal. Thus the bands bend down towards the interface as shown in the figure. Similarly, the oxide conduction band is also tilted. Again the simplest means of deciding in which direction the band bends is to consider in which direction a test electron charge would move. The positive charge on the metal surface causes the oxide band to bend downwards toward the metal as shown in the diagram. An alternative means of determining in which direction the bands bend is to consider the potential energy of an electron. Near a positive charge the electron potential energy is lowered while near a negative charge the potential energy is raised. Thus in the oxide the potential energy is lower on the metal side than on the semiconductor side and the bands bend down towards the metal.

To achieve flat band a bias must be applied to the system. The magnitude of the necessary bias is simply φ_{ms}. As mentioned above for Al–Si the work function difference is always negative. In equilibrium to align the Fermi level throughout the structure a net positive charge forms on the metal surface thus producing a positive potential on the metal. To achieve flat band the positive potential on the metal needs to be compensated by the addition of negative charge or equivalently a negative potential. Thus a negative potential needs to be applied to the metal in order to achieve flat band.

There are various sources of oxide charge that also alter the band bending of a realistic MOS system. In all cases the oxide charge is positive. Thus the oxide when impurities and defects are present acts as if it has a net positive charge. This positive oxide charge attracts electrons within the Si towards the Si–SiO$_2$ interface leading to band bending within the semiconductor. The bands bend such that a test electron will roll downhill towards the semiconductor–oxide interface. The band bending occurs in the same direction as that for the work function difference.

There are four main contributions to the oxide impurity charge. These are:

(i) mobile Na ions within the oxide, Q_m;
(ii) interface dangling bonds at the Si–SiO$_2$ interface, Q_{it};
(iii) oxide trapped charge due to imperfections within the SiO$_2$, Q_{ox};
(iv) fixed oxide charge due to ionic Si atoms near the Si–SiO$_2$ interface, Q_f.

Let us consider each one of these sources of charge in turn. The mobile Na ions arise from contaminants. Sources of contaminants are sweat from human handling, reagents used for processing, water, gases, and contaminated processing equipment. Na atoms are mobile within the SiO$_2$ and thus can move with an applied electric field. Therefore, their position within the oxide changes depending upon the bias condition. The closer to the interface the Na ions are the more they attract the negative charge within the semiconductor and thus have a stronger effect on the band bending within the structure. Fortunately, most of the ionic Na charge can be eliminated by very careful crystal growth and processing. This is one of the major reasons why virtually all integrated circuits are made without any human contact during processing. Most of the essential processing steps are fully automated requiring human interaction at most to move cassettes of chips from one station to the next. In addition, very high purity processing chemicals, gases, and de-ionized water must be used. For these reasons the cost of a fabrication plant as device size shrinks for each new generation of chips goes up substantially. In addition, mobile ion contamination can be greatly reduced using hydrogen chloride HCl and chlorine, Cl. Simply cleaning the furnace tubes used in the oxidation of the Si at elevated temperatures with HCl and Cl$_2$ reduces the Na content to negligible levels. In addition, growing thermal oxides in Cl rich environments eliminates Na ion contamination.

Interface dangling bonds formed at the oxide–semiconductor surface arise from the sudden termination of the Si at the oxide interface. The energy levels of these interface trap states lie within the forbidden gap of the Si as shown in Fig. 6.4. The interface traps below the Fermi level are generally filled while those above the Fermi level are empty. Application of a gate bias changes the band bending and as a result the number of interface traps above and below the Fermi level. Sources of interface traps can possibly be attributed to excess Si, excess O, and impurities. The net result of these sources is that unsatisfied chemical bonds at the Si–SiO$_2$ interface exist. These unsatisfied bonds are generally referred to as dangling bonds. Interface trap states can be greatly reduced by low temperature hydrogen annealing. Essentially hydrogen annealing acts to eliminate the dangling bonds formed at the surface. The hydrogen readily diffuses through the SiO$_2$ and fills the dangling bonds at the interface.

Oxide trapped charge arises from imperfections within the SiO$_2$ itself. These traps are generally electrically neutral but can be charged by electrons or holes that are injected into the oxide. As we will see in Chapter 7, as device dimensions shrink the electric field within the semiconductor in a MOSFET increases. A high electric field can produce carrier heating within the semiconductor ultimately resulting in free carrier injection into the oxide wherein it can contribute to the oxide fixed charge.

The fixed oxide charge is formed within a SiO$_2$ transition layer between the Si and the O. The most likely source of the fixed oxide charge is Si ions left near the interface

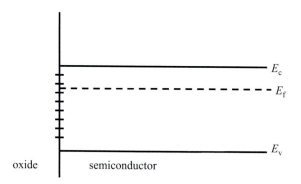

oxide   semiconductor

Figure 6.4 Interface states formed at the surface between Si and SiO$_2$. Notice that the interface states lie within the forbidden energy gap of the Si and are distributed in energy. The traps lying below the Fermi level are filled while those above the Fermi level are empty. Depending upon the bias condition, the band bending changes and thus the number of filled interface states also changes. As a result, the device threshold voltage varies with applied bias. Interface states are often passivated using hydrogen annealing.

after the oxidation process is terminated. The fixed oxide charge lies near the surface but unlike the interface charge is not precisely at the oxide–semiconductor interface. The fixed oxide is a strong function of the oxidation temperature and conditions. It decreases with increasing oxidation temperature. It can be further reduced by annealing the oxide in an argon or nitrogen atmosphere.

It is common to lump all of the above sources of oxide charge into one term called the insulator charge, Q_i; Q_i is positive and thus affects the band bending of the MIS system in the same manner as the metal work function difference. It is important to recognize that the insulator charge is generally a much greater nonideality to handle than the metal work function difference. The work function difference is completely predictable and does not vary between samples. It can be readily compensated by a fixed bias. However, the insulator charge is far less predictable. Unfortunately, it is a strong function of processing and thus can vary from one wafer to the next. Controlling the insulator charge is thus one of the most pressing and difficult problems associated with MOS device fabrication.

The combined action of the insulator charge and the work function difference requires that a net negative voltage be applied to the metal with respect to the semiconductor to compensate for the positive charge produced by the work function difference and insulator charge. At zero gate bias the bands bend down as shown in Fig. 6.3. We define the flat band voltage as the negative metal gate voltage needed to remove the band bending and make the bands flat. The flat band voltage is given then as

$$V_{\mathrm{FB}} = \varphi_{\mathrm{ms}} - \frac{Q_i}{C_i} \tag{6.1}$$

where C_i is the insulator capacitance.

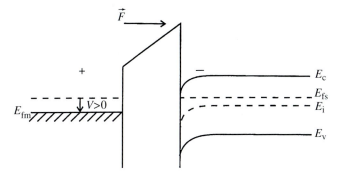

Figure 6.5 Ideal MIS system under accumulation. Notice that the surface, defined as the semiconductor–insulator interface, is more n-type than the bulk region of the device. Hence majority carrier electrons accumulate at the surface due to the positive gate bias applied to the metal (Brennan, 1999).

6.2 **MIS systems under bias**

Let us consider the operation of a MIS structure when an external bias is applied. The presence of the insulator between the metal and semiconductor ensures that there is no dc current flow in the device. Consequently, the Fermi level remains flat in the structure but no longer is coincident between the metal and the semiconductor. The application of the bias leads to a separation of the Fermi levels between the metal and semiconductor as shown in Fig. 6.5.

We will start with the ideal system and then consider the behavior of a realistic system. As an example let us consider a MIS system with an n-type semiconductor. Consider first the application of a positive voltage to the metal gate. A positive voltage on the metal gate will attract the majority carrier electrons within the n-type semiconductor to the semiconductor–insulator interface. As a result majority carrier electrons accumulate at the interface to ensure that the structure remains space charge neutral. The bands bend down towards the metal as shown in Fig. 6.5. As before, the easiest way to determine the direction of the band bending is to consider the action of a test electron charge within the semiconductor. The net positive charge on the metal, often referred to as the gate, attracts the test charge and draws it towards the interface. Since an electron rolls downhill in a potential energy diagram the bands must bend such that the test electron will roll towards the interface. Thus the bands bend down as they approach the semiconductor–oxide surface as shown in the figure. Another way to understand the nature of the band bending is to consider how the electron energies are altered by a positive potential. The positive gate potential lowers the electron energies near the interface and the bands must necessarily bend down as the interface is approached. The MIS structure under these conditions is said to be in accumulation.

The electron concentration at the interface under accumulation can be quantitatively determined as follows. Since the Fermi level is flat the interfacial electron concentration

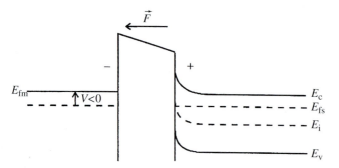

Figure 6.6 Ideal n-type MIS device biased into depletion. In this case, a negative bias is applied to the metal gate. The negative gate bias acts to repel the majority carrier electrons from the interface producing a depletion region (Brennan, 1999).

can be determined from the extent of the band bending using (1.36) as

$$n = n_i e^{(E_{fs} - E_i)/kT} \tag{6.2}$$

Inspection of Fig. 6.5 shows that the difference between the Fermi level and the intrinsic level near the interface is greater than that in the bulk. This implies that the exponent in (6.2) is larger near the interface than in the bulk resulting in a greater electron concentration at the surface than in the bulk. Since the electron concentration is greater near the interface than in the bulk, accumulation of electrons at the semiconductor–oxide interface has occurred and the system is in accumulation.

The second biasing condition of interest for the n-type semiconductor MIS structure is when a small negative voltage is applied to the metal gate. The negatively biased gate repels electrons from the semiconductor–oxide interface. In this case a test electron within the semiconductor will roll away from the interface so the bands bend upwards towards the semiconductor–oxide interface as shown in Fig. 6.6. The negative charge on the metal, corresponding to the negative gate bias, is compensated by an equal but opposite positive charge induced in the semiconductor. The positive charge arises from ionized donor atoms that form a depletion region that extends away from the interface. The electron concentration within the semiconductor at the interface can be determined using (6.2). Inspection of Fig. 6.6 shows that the difference between E_{fs} and E_i at the interface is less than that in the bulk. Thus from (6.2) the electron concentration is less at the interface than in the bulk and the semiconductor is said to be depleted. This condition is called depletion.

If the negative bias applied to the metal is sufficiently large, minority carrier holes are attracted to the semiconductor–oxide interface. As a result the semiconductor near the interface has a greater hole concentration than either the intrinsic concentration or the electron concentration. The surface of the semiconductor is said to be in inversion since the majority carrier type has changed from electrons to holes. The negative gate bias on the metal gate is balanced by an equal but opposite positive charge in the semiconductor due to the combined presence of the minority carrier holes and ionized donors within the depletion region. In this case, a test electron is repelled from the

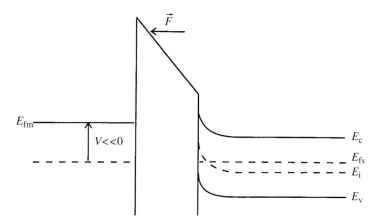

Figure 6.7 Ideal n-type MIS structure biased into inversion. In this case, the negative bias applied to the gate is sufficiently large that it attracts positively charged holes to the surface (Brennan, 1999).

semiconductor–oxide interface and thus the bands must be such that the electron rolls away from the interface as shown in Fig. 6.7. Notice that the band bending is greater in inversion than in depletion. Interestingly, the intrinsic level crosses the Fermi level at the interface while in the bulk the Fermi level is greater than the intrinsic level as shown in the figure. The crossing of the Fermi and intrinsic levels leads to inversion. This can be readily seen from the equation for the hole concentration. The hole concentration at the interface can be obtained using (1.36) as

$$p = n_i e^{(E_i - E_{fs})/kT} \tag{6.3}$$

Notice that $E_i - E_{fs}$ is positive near the semiconductor–insulator interface but negative in the bulk. Thus the hole concentration is greater than n_i near the interface but less than n_i in the bulk making the semiconductor layer near the surface p-type and the bulk n-type. Hence the name inversion.

The surface is said to be in strong inversion when the interface is as p-type as the bulk is n-type. Calling the energy separation between the intrinsic level and the Fermi level within the bulk $q\psi_B$, the magnitude of the separation between the Fermi level and the intrinsic level at the surface is also $q\psi_B$ as shown in Fig. 6.8. The voltage needed to produce strong inversion is thus $2\psi_B$.

The threshold voltage, V_T, is defined as the minimum gate voltage necessary to initiate the inversion of the semiconductor layer. As mentioned above, in inversion the negative gate potential is balanced by both the inversion hole concentration and the ionized donors forming the depletion region. For simplicity and because the most ubiquitous MIS system has an electron inversion layer and a corresponding p-type semiconductor, let us assume that the semiconductor is p-type. The magnitude of the threshold voltage can be determined as follows. The net charge on the metal gate is $+Q_m$, while the net charge in the semiconductor is $-Q_s$. The semiconductor charge

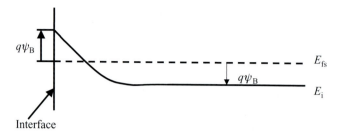

Interface

Figure 6.8 Fermi level and intrinsic level for an n-type MIS structure within the semiconductor. Inspection of the figure shows that the separation between the Fermi level and the intrinsic level has the same magnitude at the surface as in the bulk. The fact that the intrinsic level crosses the Fermi level near the interface implies that the hole concentration at the interface is larger than the electron concentration at the interface. Thus the surface is as p-type as the bulk is n-type and the system is said to be in strong inversion.

Q_s comprises both the inversion layer electrons and the ionized acceptors within the depletion region. Thus we have

$$|Q_m| = |Q_s| \tag{6.4}$$

The potential across the oxide is then

$$V_i = \frac{Q_m}{C_i} = -\frac{Q_s}{C_i} \tag{6.5}$$

At the onset of inversion the inversion charge, Q_I, is very much less than the depletion layer charge, Q_d. Thus the charge in the semiconductor Q_s can be approximated as Q_d. With this approximation (6.5) for the potential across the oxide becomes

$$V_i = -\frac{Q_s}{C_i} = -\frac{Q_d}{C_i} \tag{6.6}$$

The voltages in the MIS structure are shown in Fig. 6.9. Applying Kirchoff's Voltage Law to the MIS structure obtains

$$V_G = V_i + \varphi_s \tag{6.7}$$

But the voltage drop in the semiconductor, φ_s, necessary to produce an inversion layer is $2\psi_B$. For a p-type semiconductor layer φ_s is positive as shown in Fig. 6.9. The gate voltage needed to initiate inversion is the threshold voltage. For an ideal MIS structure the system starts in flat band and the threshold voltage is

$$V_T(\text{ideal}) = V_i + 2\psi_B \tag{6.8}$$

The voltage across the insulator is given by (6.6). Substituting (6.6) into (6.8) yields

$$V_T(\text{ideal}) = -\frac{Q_d}{C_i} + 2\psi_B \tag{6.9}$$

The depletion layer charge can be found as follows. The depletion region width is essentially the same as that for an asymmetric junction in which the depletion region

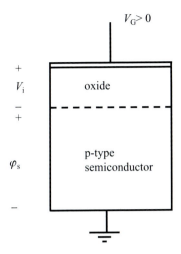

Figure 6.9 MIS structure showing various voltages within the device. Using Kirchoff's Voltage Law the sum of the voltages must be zero.

extends almost entirely into the semiconductor. In the present example the semiconductor is p-type so the depletion region extends into the p-type layer in which the charge (assuming the depletion approximation holds) is N_a. The depletion region width for a p–n junction was found in Chapter 3 to be

$$W = \sqrt{\frac{2\kappa}{q}(V_{bi} + V)\frac{(N_a + N_d)}{N_a N_d}} \tag{6.10}$$

Inspection of Fig. 6.9 shows that the voltage drop across the depletion region is simply equal to φ_s or equivalently $2\psi_B$. Replacing the quantity $V_{bi} + V$ then by φ_s yields

$$W = \sqrt{\frac{2\kappa}{q}\varphi_s\frac{(N_a + N_d)}{N_a N_d}} \tag{6.11}$$

The structure is such that the depletion region extends into the p-type layer and thus we assume that it behaves like an n^+–p junction. Consequently, $N_d \gg N_a$. With this assumption (6.11) simplifies to

$$W = \sqrt{\frac{2\kappa\varphi_s}{q N_a}} = \sqrt{\frac{4\kappa\psi_B}{q N_a}} \tag{6.12}$$

The depletion charge can be determined from use of (6.12) to be

$$Q_d = -q N_a W = -\sqrt{4q\kappa N_a \psi_B} \tag{6.13}$$

Thus the ideal threshold voltage is

$$V_T(\text{ideal}) = -\frac{Q_d}{C_i} + 2\psi_B = \frac{\sqrt{4q\kappa N_a \psi_B}}{C_i} + 2\psi_B \tag{6.14}$$

For a realistic MIS system, the threshold voltage is different from the ideal case given by (6.14). First the bands must be brought to flat band and then the necessary voltage to create the inversion and depletion regions must be supplied. Thus the flat band voltage derived above must be added to the ideal threshold voltage to obtain

$$V_T = V_{FB} + \frac{\sqrt{4q\kappa N_a \psi_B}}{C_i} + 2\psi_B \tag{6.15}$$

Upon substitution for the flat band voltage (6.15) becomes

$$V_T = \varphi_{ms} - \frac{Q_i}{C_i} + \frac{\sqrt{4q\kappa N_a \psi_B}}{C_i} + 2\psi_B \tag{6.16}$$

or more generally

$$V_T = \varphi_{ms} - \frac{Q_i}{C_i} - \frac{Q_d}{C_i} + 2\psi_B \tag{6.17}$$

The sign of each term in (6.17) bears some discussion. Let us consider each term in turn. As we mentioned above for Si the first two terms in (6.17) are both negative for all instances. The sign of the last two terms depends upon the type of channel formed and the corresponding doping type of the Si. For an n-channel device, p-type substrate, the semiconductor contains ionized acceptors which are negatively charged. Thus Q_d is negative so the resulting term, Q_d/C_i is positive. For p-type Si (n-channel) the last term is also positive. Therefore, for p-type Si, n-channel, V_T can be either positive or negative depending upon the magnitude of the last two terms in (6.17). For an n-type semiconductor with a corresponding p-type channel, all of the terms in (6.17) are negative. Q_d consists of ionized donors, which are positively charged. Hence the term Q_d/C_i remains negative. In addition, the last term in (6.17) is also negative. The threshold voltage for an n-type semiconductor, p-channel device is always negative for a Si based, Al metal gate, MIS structure.

In summary for a p-channel device it is necessary to always apply a negative gate voltage to induce the channel. For an n-channel device the threshold voltage can be either positive or negative. When the threshold voltage is positive, a positive potential must be applied to the gate to induce the channel. In this case the device is normally off until a gate bias is applied. The structure is said to be an enhancement mode device. In the opposite case, when the threshold voltage is negative for an n-channel device, this means that a negative gate bias must be supplied in order to turn the device off. A conducting channel exists with no applied gate voltage. Such a structure is said to be a depletion mode device, it is normally on and a gate bias must be supplied in order to turn the device off by shutting off the conducting channel.

The last item we will discuss in this section is the capacitance associated with a MIS structure. The presence of an insulator separating the metal gate and the semiconductor blocks dc current flow in the MIS structure. As a result, the device is essentially a capacitor. The applied voltage on the structure is dropped partially across the oxide and partially across the semiconductor as given by (6.7):

$$V_G = V_i + \varphi_s \tag{6.18}$$

The voltage drop across the insulator is simply equal to the magnitude of the electric field across the insulator times the thickness of the insulator. The electric field across the insulator is constant and is given as

$$\varepsilon_i = \frac{|Q_m|}{\kappa_i} = \frac{|Q_s|}{\kappa_i} \tag{6.19}$$

The magnitude of the corresponding voltage across the insulator is then

$$V_i = \frac{|Q_s| d}{\kappa_i} = \frac{|Q_s|}{C_i} \sim \frac{|Q_d|}{C_i} \tag{6.20}$$

and the potential drop within the semiconductor at the onset of inversion is $2\psi_B$.

In accumulation the majority carriers are collected at the semiconductor–insulator interface. Thus the capacitance is simply that of a parallel plate capacitor with the insulator as a dielectric. The capacitance under this condition is given as

$$C_i = \frac{\kappa_i A}{d} \tag{6.21}$$

As the voltage on the gate changes to drive the MIS structure into depletion, a depletion region forms within the semiconductor that acts as a dielectric in series with the insulator dielectric. As a result, the total capacitance of the device decreases. Under this condition the total capacitance is given as that corresponding to the oxide and depletion layer capacitances in series. Calling the depletion layer capacitance C_d, the total capacitance under depletion conditions is

$$C_{total} = \frac{C_d C_i}{(C_d + C_i)} \tag{6.22}$$

where the magnitude of the depletion capacitance is

$$C_d = \frac{\kappa_s A}{W} \tag{6.23}$$

The capacitance passes through a minimum value and then depending upon the frequency of the applied voltage signal increases again. At low frequencies the minority carrier concentration forming the inversion layer can follow the applied voltage. As a result, minority carrier charge has time to move from the bulk to the inversion layer and out again. Under this condition the capacitance is again due to that of the insulator. Thus at low frequencies the capacitance when the device is biased into inversion is given as

$$C = C_i = \frac{\kappa_i A}{d} \tag{6.24}$$

At high frequencies the signal changes too rapidly for the minority carriers to adjust. As a result the capacitance is due to the series combination of the insulator and depletion layer capacitances given by (6.22). The resulting capacitance vs. voltage characteristic is shown in Fig. 6.10. For simplicity we show the behavior of a MIS system with a p-type semiconductor.

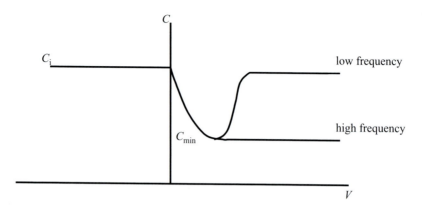

Figure 6.10 Capacitance vs. voltage characteristic for a p-type MIS structure. Under a negative gate voltage the device is in accumulation. As the voltage becomes positive first a depletion region forms and the net capacitance decreases. The frequency of the applied voltage determines whether the minority carrier electrons can follow the voltage. If the frequency is low, the minority carrier electrons have time to collect within the inversion layer formed at the interface and the capacitance increases back to its insulator value. If the frequency is high, the minority carrier electrons cannot follow the applied voltage signal and the capacitance remains at its minimum value.

Example Problem 6.1 Depletion region width for an ideal MOS capacitor

Find the maximum width of the depletion region for an ideal MOS capacitor on p-type Si in strong inversion given that the doping concentration is 10^{16} cm^{-3} and the relative dielectric constant is 11.8.

The depletion region width is given by (6.12) to be

$$W = \sqrt{\frac{4\kappa\,\psi_B}{q\,N_a}}$$

Recall that ψ_B is the potential difference between the Fermi level and the intrinsic level as shown in Fig. 6.8. ψ_B can be calculated from the hole concentration within the bulk semiconductor as follows. Recall that the bulk hole concentration is

$$p = N_a = n_i e^{(E_i(\text{bulk}) - E_{fs})/kT} = n_i e^{q\psi_B/kT}$$

Solving for ψ_B obtains

$$\psi_B = \frac{kT}{q} \ln\left(\frac{N_a}{n_i}\right)$$

Substituting in each value obtains

$$\psi_B = 0.0259\,\text{V} \ln\left(\frac{10^{16}}{10^{10}}\right) = 0.358\,\text{V}$$

The depletion region width, W, can now be found as

$$W = 2\sqrt{\frac{11.8\,(8.85 \times 10^{-14}\,\text{F/cm})(0.358\,\text{V})}{(1.6 \times 10^{-19}\,C)(10^{16}\,\text{cm}^{-3})}} = 3.05 \times 10^{-5}\,\text{cm}$$

Example Problem 6.2 MIS structures with realistic surfaces

Calculate the threshold voltage for a MIS structure given the following information. Let the semiconductor be p-type Si and the metal be Al. The doping concentration is $N_a = 10^{17}$ cm^{-3}, the oxide thickness is $d = 10$ nm, the work function difference $\varphi_{ms} = -1.05$ V, the insulator surface charge density is $Q_i = 5 \times 10^{10}q$ C/cm^2, and $\kappa_i = 3.9$. Determine the depletion region width, W, the flat band voltage, V_{FB} and the threshold voltage, V_T.

The depletion region width is

$$W = 2\sqrt{\frac{\kappa\psi_B}{qN_a}}$$

ψ_B is given from Example Problem 6.1 as

$$\psi_B = \frac{kT}{q}\ln\left(\frac{N_a}{n_i}\right)$$

Substituting in, ψ_B is

$$\psi_B = 0.0259\,\text{V}\ln\left(\frac{10^{17}}{10^{10}}\right) = 0.417\,\text{V}$$

The depletion region width can now be determined. Substituting in the values for each variable in the expression for the depletion region width obtains

$$W = 2\left[\frac{11.8(8.85 \times 10^{-14}\,\text{F/cm})(0.417\,\text{V})}{(1.6 \times 10^{-19}\,C)(1 \times 10^{17}\,\text{cm}^{-3})}\right]^{1/2} = 1.04 \times 10^{-5}\,\text{cm}$$

Next determine the flat band voltage which is given as

$$V_{FB} = \varphi_{ms} - \frac{Q_i}{C_i}$$

The surface charge density within the oxide, Q_i, is determined as

$$Q_i = (5 \times 10^{10})(1.6 \times 10^{-19})\,\text{C/cm}^2 = 0.8 \times 10^{-8}\,\text{C/cm}^2$$

The oxide capacitance, C_i, per unit area is given as

$$C_i = \frac{\kappa_i}{d} = \frac{3.9(8.85 \times 10^{-14}\,\text{F/cm})}{10^{-6}\,\text{cm}} = 3.45 \times 10^{-7}\,\text{F/cm}^2$$

Using the above results the flat band voltage is

$$V_{FB} = \varphi_{ms} - \frac{Q_i}{C_i} \qquad V_{FB} = -1.05\,\text{V} - \frac{0.8 \times 10^{-8}\,\text{C/cm}^2}{34.5 \times 10^{-8}\,\text{F/cm}^2} = -1.07\,\text{V}$$

The threshold voltage is obtained using (6.17). The threshold voltage is,

$$V_T = V_{FB} - \frac{Q_d}{C_i} + 2\psi_B$$

The depletion layer charge density, Q_d, is

$$Q_d = -qN_aW = -1.6 \times 10^{-19}\,C(1 \times 10^{17}\,cm^{-3})(1.04 \times 10^{-5}\,cm)$$
$$= -1.66 \times 10^{-7}\,C/cm^2$$

The threshold voltage is then,

$$V_T = -1.07\,V + \frac{1.66 \times 10^{-7}\,C/cm^2}{3.45 \times 10^{-7}\,F/cm^2} + 2(0.417)\,V = 0.246\,V$$

Example Problem 6.3 Threshold voltage for an n-channel MIS structure

Calculate the threshold voltage for a realistic n-channel MIS device given the following:

$$N_a = 10^{17}\,cm^{-3},\, Q_i = 10^{11}\,q/cm^2,\, d = 20\,nm,\, \text{and}\, \varphi_{ms} = -0.95\,V$$

The threshold voltage is

$$V_T = \varphi_{ms} - \frac{Q_i}{C_i} - \frac{Q_d}{C_i} + 2\psi_B$$

Find the value of each term in the expression for the threshold voltage in turn.
The work function difference is already given. The insulator charge density, Q_i, is

$$Q_i = 10^{11}/cm^2(1.6 \times 10^{-19}\,C) = 1.6 \times 10^{-8}\,C/cm^2$$

The oxide capacitance per unit area is

$$C_i = \frac{\kappa_i}{d} = \frac{(3.9)(8.85 \times 10^{-14}\,F/cm)}{0.2 \times 10^{-5}\,cm} = 1.72 \times 10^{-7}\,F/cm^2$$

The value of ψ_B is

$$\psi_B = \frac{kT}{q} \ln\left(\frac{N_a}{n_i}\right) = 0.417\,V$$

The depletion region charge density is

$$Q_d = -qN_aW$$

which is

$$Q_d = -2\sqrt{qN_a\kappa\psi_B}$$
$$= -2\sqrt{1.6 \times 10^{-19}\,C(10^{17}\,cm^{-3})(11.8)(8.85 \times 10^{-14}\,F/cm)(0.417\,V)}$$
$$= -1.67 \times 10^{-7}\,C/cm^2$$

Each of the terms is known in the expression for the threshold voltage. Substituting obtains

$$V_T = \varphi_{ms} - \frac{Q_i}{C_i} - \frac{Q_d}{C_i} + 2\psi_B$$

$$V_T = -0.95 \text{ V} - \frac{1.6 \times 10^{-8} \text{ C/cm}^2}{17.2 \times 10^{-8} \text{ F/cm}^2} + \frac{16.7 \times 10^{-8} \text{ C/cm}^2}{17.2 \times 10^{-8} \text{ F/cm}^2} + 2(0.417) \text{ V}$$

$$= 0.76 \text{ V}$$

Notice that the threshold voltage is positive so the structure is an enhancement mode device.

Example Problem 6.4 p-type MIS capacitor

Calculate the minimum capacitance for an n-channel (p-type) MIS capacitor. Assume that the capacitor is made from the Si–SiO$_2$–Al materials system. The following information is given. The p-type doping concentration within the semiconductor N_a is 5×10^{16} cm^{-3}, the oxide thickness d is 12 nm, the insulator relative dielectric constant is 3.9 and the semiconductor relative dielectric constant is 11.8.

The minimum capacitance is given as

$$C_{min} = \frac{C_i C_d}{(C_i + C_d)}$$

Determine both C_i and C_d the insulator and depletion layer capacitances respectively. The insulator capacitance per unit area is

$$C_i = \frac{\kappa_i}{d}$$

while the depletion layer capacitance per unit area is

$$C_d = \frac{\kappa_s}{W}$$

The insulator capacitance per unit area is

$$C_i = \frac{3.9(8.85 \times 10^{-14} \text{ F/cm})}{12 \times 10^{-7} \text{ cm}} = 2.87 \times 10^{-7} \text{ F/cm}^2$$

The depletion layer width, W is

$$W = 2\sqrt{\frac{\kappa_s \psi_B}{q N_a}}$$

where ψ_B is

$$\psi_B = \frac{kT}{q} \ln\left(\frac{N_a}{n_i}\right) = 0.0259 \text{ V} \ln\left(\frac{5 \times 10^{16}}{1 \times 10^{10}}\right) = 0.4 \text{ V}$$

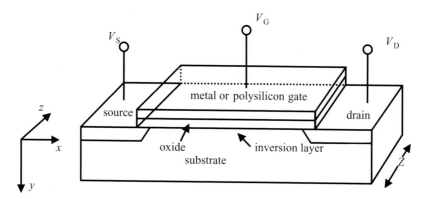

Figure 6.11 MOSFET device structure.

The depletion region width is then

$$W = 2\sqrt{\frac{11.8(8.85 \times 10^{-14}\,\text{F/cm})(0.4\,\text{V})}{(1.6 \times 10^{-19}\,\text{C})(5 \times 10^{16}\,\text{cm}^{-3})}} = 1.44 \times 10^{-5}\,\text{cm}$$

The depletion layer capacitance per unit area is

$$C_d = \frac{11.8(8.85 \times 10^{-14})\,\text{F/cm}}{1.44 \times 10^{-5}\,\text{cm}} = 7.25 \times 10^{-8}\,\text{F/cm}^2$$

C_{min} can now be calculated from C_i and C_d:

$$C_{min} = 5.78 \times 10^{-8}\,\text{F/cm}^2$$

6.3 Basic theory of MOSFET operation

A representative MOSFET device structure is shown in Fig. 6.11. The specific device shown in the figure is an n-channel structure made with p-type Si. Both the source and substrate are grounded. A positive voltage is applied to the drain with respect to the source contact. The device operates as follows. A conducting channel is induced under the gate by the application of a gate bias (enhancement mode device) such that an inversion layer is formed at the oxide–semiconductor interface. An n-type channel forms the inversion layer of the device. A conducting channel connects the source and drain regions of the device. The larger the gate voltage the greater the pileup of electrons at the surface and the higher the conductivity of the channel. This is clearly seen from the definition of the electrical conductivity, $\sigma = q\mu n$. As the concentration increases so does the conductivity. If a small positive drain voltage is applied, a current will flow from the drain to the source. Recall that electrons will flow from the source to the drain, attracted by the positive drain voltage. The current flows in the opposite direction to the electron flux. At low drain bias, the drain current, I_D, is linearly related to the drain voltage, V_D. The channel behaves as a simple resistor. As the drain voltage is increased the drain to substrate p–n$^+$ junction becomes more and more reverse biased. The depletion region at the drain end increases in size extending mainly into the p-type semiconductor region. The voltage drop from the source to the drain starts to negate the inverting effect of the gate. The potential difference between the gate and the drain end is effectively lowered which reduces the inversion effect at the drain end. Consequently, the electron carrier concentration in the channel at the drain end decreases which reduces the conductivity or equivalently increases the channel resistance. With increasing channel resistance the drain current decreases and the slope of the current–voltage characteristic decreases. Ultimately as the drain voltage increases the inversion layer completely vanishes at the drain end and the conducting channel pinches off. When the channel pinches off, the drain current (for a long channel device) saturates: I_D no longer increases with the drain voltage.

At the pinch-off point the resistance is very large since the carrier concentration is like that of a depletion region. The conducting channel no longer completely extends from the source to the drain but is interrupted near the drain end of the device. Hence, most of the drain-source voltage drop occurs in the pinched-off region and the electric field becomes substantial. The reason why the current saturates in a MOSFET is similar to that for a JFET and MESFET. For a long channel device the output conductance is essentially zero, implying zero slope for the current–voltage characteristic. Let the length of the pinched-off part of the channel be ΔL where L is the channel length. If $\Delta L \ll L$, which holds for a long channel device, then as the drain voltage increases beyond the saturated drain voltage, $V_{D,sat}$, the channel length remains essentially the same. Thus the endpoint voltages are the same and the geometry is essentially the same. As a result, the current remains unchanged from the pinch-off point value leading to saturation of the drain current.

The current–voltage characteristic can be quantitatively determined as follows. When the inversion layer exists everywhere between the source and drain contacts a conducting channel connects the source and drain. We assume that the only important current is a drift current and neglect the diffusion current. Inclusion of the diffusion current enables a full description of the current flow in the device. However, the inclusion of the diffusion current necessitates a numerical solution (see Pao and Sah (1966)). The current density neglecting diffusion is

$$\vec{J}_n = q\mu_n n\vec{\varepsilon} \tag{6.25}$$

Simplifying to one dimension, the current density is

$$J_{nx} = q\mu_n n\varepsilon_x = -q\mu_n n\frac{\mathrm{d}V}{\mathrm{d}x} \tag{6.26}$$

There are no sources or sinks between the source and the drain. The gate leakage current is neglected and no current flow is assumed within the substrate. The drain current can be determined by integrating the current density with respect to the area. The dimensions and a sketch of the active region of the device are shown in Fig. 6.12. The current is in the negative x-direction. The current flows perpendicular to the y–z plane. The current can be calculated by integrating the current density with respect to

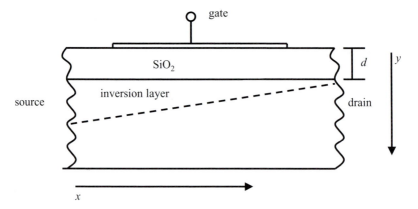

Figure 6.12 Cross-sectional view of a MOSFET showing the inversion layer. The source to drain direction is the x-direction while the y–z plane is perpendicular to the current flow.

y and z as

$$I_{D} = -\iint J_{nx}\,dy\,dz = -Z \int_{0}^{y_{c}(x)} J_{nx}\,dy \tag{6.27}$$

where Z is the device width, $y_{c}(x)$ is the edge of the channel and 0 is the interface between the Si and SiO$_2$. The negative sign enters (6.27) since the current is in the negative x-direction. Substituting (6.26) for the current density in (6.27) obtains,

$$I_{D} = qZ \int_{0}^{y_{c}(x)} \mu_{n} n \frac{dV}{dx}\,dy \tag{6.28}$$

The current flow is within the inversion layer at the Si–SiO$_2$ interface. Due to the presence of interface states and surface roughness the scattering rate is somewhat higher near the surface than in the bulk. As a result the mobility is generally lower for electrons at the surface than the bulk mobility. The mobility in (6.28) is replaced by an effective mobility that reflects the action of the interface. The effective mobility is μ_{n}' and is assumed to be constant with respect to y. The drain current is then

$$I_{D} = Zq\mu_{n}' \frac{dV}{dx} \int_{0}^{y_{c}(x)} n(x, y)\,dy \tag{6.29}$$

The total charge in the n-type channel is called $Q_{n}(x)$ (Q_{I} used earlier for the inversion charge is more general but we specialize here to an n-channel device) and is given as

$$Q_{n}(x) = -q \int_{0}^{y_{c}(x)} n(x, y)\,dy \tag{6.30}$$

Using the definition of the channel charge, (6.29) becomes,

$$I_D = -Z \frac{dV}{dx} \mu'_n Q_n(x) \tag{6.31}$$

The drain current is independent of position and can be integrated over x. The bounds on the integral over x extend over the full length of the channel, from 0 to L. The corresponding bounds for the integral over V are the potential at $x = 0$, which is assumed to be zero, and the potential at the drain end of the channel, $x = L$, which is V_D. With these bounds (6.31) can be integrated as

$$\int_0^L I_D dx = -\mu'_n Z \int_0^{V_D} Q_n(x) dV \tag{6.32}$$

The left hand side of (6.32) can be readily integrated to give

$$I_D = -\frac{Z \mu'_n}{L} \int_0^{V_D} Q_n(x) dV \tag{6.33}$$

To proceed it is necessary to specify the channel charge density in terms of the voltage. Recall that the gate voltage is given as

$$V_G = V_{FB} - \frac{Q_s}{C_i} + 2\psi_B \tag{6.34}$$

But the charge density in the semiconductor, Q_s, is equal to the sum of the inversion layer and depletion layer charge densities

$$Q_s = Q_n + Q_d \tag{6.35}$$

Substituting the expression for Q_s given by (6.35) into (6.34) yields

$$V_G = V_{FB} - \frac{Q_n + Q_d}{C_i} + 2\psi_B \tag{6.36}$$

Notice that (6.36) differs from the equation for the threshold voltage since it applies in general. In the threshold voltage expression the device is biased just at the onset of inversion, while in the above case the bias can be significantly beyond threshold thus resulting in the production of inversion layer charge at the interface. Solving (6.36) for the channel charge density, Q_n, yields

$$Q_n = -C_i \left(V_G - V_{FB} - 2\psi_B + \frac{Q_d}{C_i} \right) \tag{6.37}$$

If a drain voltage is applied, there is an additional voltage drop along x with respect to the source. This additional voltage drop is called $V(x)$ and must be subtracted from V_G to give

$$Q_n = -C_i \left[V_G - V_{FB} - 2\psi_B - V(x) + \frac{Q_d}{C_i} \right] \tag{6.38}$$

The threshold voltage, V_T, is given by (6.17) as

$$V_T = V_{FB} + 2\psi_B - \frac{Q_d}{C_i} \tag{6.39}$$

Substituting into (6.38) the expression for the threshold voltage given by (6.39) obtains,

$$Q_n = -C_i[V_G - V_T - V(x)] \tag{6.40}$$

where Q_n is the channel charge per unit area. The expression for the drain current in (6.33) can now be integrated to give

$$I_D = -\frac{Z\mu_n'}{L} \int_0^{V_D} -C_i[V_G - V_T - V(x)]dV \tag{6.41}$$

which yields

$$I_D = \frac{Z\mu_n'C_i}{L}\left[(V_G - V_T)V_D - \frac{V_D^2}{2}\right] \tag{6.42}$$

which holds below pinch-off. The conditions under which (6.42) is valid are that

$$0 \leq V_D \leq V_{D,sat} \quad \text{and} \quad V_G \geq V_T \tag{6.43}$$

Let us next examine the drain current after pinch off. In this case, the drain current saturates and is represented as $I_{D,sat}$. For $V_D > V_{D,sat}$, $I_{D,sat}$ becomes

$$I_{D,sat} = \frac{Z\mu_n'C_i}{L}\left[(V_G - V_T)V_{D,sat} - \frac{V_{D,sat}^2}{2}\right] \tag{6.44}$$

In pinch off the channel charge at L is simply zero, $Q_n(L) = 0$. The voltage at $x = L$ is $V(L) = V_{D,sat}$. Therefore, $Q_n(L)$ is

$$Q_n(L) = -C_i[V_G - V_T - V(L)] = -C_i[V_G - V_T - V_{D,sat}] = 0 \tag{6.45}$$

Solving (6.45) for $V_{D,sat}$ results in

$$V_{D,sat} = V_G - V_T \tag{6.46}$$

Using (6.46), the saturated drain current becomes

$$I_{D,sat} = \frac{Z\mu_n'C_i}{2L}(V_G - V_T)^2 \tag{6.47}$$

which holds for $V_D \geq V_G - V_T$. The corresponding current–voltage characteristic for a MOSFET is shown in Fig. 6.13.

The current–voltage relationship for a MOSFET can be derived in a different way from that above based on charge control analysis. The results obtained are the same as (6.42) and (6.47) but it is instructive to trace the derivation. The current flowing in the device is given as the ratio of the channel charge to the transit time as:

$$I_D = -\frac{Q_n}{t_{tr}} \tag{6.48}$$

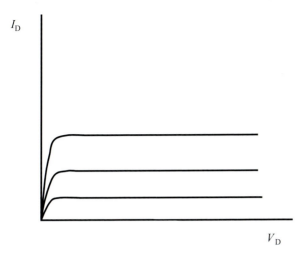

Figure 6.13 Drain current vs. drain voltage characteristic for different values of gate voltage for a long channel MOSFET device. Notice that the drain current saturates at relatively high drain voltage.

Again we assume that only a drift current exists. The diffusion current is considered to be negligible. The carrier transit time t_{tr} can be expressed in terms of the channel length, L, and the drift velocity, v_d, as

$$t_{tr} = \frac{L}{v_d} \tag{6.49}$$

The drift velocity can be written in terms of the effective mobility within the channel yielding

$$v_d = -\mu'_n F = \mu'_n \frac{V_D}{L} \tag{6.50}$$

Substituting (6.50) into (6.49), the transit time can be written as

$$t_{tr} = \frac{L^2}{\mu'_n V_D} \tag{6.51}$$

and the channel charge is

$$Q_n = -C_i(V_G - V_T - V(x))ZL \tag{6.52}$$

Equation (6.52) can be compared with (6.40). Notice that in (6.40) the channel charge is per unit area. In (6.52) we have multiplied by the area, ZL, to give only the charge. As the drain bias increases the channel bias is altered near the drain. The voltage near the drain end of the device relative to the gate decreases. As a result, the voltage difference of the gate electrode and the underlying channel becomes less, reducing the amount of inversion layer charge Q_n. To account for this voltage difference it is assumed that there exists a voltage drop along the channel of value $V(x)$. Equation (6.48) for the drain current can be rewritten using (6.51) and (6.52) to give

$$I_D = \frac{C_i Z \mu_n'}{L}[V_G - V_T - V(x)]V_D \tag{6.53}$$

The value of $V(x)$ can be approximated by its average value of $V_D/2$. With this approximation (6.53) becomes

$$I_D = \frac{C_i Z \mu_n'}{L}\left(V_G - V_T - \frac{V_D}{2}\right)V_D = \frac{C_i Z \mu_n'}{L}\left[(V_G - V_T)V_D - \frac{V_D^2}{2}\right] \tag{6.54}$$

which is precisely the same as (6.42) which holds for below pinch-off operation.

Finally, the saturated drain current can be found from (6.54) using the result given by (6.46). Substitution of (6.46) into (6.54) obtains

$$I_D = \frac{C_i \mu_n' Z}{2L}(V_G - V_T)^2 \tag{6.55}$$

In the above derivations the diffusion current has been neglected. This is generally an acceptable assumption until the device nears pinch-off. The drain-current formulations derived neglecting the diffusion current are valid below pinch-off. Above pinch-off the channel closes near the drain end and the current flow there is mainly by diffusion. If both the drift and diffusion currents are included a single expression is obtained that is valid for the full range of applied bias. In the derivations above, notice that two different equations are derived for below and above pinch off. Unfortunately, inclusion of the diffusion current requires a numerical formulation which is beyond the level of this book. The interested reader is referred to the book by Sze (1981) and the paper by Pao and Sah (1966) for a quantitative discussion of the inclusion of the diffusion current.

In addition to the drain current derived above in the presence of a conducting channel there exists a small but nonnegligible current flow that exists prior to the formation of the conducting channel. The analytical expressions for the drain current derived above were made under the assumption that the surface was inverted. When the gate voltage is below the threshold voltage of an enhancement mode device and the surface is in weak inversion a current, called the subthreshold current flows. The subthreshold current exists if the surface is not inverted and a drain voltage is applied. The subthreshold current is very important for digital circuit design. The MOSFET in a digital circuit is switched into cutoff in order to turn the drain current off. Cutoff corresponds to operation within the subthreshold region. Ideally no drain current flows with drain bias when the device is biased into cutoff. Therefore, in the ideal case there is no current flow and no power drain when the device is switched into cutoff. However, in a realistic device, the subthreshold current provides a small but nonnegligible power drain when the device is in cutoff, drawing current when the device is switched off. The presence of the subthreshold current contributes to heating of the chip from the dissipation of the power corresponding to the subthreshold current flow. As the number of MOSFETs on a chip has risen, the corresponding power dissipation has increased, producing a serious increase in the temperature of the chip. High temperature operation of an integrated circuit should be avoided since it can make the Si intrinsic (i.e. have equal numbers of holes and electrons) and lead to device and system failure. Additionally, a significant subthreshold current can cause a serious power drain that can severely limit battery powered devices. For these reasons it is important to design MOSFETs that have very small subthreshold currents. A quantitative explicit expression for the subthreshold current cannot be determined. Instead either numerical or approximate formulations are used. One such relationship given by Grotjohn and Hoefflinger (1984) is

$$I_D = \frac{\mu'_n Z C_i}{LA} \left(\frac{kT}{q}\right)^2 e^{qA(V_G - V_T)/kT} \left[1 - e^{-qV_D/kT}\right] \tag{6.56}$$

where A is given by

$$\frac{1}{A} = \frac{C_i + C_d + C_{FS}}{C_i} \tag{6.57}$$

C_{FS} is the capacitance due to fast surface states. Further discussion of the subthreshold current is beyond the level of this book. For a more detailed discussion the reader is referred to Brennan (1999, Chapter 14).

Example Problem 6.5 n-channel MOSFET operation

Consider an n-channel silicon MOSFET with the following information. The threshold voltage, V_T, is 2 V, the oxide thickness, d, is 0.1 μm, surface mobility, $\mu'_n = 500$ cm^2/(V s), device width, $Z = 10$ μm, $L = 10$ μm. Calculate the drain current assuming that the gate voltage is $V_G = 6$ V, and that the drain voltage is: (a) 1 V and (b) 6 V.

(a) To solve this problem we first need to know whether the device is operating above or below saturation. The saturated drain voltage is given by the difference between the gate and threshold voltages. If the drain voltage $V_D < V_{D,\,sat}$ the device is below pinch-off and we use the below saturation formula. If $V_D > V_{D,\,sat}$ the device is above pinch-off and we use the formula for the saturated current. The drain current is given below saturation as

$$I_D = \frac{Z\mu'_n C_i}{L} \left[(V_G - V_T)V_D - \frac{V_D^2}{2}\right]$$

and above saturation as

$$I_{D,\,sat} = \frac{Z\mu'_n C_i}{2L}(V_G - V_T)^2$$

First calculate $V_{D,\,sat}$. It is given by

$$V_{D,\,sat} = V_G - V_T = 6\,V - 2\,V = 4\,V$$

The problem states to consider two values of the drain voltage, 1 and 6 V. At $V_D = 1$ V the device is below saturation and the drain current is given as

$$I_D = \frac{Z\mu'_n C_i}{L} \left[(V_G - V_T)V_D - \frac{V_D^2}{2}\right]$$

Evaluate the coefficient in the expression for the drain current. The capacitance is

$$C_i = \frac{\kappa_i}{d} = \frac{3.9(8.85 \times 10^{-14}\,\text{F/cm})}{10^{-5}\,\text{cm}} = 3.45 \times 10^{-8}\,\text{F/cm}^2$$

The coefficient is

$$\frac{\mu'_n C_i Z}{L} = \frac{(500\,\text{cm}^2/(\text{V s}))(3.45 \times 10^{-8}\,\text{F/cm}^2)(10\,\mu\text{m})}{10\,\mu\text{m}} = 1.72 \times 10^{-5}\,\text{A/V}^2$$

The drain current is

$$I_D = 1.72 \times 10^{-5}\,\text{A/V}^2[(6-2)1 - 0.5]\,\text{V}^2 = 0.06\,\text{mA}$$

(b) In this case the drain voltage, $V_D = 6$ V is greater than $V_{D,\text{sat}}$ since

$$V_{D,\text{sat}} = V_G - V_T = 6\,\text{V} - 2\,\text{V} = 4\,\text{V}$$

and the applied drain voltage is 6 V. Therefore, the device is operating above pinch-off and the drain current is saturated. The drain current is given then as

$$I_{D,\text{sat}} = \frac{Z\mu'_n C_i}{2L}(V_G - V_T)^2$$

which upon substitution of each variable is,

$$I_D = \frac{1.72 \times 10^{-5}\,\text{A/V}^2}{2}(6\,\text{V} - 2\,\text{V})^2 = 0.14\,\text{mA}$$

Example Problem 6.6 Calculation of the drain current in an n-channel MOSFET

Consider an n-channel MOSFET with the following information:

$$N_a = 5 \times 10^{16}\,\text{cm}^{-3};\, \mu'_n = 500\,\text{cm}^2/(\text{V s});\, \varphi_{ms} = -0.96\,\text{V};\, Q_i = 5 \times 10^{10}\,\text{q/cm}^2;$$
$$Z = 50\,\mu\text{m};\, d = 30\,\text{nm};\, L = 5\,\mu\text{m};\, n_i = 10^{10}\,\text{cm}^{-3};\, \kappa_0 = 3.9;\, E_g = 1.12\,\text{eV};$$
$$\kappa_s = 11.9.$$

(a) Determine the drain current at a gate voltage $V_G = 2$ V and a drain voltage $V_D = 1$ V.
(b) Consider the case where the gate voltage is 3 V and the drain voltage in 4 V.

(a) Determine the threshold voltage of the device first. The threshold voltage is

$$V_T = \varphi_{ms} - \frac{Q_i}{C_i} - \frac{Q_d}{C_i} + 2\psi_B$$

where ψ_B is

$$\psi_B = \frac{kT}{q}\ln\left(\frac{N_a}{n_i}\right) = 0.0259\,\text{V}\ln\left(\frac{5 \times 10^{16}}{10^{10}}\right) = 0.4\,\text{V}$$

C_i is

$$C_i = \frac{\kappa_i}{d} = \frac{3.9(8.85 \times 10^{-14}\,\text{F/cm})}{30 \times 10^{-7}\,\text{cm}} = 11.5 \times 10^{-8}\,\text{F/cm}^2$$

Q_i is

$$Q_i = (5 \times 10^{10})(1.6 \times 10^{-19})\, \text{C/cm}^2 = 8 \times 10^{-9}\, \text{C/cm}^2$$

and the depletion layer charge density Q_d is

$$Q_d = -\sqrt{4\kappa_s \psi_B q N_a} = -11.6 \times 10^{-8}\, \text{C/cm}^2$$

The threshold voltage is then

$$V_T = \varphi_{ms} - \frac{Q_i}{C_i} - \frac{Q_d}{C_i} + 2\psi_B$$

$$= -0.96\,\text{V} - \frac{8 \times 10^{-9}\, \text{C/cm}^2}{11.5 \times 10^{-8}\, \text{F/cm}^2} + \frac{11.6 \times 10^{-8}\, \text{C/cm}^2}{11.5 \times 10^{-8}\, \text{F/cm}^2} + 2(0.4)\,\text{V}$$

$$= 0.78\,\text{V}$$

Next calculate $V_{D,\text{sat}}$:

$$V_{D,\text{sat}} = V_G - V_T = 2\,\text{V} - 0.78\,\text{V} = 1.22\,\text{V}$$

The drain voltage is 1 V which is less than $V_{D,\text{sat}}$ so the device is below pinch-off and saturation.

The drain current is then,

$$I_D = \frac{Z\mu_n' C_i}{L}\left[(V_G - V_T)V_D - \frac{V_D^2}{2}\right]$$

Substituting in the values for each quantity in the expression for the drain current

$$I_D = 5.75 \times 10^{-4}[(2 - 0.78) - 0.5] = 4.14 \times 10^{-4}\,\text{A} = 0.414\,\text{mA}$$

(b) The saturation drain voltage is equal to

$$V_{D,\text{sat}} = V_G - V_T = 3 - 0.78 = 2.22\,\text{V}$$

The drain voltage is greater than $V_{D,\text{sat}}$ so the device is above pinch off in the saturation region. To find the drain current we use the expression for the saturated drain current:

$$I_{D,\text{sat}} = \frac{Z\mu_n' C_i}{2L}(V_G - V_T)^2 = 1.42\,\text{mA}$$

Example Problem 6.7 n-channel MOSFET circuit

Consider the circuit shown in Fig. 6.14. The following information is given. The threshold voltage of the device is 1.42 V and the coefficient for the drain current is

$$\frac{\mu_n' Z C_i}{L} = 10^{-4}\,\text{A/V}^2$$

Neglect Q_i and assume that the work function difference between the semiconductor and metal is zero. Find the range of voltages for which the device operates in saturation.

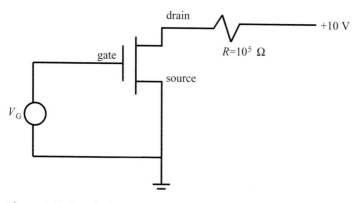

Figure 6.14 Circuit diagram for Example Problem 6.7.

Take Kirchoff's Voltage Law around the drain-source loop. This yields

$$10\,\text{V} = I_D R + V_D = 10^5 I_D + V_D$$

The value of $V_{D,\text{sat}}$ is

$$V_{D,\text{sat}} = V_G - V_T = V_G - 1.42$$

We want the device to operate in saturation. Therefore, the drain current is

$$I_{D,\text{sat}} = \frac{Z\mu_n' C_i}{2L}(V_G - V_T)^2$$

Substituting in the values for the variables, the saturated drain current becomes

$$I_{D,\text{sat}} = \frac{10^{-4}}{2}(V_G - 1.42)^2$$

Using the equation given by Kirchoff's Law and the above we have

$$10 = 10^5 I_{D,\text{sat}} + V_G - V_T = 10^5 I_{D,\text{sat}} + V_G - 1.42$$

We now have two equations in two unknowns. These are

$$10 = 10^5 I_{D,\text{sat}} + V_G - 1.42$$
$$I_{D,\text{sat}} = \frac{10^{-4}}{2}(V_G - 1.42)^2$$

Solving these equations in terms of one another yields

$$10 = 10^5 \frac{10^{-4}}{2}(V_G - 1.42)^2 + V_G - 1.42$$

Solving for $V_G - V_T$ obtains

$$V_G = 2.71\,\text{V}$$

The gate voltage is then equal to $V_G = 2.71$ V. The conditions on the gate voltage are that it be greater than the threshold voltage but less than 2.71 V. So the range of gate voltages such that the device operates in saturation is

$$1.42\,\text{V} < V_G < 2.71\,\text{V}$$

6.4 **Small signal operation of MESFETs and MOSFETs**

The small signal operation of a JFET has been presented in Chapter 5. Though JFETs are useful in many applications, for high power, high frequency device operation the most important device type is the modulation doped field effect transistor, the MODFET. We will return to the operation of MODFETs in Chapter 11. For now, we will concentrate on the operation of MESFETs and MOSFETs and for MODFETs discuss only their advantages. Typically, MESFETs are made from compound semiconductor materials. This is due to the fact that it is difficult to grow a high quality insulator on most compound semiconductors. Therefore, MIS structures are not commonly grown and used with compound semiconductor materials. Instead a Schottky barrier is used to provide gating action.

The primary advantages of compound semiconductor FETs are that many of these materials have relatively high mobility, large saturated drift velocity, and a high breakdown field. These attributes make compound semiconductors very attractive for high power, high frequency device operation. In comparison, Si has a relatively low electron mobility and a low breakdown field and is thus far less attractive for high power, high frequency applications.

The wide band gap semiconductors, specifically gallium nitride (GaN), aluminum nitride (AlN), their associated ternary compounds, and silicon carbide (SiC) are emerging as new materials for insertion in power devices. GaN and its related binary and ternary materials systems are collectively referred to as the III-nitrides. The large energy band gap of the III-nitrides and SiC makes them highly attractive for both optoelectronic and electronic devices. The large band gap energy of the III-nitrides insures that the breakdown electric field strength of these materials is much larger than that of either Si or GaAs enabling, at least in principle, much higher maximum output power delivery. Additionally, it has been found that at least for the binary compounds, GaN and InN have higher electron saturation drift velocities and lower dielectric constants that can lead to higher frequency performance of devices made from these materials.

Though most semiconductor devices are used at room temperature, power dissipation from leakage currents can result in substantial heating of the device. Given that integrated circuits are currently approaching a billion transistors on a chip, significant heating of the chip can occur. It is not unusual for chips to heat up to temperatures well above room temperature even with proper heat sinking. Therefore, modern device structures must tolerate high temperature operation. At high operating temperatures, materials with relatively small band gaps such as Si and GaAs can become intrinsic. This implies that the vast majority of carriers that enter the conduction band are

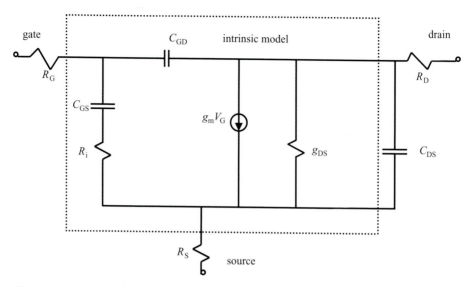

Figure 6.15 Quasi-static small signal equivalent circuit of a MESFET. The region within the dotted line corresponds to the intrinsic model of the device. The overall model includes parasitics such as the gate, drain, and source resistances. For simplicity, the parasitic inductances at each lead are neglected.

promoted thermally from the valence band rather than from dopant states. Once a material becomes intrinsic effectively its doping no longer substantially impacts the free carrier concentration. Generally, this results in device failure. The wide band gap semiconductors on the other hand can safely operate at high temperatures without becoming intrinsic. Very low thermal leakage currents can be expected in devices made from the wide band gap semiconductors if defect densities can be held at acceptably low levels. For these reasons the III-nitrides and SiC are becoming important for future high frequency, high power electronics used in wireless telecommunications systems.

To assess the ac performance of a MESFET it is useful to construct an equivalent circuit model for the device. A simplified equivalent circuit for a MESFET is shown in Fig. 6.15. The model is somewhat more complicated than that used for a JFET since the gate-source terminals are no longer assumed to be a simple open circuit. As can be seen from Fig. 6.15, the intrinsic model comprises the circuit elements within the dotted line. The extrinsic model includes the intrinsic model and external parasitics. Within the intrinsic model the gate to source is modeled as a capacitor C_{GS} and resistor R_i in series. A capacitance between the gate and drain is also included. The quantity g_{DS} is the channel conductance that is specified by the inverse of the channel resistance, R_{DS}, and there is a current source in parallel with g_{DS}. The external model contains source R_S and drain R_D resistances to take account of the finite resistance of the source and the drain regions on either side of the channel. The capacitance between the drain and the source, C_{DS}, shown in Fig. 6.15 is used to represent the substrate capacitance. In addition, there is a gate resistance, R_G, completing the parasitics for the external model.

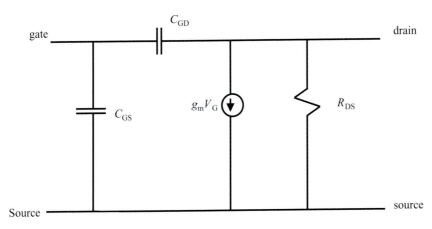

Figure 6.16 Simplified intrinsic small signal equivalent circuit model of a MESFET.

The intrinsic transistor model consists of the circuit elements shown in Fig. 6.15 within the dashed box. In many instances, the resistance R_i can be neglected and the intrinsic small signal equivalent circuit simplifies to that shown in Fig. 6.16. The extrinsic model includes the source, drain, and gate resistances and an additional capacitance between the drain and the source that represents the substrate capacitance.

The key figures of merit for microwave operation of a FET are the frequency at unity current gain called the cutoff frequency, f_t, and the maximum frequency of oscillation, f_{max}. The cutoff frequency is defined as the frequency at which the magnitude of the forward current gain is unity when the output is short circuited. In other words, the cutoff frequency is the frequency at which the device can no longer amplify the input signal. The cutoff frequency of the MESFET can be determined as follows. The output of the circuit shown in Fig. 6.16 is short circuited. Therefore, the two capacitors C_{GD} and C_{GS} are in parallel to ground. The current in general is equal to

$$i = C\frac{\mathrm{d}v}{\mathrm{d}t} \tag{6.58}$$

For the alternating current phaser form, set $v = V_G\,\mathrm{e}^{j\omega t}$ so the current is therefore

$$i = CV_G j\omega\,\mathrm{e}^{j\omega t} \tag{6.59}$$

For the circuit shown in Fig. 6.16 with the output short circuited the input current is given then as

$$i_{in} = j\omega V_G(C_{GD} + C_{GS})\,\mathrm{e}^{j\omega t} \tag{6.60}$$

since the capacitors are in parallel. The magnitude of the input current is

$$I_{in} = \omega V_G(C_{GD} + C_{GS}) = 2\pi f_t V_G(C_{GD} + C_{GS}) \tag{6.61}$$

The output current is simply equal to

$$I_{out} = g_m V_G \tag{6.62}$$

At cutoff the gain is unity so that the input and output currents are equal. Equating (6.61) and (6.62) yields

$$I_{in} = I_{out} \qquad 2\pi f_t V_G (C_{GD} + C_{GS}) = g_m V_G \tag{6.63}$$

Solving for the cutoff frequency obtains

$$f_t = \frac{g_m}{2\pi C_G} \tag{6.64}$$

where C_G is the sum of C_{GD} and C_{GS}.

The maximum frequency of operation, f_{max}, is defined as the frequency at which the unilateral power gain of the transistor rolls off to unity. The unilateral power gain is the maximum power gain achievable by the transistor. A transistor is said to be unilateral when its reverse transmission parameter is zero. In other words, for a unilateral transistor the output is completely isolated from the input: there is no reverse transmission. The maximum frequency of operation depends to a large extent upon the extrinsic elements, i.e., the parasitic resistances and capacitances that make up the extrinsic device model. The cutoff frequency is important in specifying the switching speed of the device, while f_{max} is more important in defining the device RF performance. The derivation of f_{max} is beyond the level of this book. The reader is referred to Liu (1999) for a thorough discussion. The maximum frequency of operation is given as

$$f_{max} = \sqrt{\frac{f_t}{8\pi R_G C_{GD}}} \tag{6.65}$$

where R_G is the gate resistance and C_{GD} is the gate to drain capacitance.

We now examine the small signal behavior of a MOSFET. The small signal equivalent circuit for a MOSFET is shown in Fig. 6.17. The capacitances C_{BD} and C_{BS} are the body to drain and body to source capacitances respectively. These capacitances represent the capacitance with respect to the body or substrate of the device. The equivalent circuit shown in Fig. 6.17 applies to high frequency operation. The presence of the capacitances acts to reduce the voltage gain at high frequencies. The oxide capacitance is the major contributor to the decline of the voltage gain. It is often assumed that the oxide capacitance is equal to the sum of the gate to source, C_{GS}, and gate to drain, C_{GD}, capacitances. When the device is operated in saturation the gate to drain capacitance can be neglected with respect to the gate to source capacitance. Therefore, in saturation the gate to source capacitance can be replaced by the oxide capacitance. We will utilize this result below when we calculate the cutoff frequency.

The cutoff frequency of a MOSFET can be determined as follows. The cutoff frequency is defined as the frequency at which the forward current gain is unity when the output is short circuited. Consider the small signal equivalent circuit when the output is short circuited. The input current is given by

$$i_{in} = C_{equivalent} \frac{dV_G}{dt} = j\omega(C_{GS} + C_{GD})V_G \, e^{j\omega t} \tag{6.66}$$

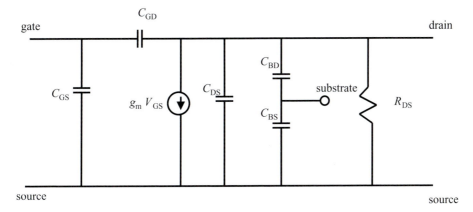

Figure 6.17 Small signal equivalent circuit model for a MOSFET. C_{BD} and C_{BS} are the substrate to drain and substrate to source capacitances respectively.

The magnitude of the output current is simply

$$I_{out} = g_m V_G \tag{6.67}$$

At unity gain, the magnitudes of the input and output currents are equal. Setting the magnitudes of the currents equal obtains

$$\omega V_G(C_{GS} + C_{GD}) = g_m V_G \tag{6.68}$$

Substituting $\omega = 2\pi f$ in (6.68) yields the cutoff frequency $f = f_t$

$$f_t = \frac{1}{2\pi} \frac{g_m}{C_{GS} + C_{GD}} \tag{6.69}$$

The usage of MOSFETs in analog or rf applications usually is as an amplifier. For amplification, the MOSFET is biased into saturation and the most useful parameters are calculated for saturation. The transconductance of a MOSFET, g_m, in the saturated region of operation is

$$g_m = \left.\frac{\partial I_D}{\partial V_G}\right|_{V_D} = \frac{C_i \mu'_n Z}{L}(V_G - V_T) \tag{6.70}$$

Equation (6.69) can be simplified using the assumption that the gate to source capacitance is dominant in saturation and can be approximated by the oxide capacitance, C_i. Under this assumption, the total capacitance at the input can be written as $C_i ZL$. Using (6.70) the cutoff frequency becomes

$$f_t = \frac{\mu'_n (V_G - V_T)}{2\pi L^2} \tag{6.71}$$

Clearly, high frequency operation occurs when the mobility is high and the channel length is very small.

Example Problem 6.8 Cutoff frequency of a MOSFET

Consider the calculation of the cutoff frequency of a MOSFET given the following information: $L = 1\,\mu m$; n-channel device with a p-type substrate; $\mu'_n = 1200\,cm^2/$ (V s); $Z = 10\,L$; $V_T = 1.1$ V; and $V_G = 5$ V.

The cutoff frequency is given in the text by (6.71) as,

$$f_t = \frac{\mu'_n(V_G - V_T)}{2\pi L^2}$$

Substituting in the given values the cutoff frequency becomes,

$$f_t = \frac{1200\,cm^2/(V\,s)(5\,V - 1.1\,V)}{2\pi(1.0 \times 10^{-4}\,cm)^2} = 7.45 \times 10^{10}\,s^{-1} = 74.5\,GHz$$

6.5 **CMOS circuits**

The question we seek to answer in this section is why is CMOS circuitry so widely used in digital computing hardware? The key issue that makes CMOS circuits highly attractive is that they dissipate very low dc power. The very low power usage of CMOS circuitry makes them important in portable applications in which battery drain is critical to the lifetime of the system (examples are in digital watches, portable computers, automobile clocks, etc.). Additionally, low power dissipation is very important in highly dense circuitry. As the number of devices per chip increases, very low power dissipation is crucial in maintaining room temperature operating conditions. As the reader may recall, as the temperature of Si increases, it ultimately becomes intrinsic, meaning that the free carrier concentration within the conduction band results mainly from interband thermal generation negating the effect of doping resulting in device failure. Even using extensive heat sinking techniques, it is difficult to completely dissipate waste heat. Therefore, usage of low power dissipation circuitry such as CMOS is essential for proper thermal management.

How though are CMOS circuits constructed such that they dissipate little dc power? Low power dissipation is accomplished by having at least one transistor in cutoff, where its drain current is extremely low, between the power source, V_{DD}, and ground for all possible logic inputs. In this way, the power source is always shielded from ground by a very high resistance path. A CMOS circuit is designed such that current only flows in the circuit during transitions or switching, with the obvious exception of leakage current.

For illustration let us consider some representative CMOS circuits. We first examine a simple CMOS inverter gate. Let us first consider the operation of the inverter using a simple switch to represent a MOSFET. Two different states of the circuit are shown in Fig. 6.18: (a) input high and output low, and (b) input low and output high. From the diagrams it is difficult initially to tell how the high and low inputs are applied, but this will become clear in the discussion below. For now, we focus on the output. In the configuration shown in Fig. 6.18(a), the output is connected to ground through the bottom switch and hence a low output is obtained. In Fig. 6.18(b), the bottom switch

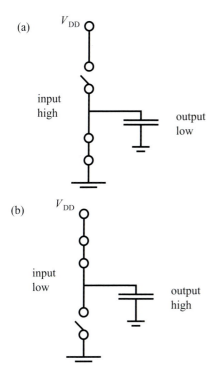

(a) V_{DD}

input
high

output
low

(b) V_{DD}

input
low

output
high

Figure 6.18 Switch representation of a CMOS inverter circuit for (a) output low and (b) output high.

is open, and the output is blocked from ground by the capacitor. This corresponds to a high output. So depending upon the bottom switch setting, the output can be either high or low.

The CMOS circuit implementation of this inverter is shown in Fig. 6.19. The top transistor is a p-channel device while the bottom transistor is an n-channel device. A p-channel MOSFET has the opposite characteristics of an n-channel MOSFET. Specifically, when the gate voltage is high, the p-channel transistor is OFF (operates in the cutoff region) while the n-channel MOSFET is ON (biased in the triode region). A conducting channel in a p-channel device is formed under a low positive or negative gate voltage. Similarly, a conduction channel in an n-channel device is formed under a high positive gate voltage. Hence, when the input voltage is HIGH, the n-channel device shown in Fig. 6.19 is ON while the p-channel device is OFF. Under these conditions, V_{OUT} is connected to ground and isolated from V_{DD}. Hence the output is LOW. Conversely, when the input voltage is LOW, the p-channel device shown in Fig. 6.19 is ON while the n-channel device is OFF. In this case, V_{DD} appears at V_{OUT}, and the output is HIGH. Therefore, the CMOS circuit behaves as follows: Input LOW – Output HIGH; Input HIGH – Output LOW, which clearly operates as an inverter.

CMOS circuitry can be used to create various logic gates. For example, NOR and NAND gates are relatively simple to implement using CMOS circuitry. Along with the inverter, NOR and NAND functions are sufficient to reproduce all the important

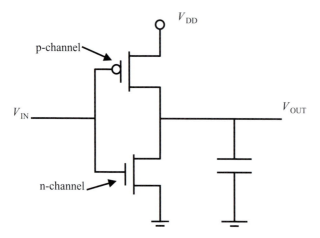

Figure 6.19 Sketch of a CMOS inverter circuit.

input		output	
A	B	OR	NOR
0	0	0	1
0	1	1	0
1	0	1	0
1	1	1	0

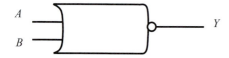

Figure 6.20 Boolean logic truth table and symbol for a NOR gate.

logic operations needed in a digital binary computer. Let us next consider NOR and NAND gates and their implementation in CMOS. The Boolean logic truth table for a NOR gate is shown in Fig. 6.20 along with a symbolic NOR gate. The CMOS circuit implementation of a NOR gate is shown in Fig. 6.21. For simplicity we consider only a two-input NOR gate. In this case, the two inputs are marked as A and B and the output is V_{OUT}. Recall that the n-channel transistors are ON when their gate voltage is HIGH, and OFF when their gate voltage is LOW. The p-channel transistors have the opposite behavior. They are ON when the gate voltage is LOW and OFF when the gate voltage is HIGH. Consider the case when both A and B are LOW. According to the truth table in Fig. 6.20, the output should be HIGH for a NOR gate. Let us see how the circuit of Fig. 6.21 behaves. If both the inputs A and B are LOW, then since M1 and M2 are

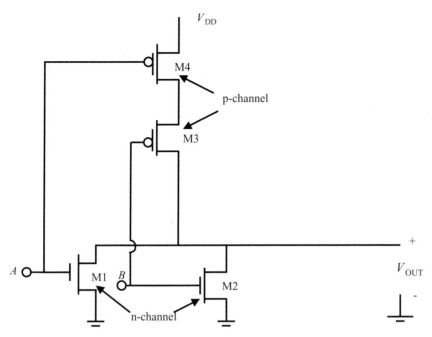

Figure 6.21 CMOS circuit implementation of a NOR gate.

n-channel transistors, they will both be OFF. Notice that V_{OUT} is connected to ground through the channels of either M1 or M2. Since both M1 and M2 are OFF, there is no conducting path connecting V_{OUT} and ground. Transistors M3 and M4 are both ON since they are p-channel devices with a LOW voltage applied to their gates. V_{OUT} is connected to V_{DD} through these devices in series. Since both M3 and M4 are on, V_{OUT} is shorted to V_{DD} and hence, the output, V_{OUT}, is HIGH. Thus the circuit yields a HIGH output for two LOW inputs. Consider what happens for any other combination, either both inputs HIGH, or either input HIGH while the other input is LOW. When both inputs are HIGH, transistors M1 and M2 are ON. When A is HIGH and B is LOW, M1 is ON and M2 is OFF. Similarly, when A is LOW and B is HIGH, M1 is OFF and M2 is ON. Notice that under any of these circumstances there exists at least one path connecting V_{OUT} to ground through transistor M1 or M2 or both. Hence, when one or both inputs are HIGH, the output is LOW. Clearly, the CMOS circuit shown in Fig. 6.21 acts like a NOR gate.

Let us next consider the operation of a NAND gate. The truth table and logic symbol for a NAND gate are shown in Fig. 6.22. Inspection of the truth table shows that only if both inputs are HIGH will the output of the NAND gate be LOW. For any other combination, the output is always HIGH. The CMOS circuit implementation of the NAND gate is shown in Fig. 6.23. Again we consider for simplicity only a two-input NAND gate. As in the NOR gate, the circuit uses two p-channel MOSFETs and two n-channel MOSFETs. The inputs are marked as A and B while the output is V_{OUT}. Consider the case when both A and B are HIGH. Then the n-channel transistors, M1 and M2, are both ON, while the p-channel transistors, M3 and M4, are both OFF.

input		output	
A	B	AND	NAND
0	0	0	1
0	1	0	1
1	0	0	1
1	1	1	0

Figure 6.22 Boolean truth table and logic symbol for a NAND gate.

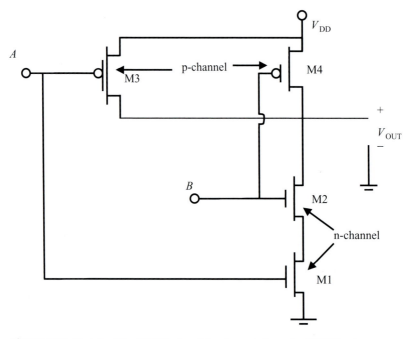

Figure 6.23 Sketch of the CMOS circuit implementation of a NAND gate.

Therefore, V_{OUT} is tied to ground through the series combination of M1 and M2 and is isolated from V_{DD} by the parallel path of M3 and M4. As a result, V_{OUT} is LOW. For any other logical input, A or B or both LOW, then the following conditions occur. If A is LOW, M3 is ON and M1 is OFF. Thus since M3 and M4 are in parallel, independent of the state of M4, V_{OUT} is connected to V_{DD}. Additionally, since M1 and M2 are in series, if M1 is OFF, then V_{OUT} is shielded from ground, independent of the state of

M2. Thus the output is HIGH. The opposite situation, B is LOW, M4 is ON and M2 is OFF, also leads to V_{OUT} connected to V_{DD} and a HIGH output. Clearly, when both A and B are LOW, the output will again be HIGH. Thus the circuit of Fig. 6.23 behaves as a NAND gate.

Finally, one might ask, why is Si CMOS the most widely used circuitry for digital logic? Although almost any complementary FET technology can be used to implement digital logic Si CMOS is almost exclusively the choice. Devices made from Si are relatively easy to fabricate. High quality semiconductor material can be inexpensively produced in a reproducible manner, important from a manufacturing viewpoint. SiO_2, the native oxide of Si, is also readily manufacturable. The resulting CMOS Si devices have few defects. Consequently, the threshold voltage across the chip is relatively uniform. In addition, both the gate leakage current and the subthreshold current are relatively small.

Problems

Information for Si CMOS: $n_i = 1.0 \times 10^{10}$ cm^{-3}, $\kappa_i = 3.9$, $\kappa_{si} = 11.8$, $q = 1.6 \times 10^{-19}$ C, and $\varepsilon_0 = 8.85 \times 10^{-14}$ F/cm.

6.1 A MOS capacitor substrate is doped with acceptors to 10^{16} cm^{-3}. Calculate the maximum depletion layer width of the device. $n_i = 1.0 \times 10^{10}$ cm^{-3}, $\kappa_i = 3.9$, $\kappa_{si} = 11.8$.

6.2 A MOS capacitor using Al–SiO_2 has substrate doping of 10^{17} cm^{-3} donors. Assume it has a gate oxide thickness of 10.0 nm, and an oxide charge of $5 \times 10^{11} q$ C/cm^3.
     Assume that the metal–semiconductor work function difference, φ_{ms}, equals -0.2 V. Determine:
     (a) the flatband voltage;
     (b) the threshold voltage.

6.3 A Si substrate is doped with acceptors to 10^{16} cm^{-3}. The thickness of the oxide is 12.0 nm. Given that the work function of a polysilicon gate is 3.95 eV, and the semiconductor electron affinity is 4.15 V, determine the threshold voltage of a polysilicon gate MOS device under:
     (a) zero oxide charge;
     (b) oxide charge of $10^{11} q$ C/cm^3.

6.4 Find the threshold voltage for a Si n-channel MOS transistor with $N_a = 10^{17}$ cm^{-3}, $\varphi_{ms} = -0.95$ V, $Q_i = 10^{11}$ q C/cm^2, and a SiO_2 thickness of 20 nm.

6.5 A standard Si MOSFET has $N_a = 10^{17}$ cm^{-3}, $L = 2$ μm, $Z = 20$ μm, oxide thickness $= 20.0$ nm, electron mobility of 400 cm^2/(V s), and $Q_i = 5 \times 10^{11} q$ C/cm^2. The metal work function of the Al is 4.1 V and the electron affinity in Si is 4.15 V. Determine:
     (a) V_T;
     (b) I_D for $V_G = 2$ V and $V_D = 6$ V;
     (c) I_D for $V_G = 2$ V and $V_D = 1$ V.

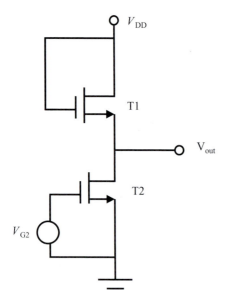

Figure 6.24 Transistors T1 and T2 are connected as shown for Problem 6.7.

6.6 A MOS capacitor uses n-type Si doped 10^{15} cm^{-3} and has an oxide thickness of 20 nm. Determine the surface charge density in the inversion layer for a gate voltage of -5 V. Assume both the metal–semiconductor work function difference and the oxide charge are zero. In this problem we are asked to find the surface charge density that is induced by the action of the gate bias. The amount of gate bias needed to produce the surface charge within the channel (holes in this case) is

$$V_G = \varphi_{ms} - \frac{Q_i}{C_i} - \frac{|Q_n + Q_d|}{C_i} - 2\varphi_F$$

where Q_n is the surface charge density, Q_d the depletion layer charge density and C_i the oxide capacitance per unit area. Simplify the above equation and solve for Q_n.

6.7 Determine the oxide thickness for the second transistor, T2, shown in Fig. 6.24. Given: $V_{G1} = 3$ V, $V_{T1} = 0.6$ V, $V_{out} = 3$ V, $V_{G2} = 5$ V, $V_{T2} = 1.2$ V, $C_{i1} = 5 \times 10^{-8}$ F/cm^2, $L_1 = 10$ μm, $L_2 = 8$ μm, $Z_1 = 15$ μm, $Z_2 = 20$ μm, $\mu_n = 500$ cm^2/(V s), $\kappa_{0x} = 3.9$, $\kappa_0 = 8.85 \times 10^{-14}$ F/cm.

6.8 A p-channel standard Si MOSFET is doped with $N_d = 10^{15}$ cm^{-3}. The MOSFET has an oxide thickness of 8 nm, $Z/L = 20$, hole mobility $= 450$ cm^2/(V s), and $Q_i = 5 \times 10^{11}q$ C/cm^2. Given $n_i = 10^{10}$ cm^{-3}, $\kappa_i = 3.9$, $\kappa_{si} = 11.8$, and $\varphi_{ms} = -0.312$. Determine:

(a) the threshold voltage of the device;
(b) the value of $V_{D,sat}$ at $V_G = -2$ V;
(c) the value of $V_{D,sat}$ at $V_G = -7$ V;
(d) the magnitude of the saturation current for $V_G = -1$ V.

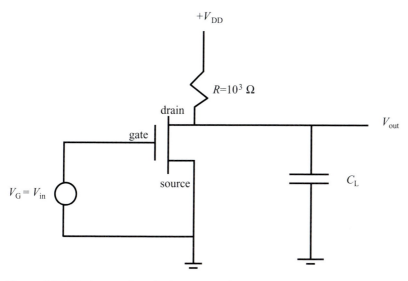

Figure 6.25 The input voltage is the same as the gate voltage.

6.9 An n-channel silicon MOSFET has a gate length L of 1 μm, $Z = 5$ μm, and $C_{ox} = 10^{-7}$ F/cm^2. At $V_D = 0.1$ V, the drain current is given as:

$I_D = 40\,\mu$A at $V_G = 1.6$ V;
$I_D = 90\,\mu$A at $V_G = 2.6$ V.

Calculate:
(a) the electron mobility;
(b) the threshold voltage.
Assume the MOSFET is in saturation and that the device has a positive threshold voltage.

6.10 Consider an n-channel MOSFET made using an Al gate with a SiO$_2$ oxide. Given: $N_a = 10^{17}$ cm^{-3}, $L = 1.0$ μm, $Z = 10$ μm, oxide thickness $= 20$ nm, $\mu_n = 400$ cm^2/(V s), $Q_i = 5 \times 10^{11}$ q C/cm^2, $kT = 0.0259$ eV, $\varphi_m = 4.1$ V, $\chi_s = 4.15$ V, $E_g = 1.12$ eV, $n_i = 10^{10}$ cm^{-3}: relative dielectric constant in oxide $= 3.9$, relative dielectric constant in the semiconductor $= 11.8$. Determine:
(a) the threshold voltage, V_T;
(b) I_D for $V_G = 2$ V and $V_D = 6$ V.

6.11 Consider an n-channel MOSFET with the following parameters: $V_T = 2$ V, oxide thickness $= 0.1$ μm, $Z = 10$ μm, $L = 5.0$ μm, $\mu_n = 500$ cm^2/(V s), $\kappa_s = 11.8$, $\kappa_i = 3.9$. Calculate the drain current under the following conditions:
(a) $V_{GS} = 6$ V, $V_{DS} = 2$ V;
(b) $V_{GS} = 6$ V, $V_{DS} = 5$ V.

6.12 An ideal n-channel MOSFET is made with $L = 7$ μm, $Z = 12$ μm, $V_T = 0.7$ V, oxide thickness of 3×10^{-7} cm, electron mobility of 500 cm^2/(V s), with $\kappa_0 = 3.9$. Determine the conductance g_d when $V_D = 3.6$ V and $V_G = 4.2$ V.

6.13 An n-channel MOS capacitor has an acceptor doping concentration of 10^{16} cm^{-3} and an oxide thickness of 10 nm. Determine:
(a) the low and high frequency capacitances under inversion conditions;
(b) the voltage where the minimum capacitance of the device occurs. Assume the ideal case for the MOS structure can be used.

6.14 Determine the transconductance in an n-channel MOSFET if the gate bias is 6 V, and the threshold voltage is 3 V with a drain voltage of 4 V. Assume that the device width is 100 μm, the mobility is 1450 cm^2/(V s), the oxide capacitance is 3.5×10^{-8} F/cm^2, and the channel length is 1.0 μm.

6.15 For the circuit shown in Fig. 6.25 determine the value of V_{DD} if the n-channel MOSFET is biased just at saturation. Assume that the device has a width, Z, of 100 μm, a mobility of 1000 cm^2/(V s), an oxide capacitance of 4.0×10^{-8} F/cm^2 and a channel length of 2.0 μm. The input voltage applied to the gate is 6 V and the device threshold voltage is 2 V. R is equal to 1.0 kΩ.

7

Short-channel effects and challenges to CMOS

Though long channel MOSFET devices are an excellent means of describing how MOSFETs work, they are rarely used nowadays. In order to increase the number of active devices on a chip and thus improve its functionality, MOSFET device structures have undergone continued miniaturization. The long channel theory developed in Chapter 6 is valid only for devices that have channel lengths greater than about 1–2 μm. Present state-of-the-art MOSFETs used in digital integrated circuits are very much smaller than this. At the time of this writing, major integrated circuit manufacturers are producing commercial products with 0.13 μm gate lengths. Devices with only 0.1 μm gate lengths are already in the design stage. Therefore, state-of-the-art devices are very different from the long channel MOSFETs discussed in Chapter 6. In this chapter we examine the processes in state-of-the-art Si based MOSFETs and discuss how reduction in the gate length influences device behavior.

7.1 Short-channel effects

There are many complications that arise as MOSFET devices are miniaturized. These can be summarized as arising from material and processing problems or from intrinsic device performance issues. As the device dimensions shrink it is ever more difficult to perform the basic device fabrication steps. For example, as the device dimensions become smaller and the circuit denser and more complex, problems are encountered in lithography, interconnects, and processing. Different intrinsic device properties are affected by device miniaturization. The class of effects that alter device behavior that arise from device miniaturization are generally referred to as short-channel effects.

There are many effects that alter MOSFET device performance as the channel length decreases. These effects can be sorted as a function of their physical origin into three different categories. The three categories correspond to three different sources as: (I) the electric-field profile becomes two-dimensional; (II) the electric field strength in the channel becomes very high, and (III) the distance between the source and the drain decreases. The most important features that arise in short-channel MOSFETs are:

  I. Two-dimensional potential profile:
    (1) Threshold voltage reduction; the gate voltage no longer controls the total gate depletion charge but depends upon the drain-source voltage.
    (2) Mobility reduction by gate-induced surface fields.
  II. High electric fields present within the channel:
    (1) Carrier-velocity saturation.

(a)

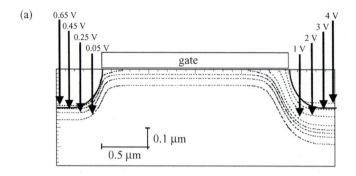

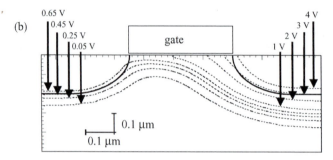

Figure 7.1 Two-dimensional potential profile for: (a) a long channel MOSFET device and (b) a short-channel MOSFET device. Notice that the profile is highly two-dimensional in the short-channel case (Brennan, 1999).

    (2) Impact ionization near the drain.
    (3) Gate oxide charging.
    (4) Parasitic bipolar effect.
III. Decrease in the physical separation between the source and the drain:
    (1) Punchthrough.
    (2) Channel-length modulation.

Let us consider each of these three main categories in turn.

One of the major differences between a short-channel and a long channel MOSFET is the fact that the potential profile becomes two-dimensional in a short-channel device. Inspection of Fig. 7.1 shows that the potential profile can be approximated using a one-dimensional model when the gate is long, but as the gate decreases in length, the potential profile becomes two-dimensional. Clearly, approximating the potential profile of a short-channel device using a one-dimensional model would be highly inaccurate. The physical consequences of the two-dimensionality of the potential can be understood as follows.

The two-dimensionality of the potential profile in a short-channel device is due to the fact that the region under the gate is relatively small because of the encroachment of the source and drain regions. The close proximity of the source and drain regions causes a fraction of the bulk charge density under the channel to have field lines terminated at the source and the drain rather than the channel. In other words, both the gate and source-drain voltages share control of the bulk charge density below the gate. This is often referred to as the charge-sharing model. Consequently, as the channel length decreases, a larger fraction of the bulk charge under the channel has field lines terminated at the source and the drain junctions. The total charge below the gate controlled by the gate voltage in a short-channel device is correspondingly less than that controlled by the gate in a long channel device. Consequently, a lower gate voltage is required to attain threshold in a short-channel device.

A similar effect to charge sharing occurs in short-channel devices. The behavior of the device can be examined by considering the potential barrier formed at the surface between the source and drain. When an n-channel device is "off," no channel exists since the gate voltage is below threshold, a potential barrier exists blocking electron injection from the source contact into the bulk semiconductor. Only a small subthreshold current flows from the source to the drain regions under this condition. In a long channel device, the potential barrier is uniform across the device and the source and drain fields are influential only near the ends of the channel. However, as the channel length decreases, the source and drain regions encroach on the channel. The source and drain fields affect the potential underneath the gate making it harder for the gate to maintain control over the channel. As a result, the potential barrier underneath the gate is lowered. The potential barrier is further lowered by the application of a drain bias. This potential barrier lowering is called drain-induced barrier lowering, DIBL. The barrier lowering facilitates electron injection under subthreshold conditions. Hence, the subthreshold current increases. An increase in the subthreshold current leads to an increase in power consumption which in turn acts to heat the chip. In addition, DIBL acts to lower the threshold voltage of the device.

There are several different means to combat DIBL in short-channel transistors. The depletion layer thickness underneath the gate can be reduced by increasing the channel doping concentration. This results in an increase in the gate control of the channel region. However, increased channel doping lowers the carrier mobility within the channel through an increase in the ionized impurity scattering rate. Increased gate control of the channel can be attained by reducing the insulator layer thickness underneath the gate contact. Unfortunately, as the gate oxide layer is reduced there is a concomitant increase in the gate leakage current resulting in greater power consumption by the transistor.

The transverse electric fields, which become of increasing importance as the channel length decreases since the gate oxide thickness is continuously reduced, alter the carrier mobility. It has been found that the action of the gate field normal to the channel in a MOS transistor degrades the carrier mobility. The mobility reduction is associated, to some extent, with enhanced surface scattering at the MOS interface. The reduction in the surface mobility can be modeled as

$$\mu = \frac{\mu_0}{1 + \theta(V_{GS} - V_T)} \tag{7.1}$$

where μ_0 is the mobility at the threshold voltage, and θ is the mobility reduction factor, which is typically determined empirically.

The second category of effects that appear in short-channel MOSFETs arise from the high electric field present within the channel. As the channel length decreases, if the voltage is not reduced, the channel electric field increases substantially. The first consequence of the high channel field we consider is carrier-velocity saturation. Both the electron and hole drift velocities in Si saturate at applied electric fields in excess of ~ 100 kV/cm. In short-channel devices, the electric field near the drain can attain values in excess of ~ 400 kV/cm. The velocity–field relationship for the carriers takes the following form (Selberherr, 1984):

$$v = \frac{\mu_0 F_y}{\left[1 + \left(\frac{F_y}{F_c}\right)^\alpha\right]^{\frac{1}{\alpha}}} \tag{7.2}$$

where F_c is the critical electric field, F_y is the channel field, and α is a parameter that depends upon the carrier type. α has a value close to 2 for electrons and close to 1 for holes.

The effect of velocity saturation on the current can be illustrated by examining the case for holes where we set $\alpha = 1$. Though the electron case is far more important, the mathematical manipulations are much more complicated. Therefore, for simplicity we examine the consequences of velocity saturation for the $\alpha = 1$ case. It is important first to recall the simplified long channel result for a constant mobility model for the drain current, (6.42):

$$I_D = \mu' \frac{Z}{L} C_i \left(V_G - V_T - \frac{V_D}{2}\right) V_D \tag{7.3}$$

where Z is the channel width, L the channel length, V_G the gate voltage, V_T the threshold voltage, C_i the oxide capacitance, and V_D the drain voltage. In the simple model, μ', the surface mobility, is assumed to be constant and equal to μ_0. The field dependent mobility is obtained assuming that the surface mobility can be approximated using (7.2) as

$$\mu' = \frac{\mu_0}{1 + \frac{F_y}{F_c}} \tag{7.4}$$

but the field in the y direction (in the direction parallel to the channel) and the critical field are

$$F_y = \frac{V_D}{L} \qquad F_c = \frac{v_{sat}}{\mu_0} \tag{7.5}$$

where v_{sat} is the saturation drift velocity. The channel charge per unit area, Q_n, can be approximated as follows. In Section 6.3, we found Q_n from the difference between the total charge in the semiconductor, Q_s, and that in the depletion region, Q_d. Neglecting $V(x)$ in the expression for Q_n in (6.40), Q_n is given as (as before, Q_n is defined per unit area)

$$Q_n = -C_i(V_G - V_T) \tag{7.6}$$

However (7.6) is a poor approximation to the channel charge when V_D is appreciable. To a first approximation, the effect of the drain bias on the channel charge can be included by considering the average voltage above threshold between the gate and the channel to be $V_G - V_D/2$. Equation (7.6) then becomes

$$Q_n = -C_i \left(V_G - V_T - \frac{V_D}{2} \right) \tag{7.7}$$

The drain current can then be found, using charge control analysis, from the ratio of the channel charge, Q_n, to the transit time, t_{tr}, multiplied by the gate area, ZL, as

$$I_D = -\frac{Q_n}{t_{tr}} ZL \tag{7.8}$$

The transit time, t_{tr}, is

$$t_{tr} = \frac{L^2}{\mu' V_D} \tag{7.9}$$

Substituting into (7.8) and (7.9) the expression for the mobility given by (7.4), and using the results in (7.5), the drain current becomes,

$$I_D = \frac{\mu_0 Z C_i}{L} \frac{\left(V_G - V_T - \dfrac{V_D}{2} \right) V_D}{\left(1 + \dfrac{\mu_0 V_D}{v_{sat} L} \right)} \tag{7.10}$$

Equation (7.10) is an approximate relationship for the drain current derived to illustrate the effects of a field dependent mobility. Notice that the drain current is less than that predicted by the constant mobility model as given by (7.3). Hence in a short-channel device the drain current is typically less within the saturation regime than that predicted by the constant mobility model. The drain current is smaller in the short-channel device as compared to the long channel device at comparable bias conditions. This behavior can be readily seen by rewritting (7.10) in terms of (7.3) as

$$I_D = \frac{I_{D0}}{\left(1 + \dfrac{\mu_0 V_D}{v_{sat} L} \right)} \tag{7.11}$$

where I_{D0} is the drain current for the constant mobility model (given by (7.3)). Notice that the drain current is always reduced below the constant mobility model when velocity saturation effects are included.

When the drain-source voltage is high, the electric field strength within the channel can act to greatly heat the carriers as they move from the source to the drain. Carrier heating can impact the device behavior in several different ways. The three most prominent effects from channel heating are impact ionization near the drain, parasitic bipolar operation, and gate oxide charging. If the drain-source voltage is sufficiently high, impact ionization of the carriers near the drain can occur. The impact ionization rate has a strong field dependence (Brennan, 1999, Sections 10.4–10.5; Brennan and Haralson, 2000). If the electric field strength near the drain exceeds the minimum

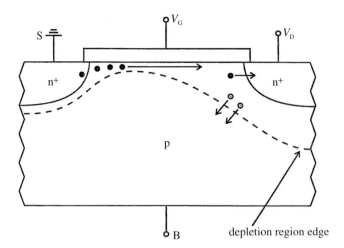

Figure 7.2 MOSFET device showing the parasitic bipolar effect. Impact ionization near the drain leads to the production of secondary holes and electrons. The secondary and primary electrons (solid dots) are swept into the drain contact while the secondary holes (open circles) are swept into the p-type bulk region thus charging the bulk. This in turn results in forward biasing the p–n$^+$ junction formed at the source resulting in additional electron injection from the source into the channel (Brennan, 1999).

needed to induce electron initiated impact ionization (the actual value within a device is somewhat geometry dependent but to a great extent can be approximated by the bulk value), carrier multiplication can become appreciable. The generated electrons are swept into the drain region, increasing the drain-source current, while the generated holes are swept into the substrate as shown in Fig. 7.2. Hence, carrier multiplication at the drain end of the device results in hole injection into the substrate. The injected holes create what is called the parasitic bipolar effect, since the holes act to forward bias the source–substrate junction. Notice that the injected holes within the substrate increase the net positive charge within the p-type substrate. This produces a positive bias on the substrate with respect to the grounded source. Thus the substrate is biased positively with respect to the source, creating a forward biased p–n junction. The forward bias acts to induce further electron injection from the source into the channel. The net result is a further increase in the drain current.

Another important consequence of the high source-drain electric field strength is the concomitant strong carrier heating associated with this field. The very high channel electric field heats the carriers to very high kinetic energy near the drain. The carriers, electrons in an n-channel device, can be heated to sufficient kinetic energy such that they can transfer from the semiconductor channel into the gate oxide. Experimental short time stress voltage tests show a significant change in the device threshold voltage with time. The threshold shift is attributed to gate oxide charging from hot electrons. The electrons can be injected into the oxide by the action of different mechanisms. These are hot electron emission over the potential barrier into the oxide and tunneling processes. The first mechanism, hot electron emission over the oxide potential barrier, is essentially a classical effect. In order for an electron to be injected into the oxide in this way, its kinetic energy in the direction perpendicular to the interface must exceed the potential barrier height. For the Si–SiO$_2$ system, this potential height is larger than 3 eV and it is likely that not many electrons survive to this energy and thus undergo hot electron emission into the oxide. Alternatively, quantum mechanical tunneling of the electrons into the oxide can occur. Tunneling can proceed either directly through band to band tunneling or indirectly with the assistance of interface traps. The magnitude of the current density associated with tunneling processes has been found to obey an empirical relationship given by (Brews, 1990)

$$J \sim F^2 e^{-K/F} \qquad\qquad (7.12)$$

where F is the magnitude of the electric field, and K has an empirical value for Si–SiO$_2$ of 19–23 MV/cm. Experimentally, oxide injection is measured by comparing the gate leakage current to the channel current. The ratio of these quantities is the injection probability, P.

Perhaps the most important consequence of hot carrier injection into the oxide is the deterioration of the device over time. This is often referred to as hot-electron aging. Though some of the electrons injected into the oxide contribute to the gate leakage current, some of the injected electrons become trapped within the oxide. The electrons generally are trapped by the oxide trapped centers, Q_{ox}, that form neutral impurities. Upon capturing an injected electron these centers become charged. As a result, the electric field beneath the oxide changes as a function of the amount of trapped injected charge. Over time with repeated hot electron stressing, the oxide charge can become appreciable resulting in a significant change in the threshold voltage of the device. This has severe consequences for long term device reliability. To summarize, gate oxide charging is a long term degradation mechanism in a MOSFET.

Typically, the most successful means of combating gate oxide charging is to introduce lightly doped regions near the source and drain n$^+$ contact regions. Such devices are called lightly doped drain, LDD MOSFETs or LDDMOS devices. In the usual MOS structures, the abrupt n$^+$–p junctions formed at the drain and source contacts result in a very high electric field in a relatively narrow region. The addition of the lightly doped n layer increases the depletion layer width thus reducing the magnitude of the electric field. Consequently, the hot carrier effects are somewhat mitigated.

Finally, we consider the last set of effects encountered in short-channel MOS devices. These arise principally from the physical decrease in the drain–source separation. The most important effects in this category are channel length modulation and punchthrough. Channel length modulation arises from the shortening of the effective channel length of the transistor because of the increase in the drain depletion region as the drain voltage is increased. The effect is sketched in Fig. 7.3. As can be seen from the figure, the effective channel length is reduced. The resulting channel length is simply equal to the metallurgical channel length minus the source and drain depletion region widths. To a good approximation, the depletion layer width at the source end of the device can be neglected with respect to that at the drain end. The effect of channel

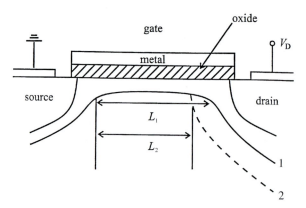

Figure 7.3 MOSFET device showing channel length modulation. Channel length modulation results from shortening of the effective channel length (L_2) from the original channel length (L_1) due to an increase in the drain depletion region (Brennan, 1999).

length modulation on the drain current can be modeled as

$$I_D = \frac{I_0 L}{L - \Delta L} \tag{7.13}$$

where L is the original channel length, ΔL is the difference in the channel length, and I_0 is the original drain current. Notice that as ΔL increases, the drain current increases. This results in an output conductance defined as the nonzero slope of the drain current vs. drain voltage for the device.

Punchthrough arises when the channel length is very small. Under this condition, the source and drain depletion regions can touch, resulting in a large increase in the drain current. The origin of the punchthrough condition can be understood as follows. For an n-channel device, both the source and drain regions are doped n$^+$ while the bulk region of the device is p-type. Therefore, the source and drain regions form n$^+$–p junctions. The drain must be positively biased for a drain current to flow in the device. Consequently, the drain n$^+$–p junction is reverse biased resulting in an appreciable depletion region formed at the junction. Since the bulk p region is much more lightly doped than the n$^+$ drain region, most of the depletion region forms in the p region. The source contact is typically grounded, but a depletion region still exists between the n$^+$ source region and the p-type bulk. Clearly, the two depletion regions formed at the source and drain contacts will ultimately touch if the source to drain separation is made small enough resulting in punchthrough.

7.2 Scaling theory

The presence of short-channel effects presents difficulties in device miniaturization. Nevertheless, devices have been successfully reduced well below the limits where short-channel effects become appreciable. Clearly, the semiconductor device industry has found a means by which MOSFET devices can be miniaturized to very small dimensions wherein it is expected that short-channel effects will dominate device behavior. The question is then, how has the industry successfully developed integrated circuits that employ submicron devices that normally would exhibit short-channel effects? The answer is through scaling. Scaling works as follows. One starts with a device that works at a satisfactory level. A miniaturized version is made by scaling the important features that dictate device performance, i.e., channel length, applied voltages, doping concentrations, etc., such that similar operation is maintained. In this way, short-channel effects can be minimized in small, submicron, structures.

There have been several different approaches to device scaling historically. It is useful to review a few of the past approaches to obtain an understanding of the basic issues involved. In any event the overall goal of scaling is to reduce the device dimensions and voltages so as to avoid device-limiting performance from short-channel effects.

Perhaps the simplest approach to device scaling is constant field scaling. As the name implies the device is scaled such that the magnitude of the electric field strength remains constant upon miniaturization. Three different variables are scaled. These are the physical dimensions, applied voltages, and doping concentrations. All dimensions are reduced by a constant scaling factor of λ. For example, the gate oxide thickness, d, scales as

$$d' = \frac{d}{\lambda} \tag{7.14}$$

where the primed variable represents the parameter for the new, scaled device and the unprimed variable is that for the original, unscaled structure. The applied voltages for the scaled device are reduced by the same factor of λ. Thus the drain voltage, for example, scales as

$$V_D' = \frac{V_D}{\lambda} \tag{7.15}$$

The doping concentration is increased by the same scaling factor. For an n-channel, p-type Si device the acceptor concentration within the substrate scales as

$$N_a' = \lambda N_a \tag{7.16}$$

The combined effect of scaling the dimensions, voltages, and doping concentration is that the threshold voltage, V_T, is decreased in direct proportion to the reduced applied voltages so the device will turn on and off as expected in circuits employing reduced supply voltages. The threshold voltage is reduced primarily by scaling the oxide thickness.

It is instructive to inspect how the threshold voltage changes with oxide thickness scaling. The threshold voltage is given by (6.20) as

$$V_T = \varphi_{ms} - \frac{Q_i}{C_i} - \frac{Q_d}{C_i} + 2\psi_B \tag{7.17}$$

The oxide capacitance, C_i, is

$$C_i = \frac{\kappa_i}{d} \tag{7.18}$$

where d is the oxide thickness. Rewriting the capacitance in terms of d (7.17) becomes

$$V_T = \varphi_{ms} - \frac{dQ_i}{\kappa_i} - \frac{dQ_d}{\kappa_i} + 2\psi_B \tag{7.19}$$

The oxide thickness scales by $d' = d/\lambda$. The scaled threshold voltage, V_T', becomes then

$$V_T' = \frac{d}{\lambda \kappa_i}(-Q_i - Q_d) + \varphi_{ms} + 2\psi_B \tag{7.20}$$

Equation (7.20) can be approximated as follows. The terms φ_{ms} and $2\psi_B$ are of opposite sign and similar in magnitude. Therefore, to a reasonable approximation they cancel one another out. The changes in doping concentration and voltage within the Q_d term also tend to cancel out. Therefore, the scaled threshold voltage is approximately given as

$$V_T' = \frac{V_T}{\lambda} \tag{7.21}$$

The drain current also scales with λ. The drain current below saturation assuming drift only is given by (6.54) as

$$I_D = \frac{C_i Z \mu_n}{L}\left(V_G - V_T - \frac{V_D}{2}\right)V_D \tag{7.22}$$

where the prime has been omitted from the mobility to avoid confusion with the scaled values. As before the mobility represents the surface mobility. Scaling each dimension and voltage by λ yields

$$I_D' = \frac{\kappa_i \mu_n \dfrac{Z}{\lambda}}{\dfrac{d}{\lambda}\dfrac{L}{\lambda}}\left(\frac{V_G - V_T - \dfrac{V_D}{2}}{\lambda}\right)\frac{V_D}{\lambda} \tag{7.23}$$

Simplifying, (7.23) becomes

$$I_D' = \frac{\kappa_i Z \mu_n}{dL}\left(V_G - V_T - \frac{V_D}{2}\right)\frac{V_D}{\lambda} = \frac{I_D}{\lambda} \tag{7.24}$$

where it is assumed that the mobility is invariant with scaling. Inspection of (7.24) shows that the drain current scales linearly with the dimensions and voltages.

Though constant field scaling is relatively simple there are some problems associated with its use. Constant field scaling requires scaling down the voltage supply to the device. Though in principle this can be done, in practice it can be very difficult. In the early days of device scaling voltage sources were not scaled down due to operating and noise margins. Constant field scaling is incompatible with noise margins. Other problems associated with constant field scaling are that the diffusion current does not scale in the same way as the drift current. Therefore, the overall behavior of the device does not remain constant upon miniaturization. Additionally, interconnection capacitances in scaled devices are dominated by fringing effects which do not scale linearly. Therefore, other scaling methodologies have been developed.

Table 7.1 Comparison of scaling law rules for the three cases: constant field, constant voltage and quasi-constant voltage

Parameter	Constant field	Constant voltage	Quasi-constant voltage
Dimensions	λ	λ	λ
Oxide thickness	λ	$\lambda^{1/2}$	λ
Doping concentration	λ	λ	λ
Applied voltage	λ	1	$\lambda^{1/2}$

There are several nonconstant field scaling approaches. The early nonconstant field scaling methods developed by Chatterjee *et al.* (1980) are constant voltage and quasi-constant voltage scaling. The manner in which each device parameter scales in the three models, constant field, constant voltage, and quasi-constant voltage is displayed in Table 7.1. Nonconstant field scaling has the important advantage that the power supply can be kept constant. Additionally, using nonconstant field scaling as opposed to constant field scaling in device design results in an increase in the available drive current in short-channel MOSFETs. The primary disadvantage of constant voltage and quasi-constant voltage scaling is that large electric field magnitudes can occur near the drain that can result in high field effects within the device.

In 1984, Baccarani *et al.* (1984) introduced a different scaling methodology. Their technique is called generalized scaling. In generalized scaling the magnitude of the field can increase, but the shape of the field and potential profiles remains the same. Basically, in generalized scaling the device dimensions and doping concentrations are adjusted such that the shape of the combined vertical and lateral electric fields remains unchanged. By preserving the field profile the deleterious short-channel effects are mitigated upon miniaturization. Generalized scaling requires that the dimensions and voltages be scaled by different factors such that the Poisson equation remains invariant and thus the profile also remains invariant. The Poisson equation is given as

$$\frac{\partial^2 \varphi}{\partial x^2} + \frac{\partial^2 \varphi}{\partial y^2} + \frac{\partial^2 \varphi}{\partial z^2} = -\frac{q}{\kappa_s}(p - n + N_d - N_a) \tag{7.25}$$

The potential is scaled as

$$\varphi' = \frac{\varphi}{\gamma} \tag{7.26}$$

The physical dimensions scale as

$$(x', y', z') = \frac{(x, y, z)}{\lambda} \tag{7.27}$$

and the doping concentration scales as

$$(n', p', N_d', N_a') = \frac{(n, p, N_d, N_a)\lambda^2}{\gamma} \tag{7.28}$$

With these scaling rules the Poisson equation, (7.25), becomes

$$\frac{\partial^2(\gamma\varphi')}{\partial(\lambda x')^2} + \frac{\partial^2(\gamma\varphi')}{\partial(\lambda y')^2} + \frac{\partial^2(\gamma\varphi')}{\partial(\lambda z')^2} = -\frac{q(p' - n' + N_d' - N_a')}{\kappa_s}\left(\frac{\gamma}{\lambda^2}\right) \tag{7.29}$$

Simplifying, (7.29) reduces to

$$\frac{\partial^2\varphi'}{\partial x'^2} + \frac{\partial^2\varphi'}{\partial y'^2} + \frac{\partial^2\varphi'}{\partial z'^2} = -\frac{q}{\kappa_s}(p' - n' + N_d' - N_a') \tag{7.30}$$

Comparison of (7.30) and (7.25) shows that the forms of the Poisson equation for the unscaled and scaled structures are identical. If the boundary conditions, namely the source, drain, and gate potentials, are proportionally scaled by γ then the solutions of (7.30) and (7.25) differ only by a scale factor. The shape of the electric field profile is the same for the two cases. However, the magnitude of the field is different between the scaled and unscaled devices. The magnitude of the field within the scaled device differs from that in the unscaled device by the factor λ/γ. Thus the magnitude of the field increases if $\lambda > \gamma$. Of course, if the magnitude of the field remains relatively high then the device will exhibit some of the high field effects discussed in Section 7.1 resulting in device degradation.

The principal advantage of the generalized scaling rule is that the applied voltages do not continuously have to be scaled within the device to preserve performance. This is highly important from a practical circuit design perspective. However, the fact that the voltage is not scaled down implies that the magnitude of the electric fields increases with miniaturization. Of course, this places different limitations on device performance, most notably in reliability and power dissipation. The higher electric field strengths present in the device can cause gate oxide charging resulting in threshold voltage variation with time. Thus the device may experience reliability problems and accelerated aging. In addition, the power density increases in the device, resulting in greater thermal management problems from enhanced resistive heating.

Though constant field and generalized scaling have contributed significantly to device reduction, for modern, state-of-the-art transistors empirical scaling rules are generally used. An example of an empirical scaling rule is that given by Brews *et al.* (1980). They proposed an empirical scaling rule for the minimum channel length, L_{min}, as

$$L_{min} = A\left[r_j d(W_s + W_D)^2\right]^{\frac{1}{3}} \tag{7.31}$$

where A is a fitting parameter, r_j is the junction depth, d is the oxide thickness, W_s and W_D are the depletion widths of the source and drain regions respectively where the abrupt, one-dimensional approximation for the junction is assumed. The criterion is that the device channel length must be greater than L_{min} as given by (7.31). As the device dimensions shrink, principally the channel length L, in order that L be greater than L_{min} as given by (7.31) and thus the device avoid deleterious short-channel effects, shallower junctions, thinner oxides, and either lower voltages or heavier doping are required. Therefore, miniaturized devices must have thinner oxides and heavier dopings if short-channel effects are to be minimized. It is interesting to consider how the constraints on these quantities affect continued miniaturization. Equation (7.31) should not be considered a rigid rule, but more of a guideline to ensure that DIBL is avoided. However, (7.31) is very useful since it clearly illustrates what constraints must be applied to several device parameters in order to effectively reduce the device dimensions leaving device performance relatively intact.

The constraints specified by the empirical scaling rule apply to conventional MOSFET device structures. These constraints can, to some extent, be used as guidelines to predict the limits of conventional MOSFET device miniaturization. Conventional MOSFET scaling dictates that the oxide thickness must be continuously decreased, and that the channel doping concentration must be continuously increased. However, there are some obvious limits to both of these trends. Clearly, the oxide thickness cannot be scaled to zero and in fact, the oxide thickness cannot realistically be made much smaller than ∼2 nm, as discussed below. Similarly, there are constraints on the magnitude of the doping concentration within the channel region since extremely high doping results in enhanced ionized impurity scattering and a concomitant reduction in carrier mobility. Even more limiting is the fact that ultimately the concentration of dopants is constrained by the solid solubility limit. The solid solubility limit is the maximum thermodynamically stable concentration of dopant atoms possible in the semiconductor. Concentrations of dopants in excess of the solid solubility limit result in the formation of clustering. Upon the onset of clustering, the free carrier concentration is no longer increased. Therefore, only a reduction in mobility occurs without any increase in carrier concentration, a most undesirable situation.

In order to mitigate short-channel effects, the gate oxide thickness needs to be scaled in accordance with the channel length. As supply voltages are scaled down, the oxide thickness must also be scaled down to maintain the same maximum electric field. The threshold voltage of a short-channel MOSFET is lower than that for an otherwise identical long channel device. The difference in the average potential drop between a long channel and a short-channel device depends upon the ratio of the change in the stored charge within the depletion region to the gate oxide capacitance. For the short-channel device the stored charge within the depletion region is less than for the long channel structure. To maintain a constant threshold voltage between the two devices, the gate oxide capacitance of the short-channel device needs to be increased. The gate oxide capacitance can be approximated using the parallel plate expression as $\kappa A/d$, where κ is the dielectric constant of the oxide, A is the gate area, and d the oxide thickness. Notice that to increase the gate capacitance the easiest quantity to vary is d, the gate oxide thickness. An increase in the gate capacitance can then be obtained by reducing d. For a MOS device with a 0.1 μm gate length, the corresponding oxide thickness required is about 3 nm, which is only about ten monolayers.

In conventional MOSFET devices, the minimum oxide thickness is considered to be about 2 nm. Further reduction in the oxide thickness is frustrated by the gate leakage current due to electron tunneling through the oxide. The tunneling current is highly dependent upon the potential barrier height and its thickness. For the Si–SiO$_2$ material system, the potential barrier height of the SiO$_2$ layer is a fixed quantity. Once the oxide thickness reaches about 2 nm, tunneling through the SiO$_2$ layer becomes strong resulting in a high gate leakage current. The power consumption resulting from a high gate leakage current is intolerable in a CMOS circuit and thus must be avoided.

Since further reduction in the oxide thickness is not feasible and the potential barrier for SiO_2 on Si cannot be altered, the only viable alternative is to find another insulator that either forms a higher potential barrier on Si or, alternatively, has a much higher dielectric constant, than that of SiO_2. If the insulator has a higher dielectric constant, then the capacitance of the gate MIS structure will be higher than that for a conventional MOS device at a given thickness. Thus a thicker insulator layer could then be used and still retain the same or potentially higher capacitance for the device. However, the identification of an alternative insulator to SiO_2 that will form as controllable an interface is not easy. It is well known that SiO_2, the oxide of Si, can be grown with exacting control to form almost perfect interfaces. In fact, it can be argued that this is the primary reason Si has emerged as the most ubiquitous semiconductor material in use today. Therefore, the replacement of SiO_2 by another insulator in MOSFET devices may not be practical. Nevertheless, there is ongoing research to identify a high dielectric constant insulator that will be compatible with future MOSFET devices. One such candidate is hafnium dioxide (HfO_2). HfO_2 has a relative dielectric constant of 22, much greater than that of SiO_2, which is 3.9. At this writing, there is some doubt that HfO_2 can be used in conventional MOSFETs since its usage apparently degrades the channel mobility. However, experimental work has shown that the usage of HfO_2 is compatible within strained layer Si–SiGe MOSFETs (see Chapter 8), leading to enhanced mobility and better device performance.

A further problem in MOSFET device miniaturization is parameter fluctuation. As device dimensions continue to shrink, random fluctuations in device parameters become more important in dictating device performance. In other words, the smaller a device becomes, the less robust its performance is to small processing fluctuations. Perhaps the most vivid example is that of random fluctuations of the dopant concentration. In many state-of-the-art MOSFET devices, the dopant atoms within the channel number only in the hundreds. Therefore, small changes in the number and location of the dopant atoms can result in significant performance fluctuations. The most sensitive parameter to dopant fluctuation is the threshold voltage. Hence, within the chip there can be a significant fluctuation in the device threshold voltage, which under most situations is intolerable in circuit design. To date, low cost mechanisms to control the dopant concentration have not been identified.

The last effect we will discuss that influences device performance upon miniaturization is the source and drain series resistances and the selection of the gate material. In long channel devices it is common to ignore the source and drain region resistances with respect to the channel resistance. However, as the device is scaled to ultra-submicrometer dimensions, the source-drain series resistance can become appreciable relative to the channel resistance. As a result, the current drive of the device is degraded.

The material used for the gate contact in many integrated circuits is polysilicon. Polysilicon gates are very highly doped such that their conductivity is nearly that of a metal. However, under bias, a narrow but nonnegligible depletion region forms between the surface of the gate and the insulator. The presence of this depletion layer acts to effectively increase the oxide thickness and consequently reduce coupling of the gate to the channel. To combat this problem, metal gates have been proposed as alternatives to polysilicon gates. The advantage of using metal gates is that due to the very high conductivity of the metal and free carrier concentration virtually no depletion region is formed at the metal–insulator interface. Thus a metal gate can control the channel region more strongly than a polysilicon gate. Identification of new metal gate materials is somewhat complicated since ideally one would like to use the same gate metal for both n-channel and p-channel devices. To date, one of the most promising candidates is a ruthenium–tantalum alloy whose composition can be varied to adjust the metal work function and thus the threshold voltage of the transistor.

In summary, scaling of CMOS devices substantially below 0.1 μm gate lengths is frustrated by fluctuations in the channel doping concentration, difficulty in reducing the gate oxide thickness below 2 nm, and supply voltage reduction. The question is then, are CMOS devices in the 0.1 μm gate length range near the limits of device scaling? In addition, can processing techniques keep pace with Moore's Law? In other words, are there limitations to processing techniques, such as lithography, resists, etching, etc., that will frustrate the practical realization of sub-0.1 μm gate length MOS devices at high levels of integration? At this writing it is unclear just how much further CMOS can be miniaturized. As we will see in the next chapter, alternative approaches to standard CMOS have been introduced that can possibly extend its rate of miniaturization. We will take up the topic of processing challenges to further miniaturization in the next section.

7.3 **Processing challenges to further CMOS miniaturization**

As discussed above, further miniaturization of conventional MOSFET devices is frustrated in several key ways. Aside from the physical limitations imposed by the gate oxide thickness and doping concentration magnitude there are limitations imposed by processing and the voltage supply. Let us next consider limitations due to processing, particularly that from lithography.

One of the key drivers in reducing device dimensions has been the continued improvement of lithographic patterning. Virtually all existing CMOS circuitry is patterned using optical lithography. The most advanced production lithography equipment presently available uses KrF excimer laser sources operating at a wavelength of 248 nm. These systems can readily provide a resolution near 0.25 μm. Further refinement in resolution can be achieved using an ArF source that operates at 193 nm. Feature size resolution near 0.1 μm has already been demonstrated with the ArF source and new projection techniques, specifically phase shifting. Phase shifting uses topological changes in the optical mask to alter the phase of the illuminating radiation. The resulting interference acts to sharpen the image at the wafer plane. However, the phase shifting technique has not yet been demonstrated to be generally applicable to arbitrary device geometries that often are encountered in advanced chip design.

Using light with a wavelength shorter than 193 nm for optical lithography presents many difficulties. The most pressing problem is the identification of suitable materials for transparent refractive optical components. Presently, there is great doubt that optical techniques can be successfully employed to define feature sizes much less than 0.1 μm.

Currently, there is no clear alternative choice for ultra-submicrometer patterning that meets the high throughput required for mass production. Competing approaches are X-ray, extreme ultra-violet (EUV), electron-beam (e-beam) lithography, and scanning tunneling microscopy (STM). X-ray is a prime candidate for high-resolution patterning. Near contact printing enables fabrication of 30 nm feature sizes, which is sufficient to pattern CMOS gates near the perceived limits of operation. The greatest challenge to X-ray lithography lies in mask fabrication. Current X-ray masks are thin membranes patterned with an X-ray absorbing material. Precise control of mechanical stress in the membrane must be maintained to preserve accuracy. X-ray proximity printing gives a 1:1 replication. Therefore, very stringent control of defects is required in order to ensure precise pattern definition.

A second approach that is considered for future device patterning is EUV lithography. Reflective optics is used at 13 nm wavelength with a four-fold reduction scheme. The major technological hurdles to the implementation of this approach are the radiation source and mask fabrication.

A third choice for future lithography is e-beam lithography, which has been the most widely used lithography tool for ultra-submicrometer device fabrication within the laboratory. Excellent patterning has been achieved down to ∼10 nm, limited primarily by the e-beam–resist–substrate interactions using current resist systems. The resolution of an optical lithography system is limited by the diffraction of the incident light. In an e-beam system, the resolution is limited by electron scattering since diffraction is very low because of the very short wavelength of the incident electrons. Due to back scatterings, the electrons can irradiate the material away from the center of the exposure beam. Thus the electron beam irradiation at one location can affect that at other locations. This is called the proximity effect. Usually, the patterns are subdivided such that the integrated exposure from several beams provides the correct dosage at a given location. In this way, the effects of the proximity effect are mitigated. The major limitation of e-beam as a commercial lithography tool is throughput. Rapid, fine patterning of chips with a high level of complexity cannot be presently accomplished using e-beam lithography. Proposed projection systems for e-beam lithography can potentially increase throughput. These systems utilize projection systems with four-fold reduction optics with a mask patterned with electron absorbing material on an otherwise electron transparent substrate. It is expected that these systems can resolve features to within 50 nm.

STM is a variation on e-beam lithography wherein a low energy beam of electrons is used. The particular advantage of using low energy electrons for high resolution imaging is that the detrimental effects of electron scattering that occur in high energy beams are mitigated. STM provides a convenient means of generating a low energy beam of electrons that can be focused into a beam with a 10 nm diameter. STM has been demonstrated to yield pattern features down to 20 nm. Again, the primary limitation of STM in commercial processing is throughput. However, it is possible that by assembling a network of independent probes that operate in parallel, a reasonable throughput can be obtained.

It is difficult to project what capability mass production lithography will have in the future. No other lithographic technology has emerged to supplant optical lithography for mass production CMOS. Projections indicate that usage of known optical lithography techniques will enable patterning down to about 100 nm. To keep pace with Moore's Law, such reduction will need to be accomplished by 2003–2005. At around that point, a significant departure from Moore's Law may occur. This is because the necessary lithographic tools, which have not even been clearly identified yet, would almost certainly need to be in place soon in order to maintain the trend given by Moore's Law.

In addition to miniaturizing the component devices, fabrication of dense circuits requires shrinking the cross-section of the interconnecting wires. It is obvious that, as the dimensions of the devices shrink, the interconnect lines between them would also reduce in length. Both the wire and insulator thickness are scaled down as well. This is because fringe capacitances and crosstalk between wires would increase disproportionally unless the wire thicknesses are scaled accordingly. However, the interconnecting wires must be capable of supporting very high current densities, almost 200 times larger than that allowed in household wiring. Such high current densities can adversely affect the wire through electromigration of the constituent atoms within the wire. As the diameter of the wire is reduced, significant electromigration can result in voids within the wire which can ultimately lead to breaks. Historically, the most ubiquitous metal used for interconnects is Al. This is because Al is highly ductile and has a very low electrical resistance. However, Al is highly sensitive to electromigration, which can subsequently lead to wire degradation.

To circumvent electromigration induced degradation, several strategies can be adopted. The most obvious approach is to limit the current flow in the wires to a value that is safely below that necessary for electromigration. This approach is not attractive since limiting the current magnitude places stringent limits on the ability to charge stray capacitances and switch other transistors. Alternatively, usage of other interconnect metals is possible. One such choice is Cu. Cu has the two favorable qualities of lower resistance and greater insensitivity to electromigration. Mass volume manufacture of integrated circuits using Cu interconnects has begun. Much progress has been made in overcoming the fabrication difficulties encountered in etching Cu. Soon Cu will be the most utilized metal in integrated circuit interconnects.

Increased speed of performance of CMOS circuitry depends critically on reducing interconnect delays between devices. There is a significant signal delay associated with the charging capacitance of the connecting wires which is specified by the resistance–capacitance, RC, product. The RC product for local wires, defined as interconnecting wires placed between devices, is invariant with scaling. Though the RC delay of local wires does not scale, the RC limit for Al based technology is well below the intrinsic delay of 0.1 μm gate length CMOS. Therefore, scaling of local wires will not limit the overall speed of a CMOS circuit.

In addition to local wires, there exist global wires. Global wires are defined as wire interconnects with lengths on the order of the chip size. These wires do not scale down with increasing circuit complexity since chip sizes typically remain the same or slightly increase. The RC delay of global wires increases by the square of the scaling factor. Consequently, global wire delay can become a serious limitation to further CMOS circuitry reduction.

Several solutions have been suggested to overcome the speed limitations imposed by global wires. The most immediate approach is to simply reduce the number of global interconnects by redesigning circuits with this in mind. Another approach is to avoid scaling the cross-sectional area of the global wires altogether. In fact, the global wires can be scaled up in cross-sectional area while the local wires are scaled down in cross-sectional area. The ultimate limits of performance of global wires are then reached when they are scaled up such that they approach the transmission line limit.

Processing challenges to the extension of CMOS involve economic issues as well as the technical issues discussed above. Through the years the cost of a fabrication facility has increased substantially. There exists a second law attributed to Moore. Moore's Second Law examines the economic issues related to integrated circuit chip manufacture. According to Moore's Second Law the cost of fabrication facilities needed to manufacture each new generation of integrated circuits increases by a factor of 2 every three years. Extrapolating from fabrication plant costs of about $1 billion in 1995, by 2010 the projected cost of a single plant would be in the vicinity of $50 billion. Such an immense cost would be unlikely to be met by a single company or even a consortium of manufacturers. Even if a company or group of companies invested enough money to build a $50 billion facility in 2010, there is clearly some limit to this process, since in 2013 the cost of a single fabrication plant would balloon to about $100 billion, etc. Thus the economics of processing plant construction will bankrupt the continuation of conventional CMOS technology unless some alternative strategy is adopted. In the next chapter we will examine alternatives to standard CMOS that may provide a means of continuing device reduction and increased integrated circuit complexity without incurring prohibitive processing costs.

Problems

7.1 Determine how the Poisson equation scales under:
   (a) constant voltage scaling;
   (b) quasi-constant voltage scaling.
7.2 Determine how the drain current below saturation scales under constant voltage scaling.
7.3 Determine how the drain current below saturation scales under generalized scaling. Also determine how the saturated drain current scales under generalized scaling.

7.4 An n-channel MOSFET is to be scaled using constant voltage scaling. If the scaling factor λ is 0.7, determine the new device parameters given that the original device has $L = 1.0$ μm, $Z = 100$ mm, d (the oxide thickness) $= 25$ nm, $N_a = 5 \times 10^{15}$ cm^{-3}, and the applied voltage is 3 V.

7.5 Using (6.50), examine how the subthreshold current behaves when a MOSFET is scaled when the device is OFF, i.e., the gate bias is zero and the drain bias is relatively high, V_{DD}. What happens to the subthreshold current if the threshold voltage is scaled down? Consider how the subthreshold current changes even if the threshold voltage is unchanged. Assume generalized scaling rules.

8

Beyond CMOS

In this chapter we consider both evolutionary and revolutionary advances that go beyond current CMOS technology. By evolutionary we mean advances that retain the standard CMOS paradigm, i.e., the devices are still FETs, consist of similar materials to that used for CMOS, and can be fabricated using current techniques. Revolutionary advances, on the other hand, break out of the standard CMOS paradigm. In these approaches novel substances are used, radically different devices take the place of CMOS FETs, and even the nature of computation is different. The revolutionary approaches to replacing CMOS may or may not prove successful. It is the purpose of this chapter to introduce the student to potential replacements to CMOS keeping in mind that all, some, or none of these approaches may supplant CMOS for computing applications. Finally, the list of revolutionary approaches that we will address is not exhaustive. Specifically, most of the quantum effect methods will not be discussed here, in particular, single electron transistors, spin based devices (spintronics), resonant tunneling devices, and quantum computing. The interested reader is referred to the book by Brennan and Brown (2002) for a discussion of these topics. We will restrict ourselves to discussion of devices that do not require knowledge of quantum mechanics.

8.1 Evolutionary advances beyond CMOS

In this section we examine three different evolutionary approaches that go beyond standard CMOS devices. These are Si on insulator, SOI, dual gate FETs, and silicon–germanium, SiGe, structures. Let us first examine SOI.

As the name implies an SOI device consists of a Si FET grown on a buried SiO_2 layer which is in turn grown on a Si substrate. The FET structure is typically quite thin. The same basic technology is used to make an SOI device as is used for CMOS. Therefore, SOI devices can leverage the great technological strides that have been realized in Si based integrated circuits. SOI utilizes the same substrate material, oxide, and fabrication techniques as conventional CMOS but offers substantial improvement in device level performance since it has the added advantage of nearly complete electrical isolation. In addition, SOI devices are less susceptible to radiation induced failure, such as soft errors, than conventional CMOS devices and typically have better immunity to short-channel effects. A soft error is defined as a change in the logic state of a device due to the accumulation of charge resulting from carrier generation following the transit of a high energy particle or ionizing radiation through the device. The enhanced radiation hardness of SOI devices makes them suitable for space flight applications. However,

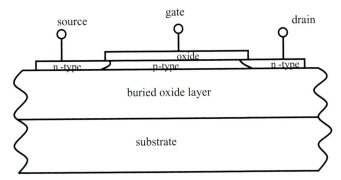

Figure 8.1 Sketch of an SOI MOSFET structure. Notice that the active region of the device is relatively thin and is grown onto the buried oxide layer which results in excellent device isolation.

radiation tolerance is also important in terrestrial applications to avoid alpha particle induced errors. Consequently, SOI devices are possibly better suited to miniaturization than conventional devices. A representative SOI structure is shown in Fig. 8.1.

The principal advantages of SOI devices can be understood as follows. One of the problems of standard CMOS circuits is caused by latch-up between adjacent devices. As the density of CMOS devices increases, the electrical isolation of each device becomes less effective. As a result, the electrical behavior of one device affects that of a neighboring device. The major reason for an interaction between CMOS devices is that the transistors are not physically isolated from one another and an electrical interaction can occur between neighboring tubs and/or the substrate. SOI overcomes this problem since each device is made on top of an insulating layer, SiO_2, in which no current can flow. Consequently, nearly complete electrical isolation can be obtained between neighboring devices eliminating latch-up problems. This advantage of SOI is particularly important at high device packing densities. SOI integrated circuits can be packed more closely than standard CMOS circuits.

As mentioned above, SOI devices are highly immune to radiation induced failure. This feature is called radiation hardness, or rad-hard. The device structures in SOI are made on a thin Si film deposited onto the buried oxide. Since the active layer of the device is very thin it has a very small effective cross-section for radiation damage. In other words, the device volume in which radiation induced carrier generation can occur is very much smaller in an SOI device than in conventional CMOS. Incident radiation will likely penetrate through the thin Si layer and produce electron–hole pairs instead within the buried oxide. The buried oxide is a good insulator so that any carriers produced from incident radiation fail to move through the oxide and enter the active device. As such, the device is relatively unaffected by radiation. The enhanced radiation hardness of SOI devices makes them suitable for space flight applications. However, radiation tolerance is also important in terrestrial applications to avoid alpha particle induced errors. Consequently, SOI devices are possibly better suited to miniaturization than conventional devices.

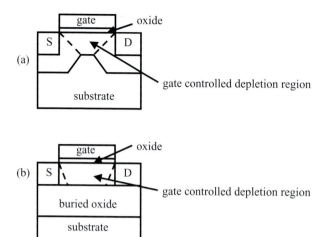

Figure 8.2 Comparison of bulk MOSFET and SOI MOSFET devices showing the difference in the gate controlled depletion regions: (a) bulk device; (b) SOI device. Notice that in the bulk device the source and drain contacts control more of the charge underneath the gate than those of the SOI device control. As a result the bulk device is more susceptible to short-channel effects than the SOI device is.

Isolation of the source and drain from the substrate in an SOI device can reduce the parasitic capacitances of the device by the elimination of the lateral junction area. SOI devices are also more immune to short-channel effects than standard CMOS. Improvement in short-channel effects immunity arises from the fact that the source and drain depletion region widths are restricted and thus do not have as great an influence on the threshold voltage. Comparison of charge sharing between the gate and source and drain regions of a bulk device and an SOI device is shown in Fig. 8.2. As can be seen from the figure, usage of an SOI device restricts the encroachment of the source and drain depletion regions under the gate. As a result, the gate controls more of the charge beneath it than in the bulk device for comparable channel lengths. The SOI device is thus more immune to short-channel effects than the bulk structure. The degree to which an SOI device is affected by short-channel effects depends upon the thickness of the Si layer and whether the device is operated as fully or partially depleted.

In general there are two different manners in which SOI devices are operated. These are partially or fully depleted. Depending upon the thickness of the Si layer and its doping concentration, the Si layer can be either completely depleted of free carriers or only partially depleted. Partially depleted SOI devices typically are made with thicker Si active layers. In the fully depleted SOI devices, the active Si layer is sufficiently thin and the doping concentration sufficiently low that it is depleted of free carriers. In either case, partially or fully depleted SOI, the benefits of SOI devices over conventional CMOS devices are that the source and drain regions are grown onto an insulating layer resulting in a substantially lower parasitic capacitance, better device isolation, and excellent immunity to radiation. However, partially and fully depleted

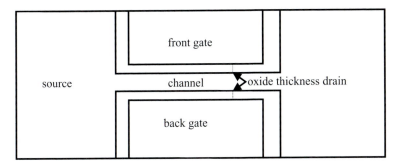

Figure 8.3 Sketch of the cross-section of a dual gate MOSFET device. Notice that the channel is controlled by both the front and back gates.

SOI devices have different advantages and disadvantages relative to one another. The principal disadvantage of a partially depleted SOI device is body charging. In the partially depleted SOI device the substrate is electrically floating. Since the region of the device beneath the conducting channel, usually referred to as the body, is not depleted or contacted, it can charge up with charge generated through impact ionization events near the drain of the device. The concomitant increase in the body potential lowers both the threshold voltage and the source potential barrier leading in turn to an increase in the drain current. This acts as a feedback mechanism producing a sudden increase in the drain current called the kink effect. Additionally, charging of the floating body can result in premature breakdown as well as an increase in the subthreshold slope at high drain bias. Since the degree to which the floating body of the device is charged is variable, different devices may exhibit different behaviors thus complicating circuit performance. One particular problem a floating body SOI device experiences occurs during switching. The body potential rises with the gate potential, which acts to reduce the threshold voltage resulting in an increase in the transient current. This effect is called drain current overshoot and its magnitude is a function of the earlier states of the device. Thus the device performance is history dependent, i.e., its performance depends upon how recently and how often the device has been switched through conditions under which impact ionization can occur.

The fully depleted SOI devices on the other hand have no body effect since they have no floating body. Unfortunately, the fully depleted devices are more susceptible to short-channel effects than partially depleted SOI. The threshold voltage of fully depleted devices is a function of the Si and buried layer thickness whereas that of the partially depleted SOI device is not. In a fully depleted SOI device there is a stronger coupling between the source and drain into the channel, thus aggravating the short-channel effects.

Another approach that goes beyond CMOS is a dual gate FET. A representative dual gate device is shown in Fig. 8.3. As can be seen from Fig. 8.3, both the front and back gates control the channel. The two gates act to shield the source region of the channel from the action of the drain, providing a high degree of isolation of the source and drain. As a result, short-channel effects are reduced in this structure. Some of the advantages of dual gate devices relative to single gate MOSFETs are as follows. As mentioned, short-channel effects are reduced in dual channel devices. The dual gates reduce the drain-induced barrier lowering and minimize the threshold sensitivity to the channel. The dual gate device has a relative scaling advantage of about a factor of 2 over conventional CMOS devices. One of the major disadvantages of dual gate MOSFETs is that the presence of two Si–SiO$_2$ interfaces from the two gates increases the surface roughness scattering. As a result the carrier mobility is reduced thus lowering the frequency of operation of the device.

There are, in general, two different types of dual gate MOSFET structures. These are symmetric and asymmetric gated devices. The symmetric type is made such that both gates have the same work function. This results in the surface channels turning on at the same gate voltage. For asymmetric gated devices the work functions of the two gates are different. Thus the surface channels turn on at different voltages. The asymmetric devices have roughly the same current drive that the symmetrical devices have. The asymmetric structures have the additional features that they are more easily fabricated, have more design flexibility, and have potentially better immunity to short-channel effects.

In practice there are three different ways in which dual gate MOSFETs can be made. In all three approaches the device is no longer planar but is essentially three-dimensional instead. As a result, the relative ease of planar fabrication is lost in dual gate devices. Though the dual gate devices are more difficult to manufacture, there has been much progress in developing viable structures. The simplest dual gate scheme is made with top and bottom gates formed parallel to the plane of the substrate. In this configuration the source and drain regions straddle the two gates and the current flow in the channel runs parallel to the substrate. Alternative dual gate FETs are made vertically. In the vertical dual gate FETs the gates are perpendicular to the substrate and the current flows parallel to the substrate. Perhaps the most popular design is called the FinFET. In the FinFET a narrow vertical fin sticks up from the substrate that is formed via etching. The source and drain regions are formed on either end of the fin with the gate electrodes on either side of the broad face of the fin. One advantage of this design is that the fin can be made very thin such that the gate electrodes control the entire region everywhere beneath the gates. Hence, the drain-source subthreshold current can effectively be totally eliminated thus improving device performance.

The last evolutionary advance we consider is the usage of Si and Si–Ge together to form a heterostructure. As was discussed in Section 5.2, a heterostructure results from the sequential growth of two dissimilar semiconducting materials on one another. For example, one can grow a layer of Si$_x$Ge$_{1-x}$ on a Si substrate to form a SiGe–Si heterostructure. The lattice constants of Si and Si$_x$Ge$_{1-x}$ are different. As a result, when the two are grown on top of one another, they do not readily "fit together." Two possibilities can result. If the top layer is sufficiently thin, it can adopt the lattice constant of the underlying layer. As a result the thin top layer can be in either compressive or tensile strain as shown in Fig. 8.4. Notice that if the lattice constant of the top layer is less than that of the underlying substrate, then the atoms forming the top layer must

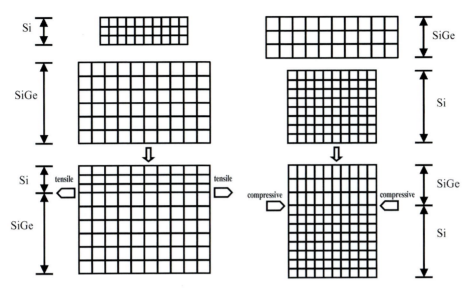

Figure 8.4 Illustration of the formation of a heterojunction when the lattice constants of the constituent semiconductors are not equal. If the top layer is grown sufficiently thin then it will adopt the lattice constant of the bottom layer. If the top layer has a smaller lattice constant than the substrate then the top layer will be in tensile stress as shown in the diagram. Alternatively, if the top layer has a larger lattice constant than the substrate the top layer will be in compressive stress.

move further apart to match the lattice constant of the substrate material and the system is in tension. Alternatively, if the lattice constant of the top layer is greater than that of the underlying substrate, then the atoms forming the top layer must move closer together to match the lattice constant of the bottom layer resulting in compressive stress. It is important to recognize that the top layer must be quite thin, less than a critical thickness. If the top layer is thicker than the critical thickness the strain relaxes and the two layers each retain their inherent lattice constants. When this occurs defects are necessarily formed at the interface, since the two lattice constants can no longer be accommodated.

Si_xGe_{1-x} has a greater lattice constant than Si. Either SiGe can be grown on top of a Si substrate producing a system under compressive stress or Si can be grown on top of SiGe producing a system under tensile stress. The system of greatest importance to MOSFETs is Si on top of SiGe. In this case the thin Si top layer is strained and in tension. The resulting tensile strain within the Si layer modifies the conduction band structure of Si. In unstrained bulk Si the minimum energy of the conduction band occurs along the [100] crystallographic direction, which means along the x axis. Si has cubic symmetry and thus all of the 100 directions are equivalent. This means that [100], [010], [001], and their corresponding negative quantities are all equivalent directions. In other words, the energy band structure remains the same whether one moves along the x, y, z, or negative x, negative y, and negative z directions in k-space. Therefore, there are six equivalent minima for the first conduction band of Si. These minima, or valleys, are referred to as the X points. Two of the valleys, called the normal valleys, have their longitudinal axis normal to the growth plane. The other four valleys, called the parallel valleys, lie within the growth plane. When the Si layer is under tensile strain, the normal valleys are shifted downwards in energy while the parallel valleys are shifted up in energy. The degree to which the valleys shift in energy depends upon the Ge mole fraction in the SiGe layer upon which the Si layer is grown. The lower lying normal valleys are more readily occupied than the higher lying parallel valleys. Depending upon the Ge concentration the energy separation between the normal and parallel valleys can be made significantly larger than the mean thermal energy, thereby ensuring excellent confinement of the electrons within the normal valleys over a wide range of applied field strengths. Transport parallel to the interface is governed by the smaller transverse effective mass within the normal valleys. The electron mobility then is greater for the strained Si layer as opposed to that of bulk Si. Therefore, in the strained Si layer at relatively low applied electric fields the electrons move within the normal valleys and experience a high mobility.

The enhanced electron mobility within the strained Si acts to increase the speed of the SiGe MOSFET. Therefore, strained Si–SiGe MOSFETs have an intrinsically higher transconductance and thus higher frequency of operation than conventional MOSFETs.

Example Problem 8.1 Partially depleted SOI device

Determine the thickness of the Si active layer in a partially depleted SOI device if the layer is only 50% depleted in equilibrium. Assume that the Schottky barrier height is 0.72 eV, the effective density of states within the conduction band is $N_c = 3.22 \times 10^{19}$ cm^{-3}, and the doping concentration is 10^{16} cm^{-3}. Assume that all of the donors are ionized.

From Example Problem 3.6, the built-in voltage of the Schottky barrier gate is

$$q V_{bi} = q \Phi_B - kT \ln \left(\frac{N_c}{n} \right)$$

Substituting in, the built-in voltage is

$$q V_{bi} = 0.72 \text{ eV} - 0.0259 \text{ eV} \ln \left(\frac{3.22 \times 10^{19}}{1 \times 10^{16}} \right) = 0.51 \text{ eV}$$

The width of the depletion region for a Schottky barrier is

$$W = \sqrt{\frac{2\kappa}{q N_d} (V_{bi} - V)}$$

V is zero in equilibrium and the depletion region width is then

$$W = \sqrt{\frac{2\kappa}{q N_d} (V_{bi})}$$

Substituting in values for each variable, the depletion region width is given as

$$W = \sqrt{\frac{2(11.8)(8.85 \times 10^{-14}\,\text{F/cm})}{(1.6 \times 10^{-19}\,\text{C})(1 \times 10^{16}\,\text{cm}^{-3})}0.51\,\text{V}} = 2.59 \times 10^{-5}\,\text{cm}$$

If the device is a partially depleted SOI device such that only half of the active region is depleted, the width of the active n-type semiconductor region is $2W$. Thus the active region width is 5.18×10^{-5} cm.

8.2 **Carbon nanotubes**

The most obvious approach to increased reduction in device size is to continue to make FETs but make them at nanoscale dimensions, of the order of 10^{-9} m or less. As discussed in Chapter 7, the continued miniaturization of standard Si based MOSFETs to these dimensions faces many daunting challenges. These challenges may prove technically too difficult or become too costly to overcome. Therefore, to maintain progress in accordance with Moore's First Law, alternative revolutionary approaches to CMOS may have to be taken, wherein CMOS and its evolutionary derivatives are abandoned. In this section we discuss very small FET structures that are made using a radically different technology than Si, that of carbon nanotubes. Though the growth and behavior of carbon nanotubes is vastly different from that of Si, the basic device structure made with these materials is still a FET. In later sections of this chapter we will discuss approaches that not only use different materials with different properties from Si, but employ totally different device strategies as well.

Carbon nanotubes are essentially very small graphite sheets rolled into a seamless cylinder. In most respects a nanotube can be defined as a tubular nanocrystal of graphite. Nanotubes are very thin and are long macromolecules of carbon. Nanotubes can be either multi-walled or single walled. Multi-walled nanotubes consist of concentric cylinders wrapped around each other while single walled nanotubes consist only of one layer. These nanotubes as the name suggests have dimensions within the nanometer range. Nanotubes are relatively easy to make, though as we will see below they are somewhat more difficult to manipulate and deposit in a controlled way. They arise naturally as components of soot deposits following high energy arcing of carbon at high temperature. Electrically carbon nanotubes can act as either a metal or a semiconductor depending upon the diameter of the tube and its helicity, or degree to which the carbon spirals around the tube. The energy band gap of a carbon nanotube depends mainly upon the diameter of the tube; the larger the diameter the smaller the energy gap. A nanotube can act like a semimetal if it has a sufficiently large diameter. Conversely, a narrow width nanotube is semiconducting. Electrical transport at very low temperature near that of liquid helium in a nanotube is mainly ballistic. In this regime the electrons flow without collision from one end of the tube to the other. As the temperature of the nanotube is raised the transport becomes collision dominated and is mainly diffusive.

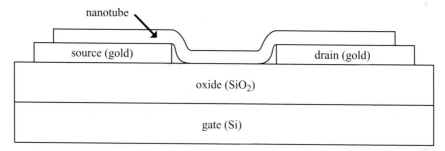

Figure 8.5 Nanotube FET device. A single nanotube connects the source and drain gold contacts as shown in the figure. The back gate comprises Si with a SiO_2 insulator.

Mechanically, carbon nanotubes have one of the largest strength to weight ratios of any material. Thus potentially they can form a very strong fiber.

A typical FET structure made with a carbon nanotube is shown in Fig. 8.5. The nanotube connects two gold (Au) contacts that form the source and drain respectively. The structure is formed on top of a SiO_2 layer which in turn is grown on top of a Si layer. The bottom Si layer acts as a back gate. The drain current vs. drain voltage curves are linear for applied gate voltages less than zero and at a gate voltage of zero. As the gate voltage becomes greater than zero, at some point the current within the nanotube becomes immeasurably small. Since the nanotube loses its conductivity at high positive gate voltages the charge carriers are dominated by holes. Thus the device works in a similar manner as does a p-channel MOSFET: the conducting channel becomes depleted at a high positive gate bias. A conductance modulation of about five orders of magnitude has been observed in this type of nanotube FET (see Martel *et al.* (1998)), thus making it an effective nonlinear device. The conductance modulation of the carbon nanotube FET is comparable to that of a Si MOSFET. Thus the nanotube device can operate at room temperature with characteristics similar to those of a standard Si MOSFET. However, the nanotube device can be made very much smaller and because of its small size be able to switch using far less power than a Si CMOS MOSFET.

An additional advantage of carbon nanotubes for microcircuitry is their high thermal conductance. Carbon nanotubes have thermal conductivities close to that of diamond, making them efficient heat conductors. For this reason, nanotubes could be used to cool very dense arrays of devices.

Though nanotubes offer some intriguing possibilities for nanoscale FETs, they have some serious shortcomings that can affect their ultimate usefulness in circuits. The nanotube FETs presently made require manipulation of the nanotube by atomic force microscopes. Not only does it require an atomic force microscope to examine the nanotube but its manipulation, i.e., placement onto the Au contacts, is often performed with such microscopes. Therefore, the fabrication of nanotube FETs presently is done one at a time with great effort, making them unattractive for high density circuits. In addition, it is presently difficult to grow nanotubes where they are wanted. Some researchers have found that using nickel (Ni), iron (Fe), or other catalysts nanotubes can be selectively deposited onto a substrate.

Nanotubes may find usefulness as interconnects. Their very small diameter and high conductivity make them attractive candidates for wires in an integrated circuit. One of the problems encountered in dense CMOS circuits is that of interconnects. As the packing density of devices increases it is essential to scale down the width of the interconnects. However, continued shrinkage of metal interconnects is challenged primarily by electromigration, as was discussed in Chapter 7. As the width of an interconnect decreases, the current density flowing within the wire increases. Electromigration occurs when the atoms of the metal interconnect are displaced by the high current densities. This can result in wire breakage leading to failure of the chip. Many chip manufacturers have changed from Al to Cu (see Section 7.3). Cu has been introduced as a new interconnect metal due to its greater immunity to electromigration. However, ultimately even Cu wires will fail if sufficiently reduced in width. Heat removal is also a consideration. As the packing density of devices increases, removal of waste heat from the devices becomes much more difficult. Therefore, packing devices in more tightly results in thermal problems that adversely affect the operation of the chip. Nanotubes may provide an alternative interconnect material that has a high thermal conductivity, and can thus help in dissipating heat in a highly packed circuit. Additionally, nanotubes can support extremely high current densities without any electromigration problems.

Carbon nanotubes can also be used to make emitters for compact, flat-panel displays. Electrons can be emitted from the narrow tip of a metal rod in what is termed field emission. Essentially, the application of a high electric field can supply sufficient energy to the electrons that they can be emitted from the surface of the metal tip. The electrons then travel through vacuum and strike a phosphor screen, selectively illuminating a portion of the screen. To make a display a two-dimensional array of field emitters is used. Though metallic field emitters have been around for some time they generally require a very high applied voltage for proper operation. Alternatively, nanotubes can emit electrons at a much lower applied voltage thus greatly reducing the power requirements of an array of nanotube emitters. Nanotube displays have already been made and have shown very good lifetime and operating characteristics. However, the major drawback to nanotube displays is their orderly operation. Typically, each pixel (which consists of at least one field emission tip) must be separately contacted. A pixel is essentially a unit cell within the flat-panel display that provides a small portion of the final image. The intricacies of wiring such a vast array of pixels present an ongoing challenge.

8.3 Conventional vs. tactile computing, molecular and biological computing

Conventional computers are machines that follow a well described set of instructions to process data. Basically, a set of instructions is read into the machine and it works sequentially in an ordered way to execute a task. Such machines are often referred to as Von Neumann computers or classical machines. The major components of a Von Neumann computer are memory, processing, and bandwidth. Using these ingredients classical computers are very well suited to number crunching, communication, and data manipulation. However, there are problems that classical digital computers are not well suited for. The most important of these involve pattern recognition. Specifically, it is very difficult to program a classical computer to recognize a complicated molecule or distinguish between different microorganisms. In the chemical and biological world pattern recognition is highly efficient and readily accomplished. For example, the immune system in the human body is adept at identifying foreign organisms such as viruses and bacteria. When the human body is attacked by foreign microorganisms, the immune system is able to identify that the invader is foreign, decide if antibodies should be made to combat the invader, decide if and when these antibodies should be released and which of a variety of other responses should be activated. This behavior is an example of a different type of computing, which we will refer to as tactile computing.

The primary difference between conventional and tactile computing platforms can be summarized as follows. Conventional electronic digital computers can be thought of as structurally programmable machines. This means that the program controls the behavior of the machine. A compiler translates the input code into machine language that is expressed in terms of the states of simple switching devices and their connections. The machine computes symbolically and the result depends to some extent upon the human input.

Tactile computing machines are quite different. These machines are not structurally programmable. In biomolecular computing, pattern processing is physical and dynamic as opposed to the symbolic and passive processing in a conventional machine. Programming depends upon evolution by variation and selection. Such a molecular computer is called a tactile processor. A tactile processor can be thought of as a computer driven by enzymes; the inputs are converted into molecular shapes that the enzymes can recognize. An enzyme thus performs sophisticated pattern recognition since it scans the molecular objects within its environment, interacting with only those molecules that have a complementary shape. Recognition is thus a tactile procedure.

An example of a tactile computer would be strains of bacteria bred to dissolve oil spills. Basically, the bacteria would be "programmed" by altering its DNA such that it would bloom in the presence of an oil spill, eat the spilt oil, and then die off after the oil has been eliminated. The bacteria would act like a type of computer in the sense that they have been programmed to identify and respond to a given problem, albeit its programming and response are far different from those of a classical computer. In this sense the unique and powerful information processing capabilities of life, pattern and object recognition, self-organization and learning, and effective use of parallelism are harnessed. It is just these properties that are not well reproduced by a classical, conventional computing machine. Tactile computing provides an entirely different approach to computing than conventional computing.

Another example problem that a tactile computer can attack is that of sensing bioagents or toxins. Molecules can be used to perform complicated pattern recognition of dangerous toxins released into the air or water. Very dilute amounts of toxins can be identified using tactile computing. Identification of toxins in the external environment would present an almost overwhelmingly difficult task for a binary Von Neumann computer. A tactile computer on the other hand can readily make an identification. Thus such an approach can be used to fight terrorist bioattacks, or chemical warfare.

It is important to recognize that biomolecular or tactile computing proceeds along an entirely different line than conventional computing. Essentially, tactile computing provides a high degree of pattern recognition that can work relatively quickly; an enzyme encounters and interacts with a substrate molecule providing the "molecular switching action." The "program" is in the molecule itself; the computation occurs by the recognition of one molecule by an enzyme and their subsequent chemical reaction. Conventional computing requires massive amounts of code to instruct a machine to recognize a single molecule. Therefore, tactile computing provides a different yet complementary approach to computing than conventional machines. Below we will examine tactile computing using biological systems, DNA, and molecules.

It is useful to examine the advantages of molecules as computing engines. Biological systems are a special case of molecular computing. Therefore, we will consider the broader context of molecules and then refocus on biological molecules and biological systems.

The major qualities of molecules that make their potential usage in future applications attractive are that they offer simplified fabrication, low cost and relatively unlimited availability, and devices made from molecules are very much smaller and compact than existing CMOS structures. For example, a CMOS OR gate requires anywhere from 4–6 orders of magnitude more surface area than a comparable molecular logic gate. Molecular electronics offers the potential then for extremely compact computing capability, i.e., a microprocessor on a pinhead. Coupled with their expected lower power consumption, molecular devices offer a revolutionary trend in electronics: highly miniaturized, low power devices that are relatively inexpensive to fabricate and mass produce. The fact that molecular materials can be readily engineered and occur in many different forms with a variety of properties enables the potential realization of a vast number of possible structures.

The key advantages of molecular materials include:

(1) Size: the dimensions of individual molecules are in the nanometer range; on average they are about 4 nm. These device dimensions are about two orders of magnitude smaller than that which can be obtained using Si CMOS technology. Quantum dots and spin systems are also typically larger than these dimensions.
(2) Three-dimensional structures: molecular materials are inherently three-dimensional. In contrast, in Si based technology, much fabrication effort and cost is required to produce three-dimensional geometries.
(3) High packing density: the combined features of small size and three-dimensionality make very high packing densities possible. It is estimated that the packing density can be increased by 6–9 orders of magnitude over CMOS.
(4) Bistability and nonlinearity: bistability and nonlinearity can be utilized to perform switching functions. Both of these properties are commonly available in molecules.
(5) Anisotropy: it is important to recognize that the electronic and optical properties of a molecule are inherent in the molecular structure instead of being fabricated by the processing technologies as in CMOS.

(6) Upward construction: organic synthesis enables growth of microstructures from the small upward. In standard CMOS, device and circuit functionality is sculpted from a relatively large piece of material.

(7) Self-organization: molecules present some new important functionalities not found in other materials. These are: self-organization, self-synthesis, and redundancy factors well known in organic and biological molecules that could potentially be applied to molecular electronic devices.

(8) Low power dissipation: biological molecules require much lower amounts of energy to switch than CMOS circuits. Estimates are that the total power requirements for molecular switches will be about five orders of magnitude lower than for CMOS switches.

(9) Molecular engineering: it is possible that molecules can be tailored or engineered to perform specific tasks or have specific properties. In other words, molecules can be selectively grown and made to inherently possess desired qualities to perform a task.

The above features of molecules also apply to biomolecules. The most important attributes of biomolecules can be summarized as follows. Biomolecules essentially provide a parallel environment for computation, they occur abundantly in nature, and their implementation due to self-organization and upward construction is relatively simple compared with the difficult and costly fabrication of CMOS circuitry. As mentioned above, size is a key advantage of molecules. For example, the entire genetic code for humans is contained in the nucleus of most cells. The DNA code consists of over three billion nucleotide pairs and fits into a few double helices about 3.4 nm in width and measuring micrometers in length.

Researchers are developing what are called applets for biological systems. An applet is generally defined in computer science to be a short piece of code that is often employed on web pages to perform a specific task. In the biomolecular world one creates genetic applets. These applets enable the system to respond to an external event, "program" a cell to produce a desired chemical or enzyme, or enable a cell to identify a reagent. Some potential future applications of biological applets are in gene therapy for treating diseases such as hemophilia, anemia, etc. Another potential application of biological applets is in the treatment of diabetes. A genetic applet can be used that senses the glucose level in the blood and another applet can then direct the production and release of insulin if needed.

The primary biomolecule chemistry of relevance is that of DNA. DNA molecules are composed of four basic nucleic acids called adenine (A), guanine (G), cytosine (C) and thymine (T). A and T, and C and G naturally bond together to form pairs. To compute with DNA there are three basic steps. These are: encoding that maps the problem onto DNA strands, basic processing using a chemical process called hybridization that connects two complementary DNA strands into a double strand, and outputting the results. Progress has been made in programming DNA to perform computational tasks. The technique involves the usage of what are called DNA tiles. These tiles consist of multiple strands of DNA knotted together. The ends of each tile are created such that the tile will recognize and attach to other pre-designed tiles to make self-assembled structures. These tiles can be used to add or multiply numbers, thus providing a new approach to numerical computation. In addition, it has been shown that DNA tiles can be utilized to perform any operation that can presently be done by an electronic computer.

Molecules can also be used to build devices that mimic CMOS functionality. Molecular diodes and other quantum based devices have been designed. The primary components of these molecular structures are conducting groups called polyphenylenes and insulating groups called aliphatic molecules. By arranging these molecules in various orders similar device action to that found in semiconductors can be attained. These structures can be combined to make various logic gates such as NOR and NAND gates. This approach to doing molecular computing is often referred to as moletronics. The workings of some moletronic devices requires some understanding of quantum mechanics (see Brennan (1999) Chapters 1–4, for an introductory discussion of quantum mechanics). However, there are some diodelike devices that can be made from molecules. In the next section, we present a short discussion of these molecular devices and illustrate how they can be used to perform digital logic functions.

Though biological and more generally molecular tactile computing offer a wide range of new approaches, there are some important drawbacks that currently restrict their application. These are basically related to the difficulty encountered in regulating or controlling the basic chemistry. Implicit to the challenging chemistry are reliability, efficiency, and scalability. In this context, reliability means the degree of confidence in correctly solving a problem. Efficiency is related to the effective manipulation of the molecules used to perform the computation. Scalability is the successful reproduction of the desired event many times. For example, performing one successful experiment instead of many is inadequate. To be truly useful biological computing must be able to reproducibly perform many operations in an efficient manner. Currently, this is not the case.

In conclusion, it is important to recognize that molecular and its special case of biomolecular computing offer an entirely new approach to computation that cannot be readily performed on an electronic, Von Neumann machine. Additionally, molecules can be used to mimic CMOS functionality providing a much lower power, greater compaction alternative to CMOS devices. An example of this approach is presented in the next section. Presently, the challenges of successfully efficiently controlling the chemistry of molecules in a reproducible manner have not been fully realized. Nevertheless, much progress has been made and this area remains a fruitful and exciting avenue of research.

8.4 **Moletronics – molecular diodes and diode–diode logic**

We begin our discussion of moletronics by examining the types of molecules used and their properties. Generally, both conducting and insulating molecules are needed to form devices. Presently, the two most promising conducting molecular species are polyphenylenes and carbon nanotubes. The polyphenylenes are essentially

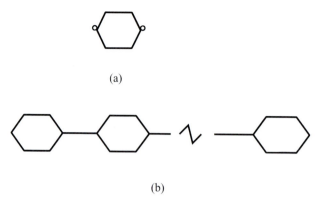

(a)

(b)

Figure 8.6 Sketch of (a) a phenylene group and (b) polyphenylene. The two free binding sites on the phenylene group are represented by the open circles on the ends of the ring. Notice that with two free binding sites, each phenylene group can be bound to two others. In the polyphenylene sketch, each ring represents one phenylene group.

chain-like molecules formed by linking a basic molecular unit, C_6H_4, phenylene, which is a derivative of the benzene ring but with two free binding sites (see Fig. 8.6). Given that each phenylene has two free binding sites it can be linked to two other phenylenes to form a chain as shown in Fig. 8.6(b). These chains can then be used to form molecular wires. Polyphenylene wires are fairly conductive. Conduction in these molecules proceeds by electrons moving through extended molecular orbitals that span or nearly span the length of the entire molecule. The extended molecular orbitals are called π-type and lie above and below the molecular plane when the polyphenylene is in a planar conformation. Basically, the extended π-orbital states occur in much the same way that extended states in a solid occur, i.e., when the atoms become close enough spatially, the wavefunctions overlap leading to extended states. The resulting extended states are no longer degenerate in energy. As in a superlattice (Brennan, 1999), if many atoms contribute π-orbitals, corresponding to a relatively long polyphenylene chain, a range of extended states is formed producing a band. If the chain is relatively small containing few atoms contributing π-orbitals, only a relatively small set of extended states is produced. In any event, these extended states provide a conducting path through the molecular chain. Hence, the polyphenylene chains are conducting.

Alternatively, chains formed with aliphatic organic molecules can produce an insulator. Aliphatic molecules are singly bonded molecules that do not contain π-bonds, but have what are called σ-bonds instead. The σ-bonds lie along the axes of the molecules and are not easily extended between atoms. This is because at each end of a σ-bond there is a positively charged nucleus and thus its spatial extent is "interrupted" by the nucleus. Chaining together methylene, CH_2, molecules can form an aliphatic molecule. Inserting aliphatic molecules into a polyphenylene chain can also form insulating molecules. The aliphatic molecules interrupt the extended states formed by the π-bonds thus breaking the conducting pathway through the polyphenylene chain.

The primary molecular device type we will consider is a molecular rectifying diode. As in a p–n junction diode, two different doped regions are required: a p-type region and

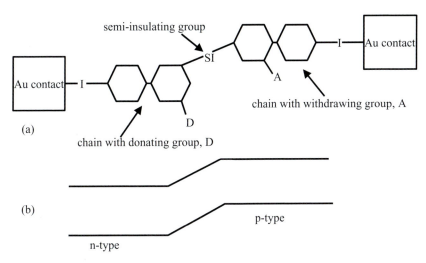

(a)

(b)

Figure 8.7 Sketch of (a) a molecular diode and (b) a solid state diode. In (a) the molecular arrangement for the rectifying diode structure is shown. The insulating groups I provide potential barriers for tunneling into and out of the diode. The semi-insulating group, SI, in the center of the device plays the role of the built-in potential of the diode maintaining the electron density imbalance within the junction. Note that the drawing does not represent the energy of each side of the junction. The side with the donating group is at a higher energy than the side with the withdrawing group. In (b) the p-type layer is on the right hand side and the n-type layer on the left.

an n-type region. As in semiconductors, the electron concentration can be altered by the introduction of foreign agents. In semiconductors, one adds dopants. In molecular systems, one adds molecular groups that attach themselves at specific places within the chains altering the electron concentration. Groups that add electrons to the system are called electron donating groups, while groups that remove electrons are called electron withdrawing groups. Electron donating and withdrawing groups are analogous to donors and acceptors in semiconductors. As in a semiconductor, a p–n junction can be formed by placing a chain with withdrawing groups together with a chain with donating groups. A semi-insulating group is inserted to form a potential barrier between the donating and withdrawing group chains. The potential barrier formed by the semi-insulating group serves to maintain the charge imbalance between the two sides of the junction. In other words, the differing electron densities between the two sides of the junction would equilibrate if the potential barrier did not exist, thus removing any diode action. In a p–n junction diode, the built-in potential plays the role of the semi-insulating group. The basic scheme is shown in Fig. 8.7(a) along with a solid state p–n junction for comparison (Fig. 8.7(b)).

Before we discuss the current flow in the molecular diode it is important to understand the energy level diagram under equilibrium. In equilibrium the Fermi levels are of course aligned as shown in Fig. 8.8. The lowest unoccupied molecular orbital (LUMO) levels in the donating chain are at a higher energy than those in the withdrawing chain as shown in the diagram. This is because the electron density is increased

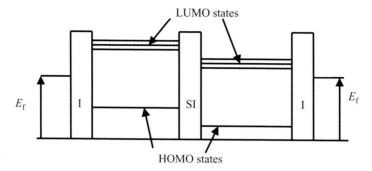

Figure 8.8 Sketch of the molecular diode energy levels under equilibrium. The center potential barrier, labeled SI, corresponds to the semi-insulating layer. The two end potential barriers, labeled I, are the insulating layers between the two chains and the contacts. In equilibrium notice that the Fermi levels are aligned. The LUMO levels on the donating chain lie at a higher energy than the LUMO levels on the withdrawing chain.

within the donating chain resulting in an increased electron–electron repulsion and hence a higher total electron energy. Similarly, the lower electron density within the withdrawing chain results in a lower total electron energy. Consequently, there exists an energy difference between the two sides of the molecular diode, i.e., the levels within the donating chain lie at a higher energy than the levels within the withdrawing chain.

Current flow within the molecular diode can be understood as follows. A molecular diode has two bias conditions, forward and reverse, similar to the case of a solid state diode rectifier. As in a solid state diode, the current flow is asymmetric with respect to the applied bias. Under forward bias, the current flow is high while under reverse bias the current flow is much lower. The key to understanding the asymmetry in the current flow is the fact that the difference in energy between the Fermi level and the LUMO levels on the side with the withdrawing group is less than the energy difference between the Fermi level and the LUMO levels on the side with the donating group. The bias required to align the Fermi level with the LUMO levels on the withdrawing side is less than that required to align the Fermi level with the LUMO levels on the donating side. Therefore, for the same magnitude of bias, the current flow within the device will be highly nonlinear and asymmetric.

Let us consider each bias in turn. The insulator groups, I, and the semi-insulating group, SI, act as potential barriers in the structure. Therefore, in the band diagram for the device they are shown as barriers. The donating chain (left hand side of the diode) and withdrawing chain (right hand side of the diode) are conducting with HOMO (highest occupied molecular orbital) and LUMO states. Forward bias occurs when a high potential is applied to the left hand contact with respect to the right hand contact. A positive potential acts to lower the electron energies. Hence, the occupied energy levels within the left hand contact are lower than those within the right hand contact as shown in Fig. 8.9(a). Since the contacts are metals, the uppermost occupied levels in the contacts are the Fermi levels. When the Fermi level in the right hand contact

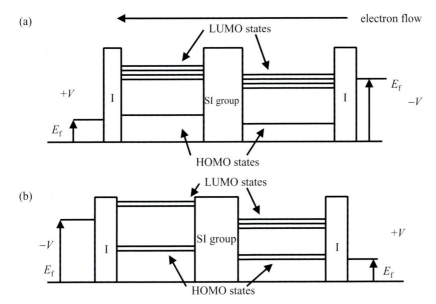

Figure 8.9 Sketch of the electronic structure of the molecular diode under (a) forward and (b) reverse bias. In both cases the same magnitude of bias is applied. In (a) the intended electron flow is from the right to the left. In (b) the electron flow is from left to right. Notice that resonant alignment does not occur at the same magnitude of bias as in (a).

is sufficiently raised so as to align with the LUMO levels in the withdrawing chain, electrons can tunnel from the contact into the LUMO levels (see Fig. 8.9(a)). These electrons can then tunnel through the SI potential barrier into the donating chain LUMO levels. Finally, the electrons tunnel through the last potential barrier into the positively biased left hand contact. It is important to note that the electrons can tunnel appreciably when the levels are aligned in terms of energy. When the levels are not aligned a much smaller current flows called the nonresonant current. Notice that under what is called forward bias only a relatively low bias is necessary in order for a current to flow. As shown by Fig. 8.8, the Fermi level lies closer to the LUMO levels in the withdrawing chain than in the donating chain. Hence, only a relatively low bias is necessary to align the LUMO levels in the withdrawing chain with the Fermi level in the contact.

The reverse bias conditions can be understood as follows. Under reverse bias, a high voltage is applied to the right hand contact with respect to the left hand side contact as shown in Fig. 8.9(b). The Fermi level is lowered on the right hand contact and raised on the left hand contact as shown in the diagram. For the same magnitude of bias as in the forward case, the Fermi level on the left hand contact is *not* aligned with the LUMO levels. As such no resonant alignment occurs and little current flows in the device. The principal current flow mechanism under reverse bias is the nonresonant current which is much smaller in magnitude than the resonant tunneling current.

Given that molecular diode rectifiers can be made, the next question is whether logic gates can be constructed. It is well known that simple AND and OR gates can be constructed using rectifying diodes. A simple example of a two-input AND gate using

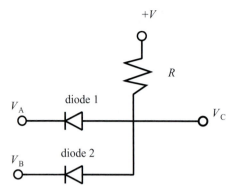

Figure 8.10 Sketch of a two-input diode AND logic gate. The inputs are V_A and V_B. The output is V_C.

solid state diodes is shown in Fig. 8.10. The operation of the gate can be understood using the ideal diode model. A positive voltage is applied across the resistance, R, as shown in the diagram. If either V_A or V_B is low, or both are low, then the diodes 1 and 2 are forward biased. As a result, there is little voltage drop across the diodes, and in the ideal approximation no voltage drop across the diodes. Therefore, the point, V_C is also low. Thus for either input low or both inputs low, the output is low. Alternatively, if both of the inputs, V_A and V_B, are high, then both diodes 1 and 2 are reverse biased. Under the ideal approximation the two diodes are open circuits. Therefore, V_C is high. Hence, if both inputs are high the output is high. This constitutes AND gate performance. An OR gate can be made in a similar manner.

8.5 **Defect tolerant computing**

Another issue that impacts molecular computing is the tolerance of defects. Molecular computers have been envisioned that will be something like the human brain, containing many orders of magnitude computing elements, far more than are currently present in electronic digital computers. As in the brain, it is virtually guaranteed that not all of the components will operate correctly. There necessarily will be many defective devices and pathways. Nevertheless, the machine must still function and be able to "program" itself around the defects. In this section we examine defect tolerant computing wherein the operation of a computer can be maintained even if part of its hardware is defective.

The standard paradigm for computing hardware is that the microprocessor chip needs to be perfect and free of defects. Perfectly working chips are necessary to ensure that the processing power of the computer remains flawless. Additionally, chip operation must be perfect to ensure a high level of versatility such that the chip can perform a wide variety of functions without error. As the complexity of the chip grows the number of active device elements on the chip increases. Ensuring that all active device elements work properly and that there are no shorts or opens in the connecting circuitry makes the fabrication process highly demanding. The subsequent yield of the chips, defined as the percentage of perfectly operating chips vs. the total number made on the die, decreases with increasing chip complexity. As the yield decreases the cost increases. Substantial cost savings in manufacture can be realized if the requirement of near perfect operation of the chip is alleviated. Consequently, computer designs that utilize chips that contain some defects may prove to be far less expensive than those that require nearly perfect chip performance. The ability to configure a chip to operate around defects and imperfections would greatly improve the manufacturing yield and consequently the cost of the chips.

Though some error correction can be made to overcome defects, most microprocessor chips need to be nearly flawless in performance. In fact the consumer has come to expect nearly perfect performance so that when a chip shows up with a defect, a major product recall occurs. For example, a few years ago, Intel marketed a chip that had a minor defect in it that resulted in a slight error in the calculation of a built-in function. Though the vast majority of users would never need to use this function and its occurrence was very rare, the product ultimately had to be recalled and replaced by fully functioning chips simply because the prospect of an incorrect calculation, as rare as it was, was unsatisfactory to the consumer. Clearly the existing computing hardware paradigm is that the chip must function in all applications perfectly.

The perfectly operating chip paradigm is not the only possibility for computing hardware. It is conceivable that flawed chips can still be used provided that the defective components in the chip are known in advance and are thereby avoided during the chip's operation. Such an approach is called defect tolerant computing. As mentioned above, for highly dense circuitry like that possible using molecular based hardware, it is essential that the machine be able to function in the presence of defects since the probability of creating a machine with about 10^{18} or more elements that work perfectly is nearly zero. Below we discuss some of the features of defect tolerant architectures and show how such a scheme can be and has been used to create a computer.

As discussed above, the manufacturing cost of integrated circuit chips is a strong function of their degree of perfection. If nearly perfect performance is required, as the complexity of the chip increases the yield decreases and the resulting cost escalates. However, if a defective chip can still be used, the yield would then greatly increase and the cost of the chips would decrease substantially. In addition, it is likely that the manufacturing costs associated with fabrication plant construction and operation would also be mitigated. The question is then, can a computing machine be made that utilizes defective chips? A group of researchers at Hewlett-Packard Laboratories and UCLA have announced a defect tolerant computer design (Heath *et al.*, 1998). These researchers have developed a machine that uses conventional Si integrated circuit technology. The machine is called "Teramac," an acronym that stands for *tera multiple architecture computer*. The machine is designed to operate at 10^{12} operations per second. The teramac was constructed using components that were defective but inexpensive. Many of the chips used in the machine were "throw-aways," chips that did not meet specifications and thus were disposed of rather than sold. Due to the manner in which the chips were interconnected and the utilization of configurable computing algorithms, the defects within the chips were mapped out in advance and

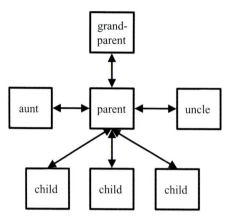

Figure 8.11 Family-tree-like branching architecture. Notice that in the present scheme the box labeled parent is a critical node in that it serves to connect to all of the other nodes. To pass a message from one child to a grandparent, the message must first pass through the parent. To pass a message between any of the children, the message must pass first to the parent and then to the other child. If the parent becomes disabled communication between all of the other boxes is lost.

suitable computational pathways were identified. The teramac configured itself to avoid defective devices or zones in each chip, thereby providing reliable computing pathways. In this way, many defective chips could be connected to produce a powerful, yet very inexpensive parallel computer.

One of the key issues in the design of the teramac is the manner in which the chips are connected. The basic problem encountered in configurable computing is how to maximize the number of interconnections such that the machine can have the greatest number of configurations, and hence have the greatest flexibility. The manner in which the logical units of the teramac were linked is called a "fat-tree" architecture. The fat-tree architecture can be understood best by comparing it to a family tree architecture. In a family-tree-like architecture, each child branches from one parent. Easy communication between the parent and child then results as shown in Fig. 8.11. However, communication between children of the same parent is not direct. Instead, a message must be passed to the parent first and then to the other child. Communication to a grandparent also must proceed through the parent. Not only does this slow communication, but it makes it extremely vulnerable to disruptions. For example, if the pathway between the parent and grandparent becomes disabled, then communication between a whole branch of the tree and the rest of the family is destroyed. Such a scenario is not likely to unfold in a computer in which nearly perfect chips are employed. However, if one utilizes defective chips, it is highly likely that important pathways between parts of the structure will be blocked. Consequently, a different architecture from that of the family tree must be employed in a machine like the teramac that employs defective chips.

The fat-tree architecture avoids the vulnerability of the family-tree-like approach by interconnecting the nodes as shown in Fig. 8.12. As can be seen from the figure each node is connected in turn to several other nodes. For example, each child is

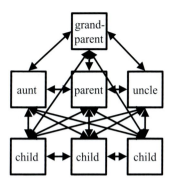

Figure 8.12 Fat-tree architecture showing the many interconnects between each node. Notice that each child is connected to the aunt, uncle, parent, and grandparent directly. Thus if the parent node becomes disabled communication between the other nodes continues.

interconnected in addition to the parent to the uncle, aunt, at least one other child and in some cases to the grandparent. Thus there exists a high degree of connectivity between the nodes such that any disruption does not lead to system failure. If one of the pathways is disabled in a fat-tree architecture, this would have little effect, since many other routes and connections exist. Therefore, the fat-tree architecture is relatively immune to pathway interruptions and hence is capable of working even when many defects are present. For these reasons, the teramac utilizes a fat-tree architecture.

An important part of the operation of the teramac machine is the location and identification of the defects in the chips. In teramac, all "repairs" are made using software. A software routine locates and maps the defects. Once all of the defects are identified, the software also maps configurations that avoid the defects providing computational pathways.

Teramac has demonstrated that a computing machine can operate reliably even in the presence of many defects (see Heath *et al.* (1998)). There are several useful lessons that the successful operation of teramac illustrates. These are:

(1) It is possible to build a powerful computer that contains a high level of defects provided that there is sufficient bandwidth communication between nodes.
(2) A high degree of connectivity is more important than regularity in a computing machine. It is not essential that each component be connected in an orderly manner, as long as each component can be located and characterized. If the testing algorithm is sufficiently robust, then it can determine after the fact how the machine is connected and which parts can contribute to its execution. Therefore, the architecture of the machine need not be highly organized.
(3) The most essential components for a nanoscale computer are the switching and interconnect mechanisms.
(4) The teramac computer has introduced a new algorithm in computer manufacture. It is: build the computer, find and map the defects, configure the resources with software, compile and execute the program. In this way, imperfect hardware can be utilized, thus greatly reducing the cost of the machine.

8.6 **Quantum dot cellular automata**

One of the principal limitations to increased computing capability is the difficulty encountered in interconnecting devices in a high density array. In the standard computing algorithm, each individual device must be contacted to enable independent switching of its state. As the density of devices increases, an attendant increase in the interconnect complexity occurs. It is expected that the interconnect complexity ultimately will circumvent further reduction in device size and packing density. It is likely then that an ultimate limit in packing density will occur in standard CMOS circuitry.

One approach by which this packing density limit can potentially be overcome is through the exploitation of a different computing paradigm. If computing can be accomplished without the necessity of contacting each individual device, it is conceivable that much greater device packing densities can be achieved. In this section, we discuss an approach in which each device does not need to be interconnected; contacts are made to only those devices on the perimeter of the system. This scheme is called a quantum dot cellular automata. There are other approaches in which interconnects to each device can be avoided. The most prominent method is to utilize the spin of the electron. This approach will not be discussed here but can be found in Brennan and Brown (2002) to which the interested reader is referred for a full discussion of spin based computing devices.

The most common architecture for a quantum dot computing machine is a two-dimensional array (Porod, 1997; Lent and Tougaw, 1997). The fundamental principle behind the operation of a quantum dot array is that information is contained in the arrangement of charges and not in the flow of charges, i.e., current. This means that the devices interact by direct Coulomb coupling and not by currents flowing through wires. In this approach the actual dynamics of the computing operation proceeds by direct Coulomb coupling between neighboring cells without introducing wired interconnections between each cell. As such, a distinct advantage over conventional CMOS architectures is achieved since complex interconnections can be avoided.

Before we discuss how quantum dots are arranged into an array, it is necessary to first describe the operation and fabrication of quantum dots. Exacting materials growth capabilities have enabled the realization of atomically thin layers of material. As is well known, when the device dimensions are comparable to the electron de Broglie wavelength, spatial quantization effects occur. In a system comprising an atomically thin layer of a narrow band gap semiconductor sandwiched between two layers of a wider band gap semiconductor, spatial quantization levels appear in the narrow gap semiconductor material. Such a system is said to be a two-dimensional quantum well system. The electron motion is constrained only in one direction, that parallel to the well dimension. The electron motion in the other two directions is not quantized since the electron motion is not constrained in either of those two directions. Etching the resulting layers to form either wires or dots as shown in Fig. 8.13 can further constrict electron motion. As can be seen from Fig. 8.13, quantum wires are structures in which the electron motion is quantized along two directions leaving one direction free. A quantum wire is thus a one-dimensional system. Further constriction can be obtained

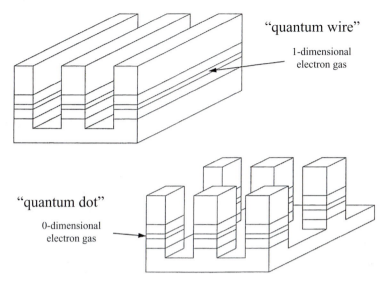

Figure 8.13 Sketches of a quantum wire and a quantum dot. The quantum wire is a one-dimensional system while the quantum dot is a zero-dimensional system. (Reprinted from W. Porod (1997). Quantum-dot devices and quantum-dot cellular automata, *Journal of Franklin Institute*, **334B**, 1147–1175, Copyright (1997), with permission from Elsevier Science and W. Porod.)

by etching the wires to form dots as shown in Fig. 8.13. In a quantum dot the electron is quantized in all three dimensions. Such a system is said to be zero-dimensional.

The simplest implementation of the quantum dot cellular automata scheme is to arrange the dots into a four-site cell with each dot at the corner of a square. Each cell contains two additional mobile electrons that can tunnel between the four different sites in the cell. The tunneling rate depends upon the spatial separation of the dots. As the spatial separation increases the tunneling probability decreases exponentially. In order to produce tunneling between neighboring quantum dots it is essential that these dots be spaced relatively closely together. If the spatial separation between each cell is sufficiently large tunneling of the mobile electrons between different cells is prohibited. Therefore, the electrons are localized within each cell. The spacing between neighboring quantum dots within a cell is much closer than dots within different cells. Consequently, tunneling can occur between the four neighboring dots in a single cell enabling the electrons to rearrange themselves depending upon the potential energy of the system. The minimum energy configuration of the cell is such that the electrons occupy antipodal sites (diagonally apart sites) as shown in Fig. 8.14 (see Example Problem 8.2). As can be seen from Fig. 8.14, if the two electrons are placed into adjacent dots, the Coulomb repulsion is relatively high leading to an unstable arrangement. The lowest energy configuration occurs when the two electrons occupy opposite corners of the cell as shown in Fig. 8.14. Two such arrangements are possible. These are shown in Fig. 8.15, where the shaded dots represent dots that hold an additional electron. As is shown in Fig. 8.15, the different arrangements can be referred to as $P = +1$ or

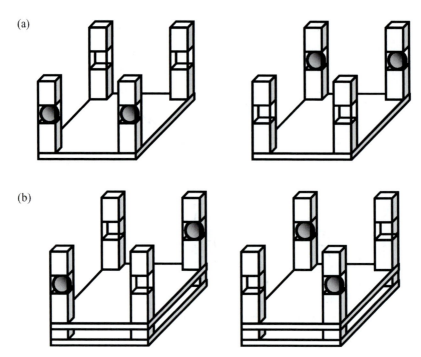

Figure 8.14 (a) Unstable states for a four-quantum-dot cell. Notice that the close proximity of the two electrons leads to a high repulsion and a high energy state and is thus unstable. (b) Stable states for a four quantum dot cell. Notice that there are two stable states, each with the electrons at two of the four corners. The larger spatial separation of the two electrons reduces the Coulomb repulsion thus leading to a lower energy state (Brennan, 1999).

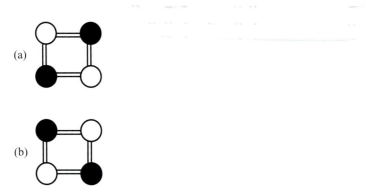

Figure 8.15 Two different polarization states in a four-site cellular automaton: (a) $P = +1$ (b) $P = -1$. Both sites have the same energy of configuration when independent.

$P = -1$. Therefore, two different states of the cells exist which correspond to two different logic states, $+1$ or -1.

Cells that are adjacent to one another tend to align. Cells that are arranged diagonally to one another tend to antialign. From these simple rules, the cell polarization can be used to encode binary information. Let us consider how information propagates from

(a)

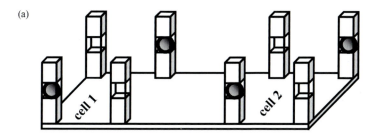

(b)

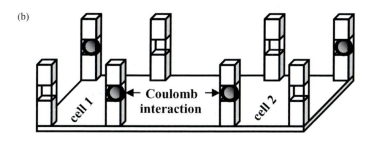

(c)

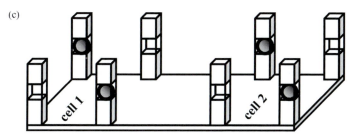

Figure 8.16 Sketch of a two-cell quantum dot cellular automaton showing: (a) the initial condition, (b) resetting cell 1 from an external stimulus, and (c) the final condition of the two cells after the Coulombic interaction causes cell 2 to reset (Brennan, 1999).

one cell to another. Consider the system shown in Fig. 8.16(a). Initially, both cells have electrons at the same opposite corners as shown in the diagram. If cell 1, called the driver cell, is reset by an external agent as shown in Fig. 8.16(b), then a Coulomb repulsion is produced between cells 1 and 2. This repulsion forces cell 2 to reset as shown in Fig. 8.16(c). Therefore, the information fed into the driver cell, cell 1, propagates to cell 2. The propagation would continue then from cell 2 to its next nearest neighbor cell and so on along a linear array of cells. Several simple cell arrays are shown in Fig. 8.17. In each case, the cell furthest to the left in the diagram is the driver cell. The state of the driver cell is fixed and forms the input to the array. The driver cell determines the state of the entire array in the manner shown in Fig. 8.16.

The simplest array is that shown in Fig. 8.17(a), that of a simple cellular wire. Notice that the state of the driver sets the state of each of the following cells within the wire.

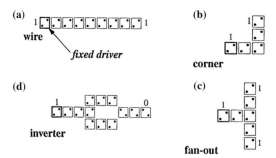

Figure 8.17 Arrangements of quantum dots showing different functionality: (a) wire, (b) corner, (c) fan-out, and (d) inverter. In all situations the left most cell provides the input and is called the fixed driver. Notice that in the wire, a "1" input results in a "1" at the output. For the corner, a "1" input leads to a "1" at the output. In the fan-out configuration, a "1" at the input leads to "1"s at both outputs. Finally, for the inverter, an input "1" is inverted to a "0" at the output. (Reprinted from W. Porod (1997), Quantum dot devices and quantum dot cellular automata, *Journal of Franklin Institute*, **334B**, 1147–1175, Copyright (1997) with permission from Elsevier Science and W. Porod.)

An input state of "1" is transferred to the output as a "1" as shown in the diagram. It is important to recognize that no charge is transferred between cells, only information. Hence, there is no need for interconnect circuitry in this scheme. The propagation of information occurs by the arrangement of the charge configuration in each cell. The driver cell acts to set all of the succeeding cells by direct Coulomb interaction. A simple inverter is shown in Fig. 8.17(d). In this case, the input state "1" is switched to a "0" at the output. Notice that because the cells are arranged diagonally, as can be seen from Fig. 8.18 the charges antialign since this leads to the lowest energy configuration of the cells in accordance with the second rule stated above. Conventional AND and OR logic gates can be obtained using the quantumdot logic functions. Therefore, all of the important logical operations can be made using quantum dot cellular automata (QCA).

The question is though, how can computations be performed using this scheme? The basis for computation in the cellular automata scheme lies with two key concepts. These are: (1) computing with the ground state and (2) edge-driven computation. The basic structure of the quantum dot computation scheme is outlined in Fig. 8.19. As can be seen from the figure, the QCA is connected to the external world through input and output cells. It is important to notice that no interconnects need be made to any cells within the QCA array. The only interconnects are made to the input and output cells, thus greatly reducing the interconnect complexity of the system. Further inspection of Fig. 8.19 reveals that the input to the computation is addressed by setting the edge cells to the QCA array. The solution of the computational logic problem is obtained physically from the collective ground state of the QCA array.

The basic workings of the QCA computer can be summarized as follows. First, the polarization state of the input cells is fixed in accordance with the input logic. This is what is meant by edge-driven computation. Second, since the internal cells within

(a)

(b)

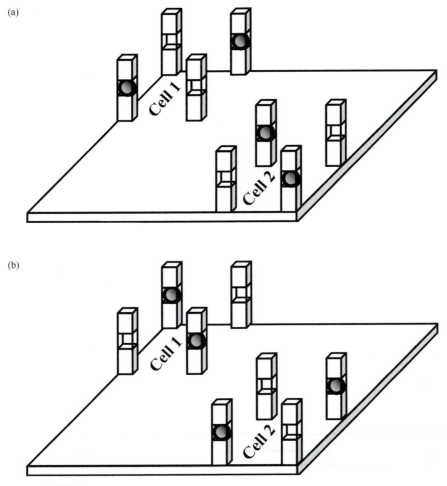

Figure 8.18 (a) Two cells arranged diagonally. Since the spatial separation of the electrons is maximized, the two cells are at their lowest energy. (b) An alternative arrangement of the electrons in the two diagonal cells. Notice that again the electrons are furthest apart leading to the lowest energy configuration of the system, which is degenerate with that shown in (a). In both (a) and (b) the cells are antialigned (Brennan, 1999).

the QCA array are not independently contacted, they cannot be supported indefinitely in a high energy, nonequilibrium state. Instead, the array must eventually collapse into a ground state that is uniquely specified by the input cells. In other words, the QCA array relaxes to some stable ground state consistent with the condition of the input cells. It is important to note that many ground states are, in general, available for the array, but for each initial state condition on the input cells, the resulting ground state of the array is unique. Therefore, the array will be found in only one unique ground state for any given initial state of the input cells. This resulting ground state is "read" by the output cells by sensing the polarization of the periphery cells of the QCA array.

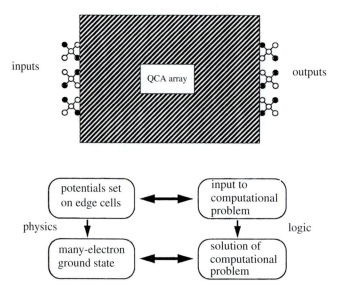

Figure 8.19 Format of a quantum dot cellular automata. The only interconnections to the external environment required are to the input and output cells that are connected to the array. (Reprinted from W. Porod (1997), Quantum-dot devices and quantum-dot cellular automata, *Journal of Franklin Institute*, **334B**, 1147–1175, Copyright (1997) with permission from Elsevier Science and W. Porod.)

The crucial aspect of the QCA computing engine is that the computational result resides in the ground state of the array obtained following its excitation by the input cells. The particular ground state that the array relaxes into depends uniquely upon the input cell states and not upon the dynamics by which the system relaxes. The actual mechanism by which the array relaxes plays little role in determining the result. Therefore, no external control of the array is needed, hence no interconnects are required.

It is important to recognize that ground state computing is a fundamental requirement of a computing system that has an interconnectless architecture. Since the internal devices within the array are not connected to an external power source, there is no mechanism to retain them perpetually in excited states. As a result, each device will ultimately decay into its ground state. The ground state computing algorithm simply utilizes this concept to construct a computing machine.

An additional key issue in understanding the operation of a QCA array as a computing machine is that the information propagates from the input cells to the output cells not by current flow, as in conventional CMOS circuitry, but by an electrostatic interaction between cells. The electrostatic interaction occurs between electrons in a spontaneously polarized system of quantum dots. Through this interaction, if the polarization condition on the input dot is initially set, that input will propagate through the array flipping dipoles in the following cells, much like a soliton wave. Such a propagation would result in a rapid transfer of information from the input cells into the array.

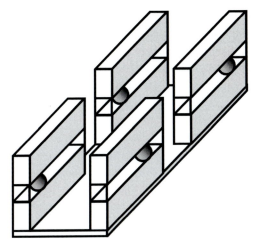

Figure 8.20 Sketch of a single cell of a quantum dash structure. Notice that in the two dashes in the foreground, the charges are repelled forcing them spatially apart. As a result, the charge in the lower right cell then repels the charge in its nearest neighbor dash lying above it. The resulting configuration of the charges is given then as shown (Brennan, 1999).

Two different strategies have been proposed for switching in a QCA array. These are by tunneling, wherein electron transfer occurs between dots, and by a long range electrostatic interaction, wherein the information transfer occurs from dipole interactions. Tunnel transfer of electron charge between dots can occur if the dots are made sufficiently "leaky" such that a nonzero probability exists that an electron can tunnel from one dot to another. As discussed by Porod (1997), the electrons within the QCA array are allowed to tunnel between dots within a single cell but not between cells. Since the tunneling probability is a function of dot separation, this is accomplished by placing the dots in relatively close proximity within each cell and spacing the cells somewhat further apart. Therefore, the long range Coulomb interaction affects the polarization state between cells. To minimize the energy of the cell consistent with the polarization, electrons will tunnel from one dot into another to produce the minimum energy dot configuration. Notice that within this model, electron tunneling occurs between dots. A second strategy that avoids tunnel transfer was first proposed by Bakshi *et al.* (1991). In their approach they employ "quantum dashes." A quantum dash is a quantum structure in which each dot is elongated along one dimension as shown in Fig. 8.20. The Coulomb force between electrons within adjacent dashes is greater than the confining force along the dash direction of elongation. Therefore, the interdash Coulomb force can overcome the confining force shaping the electron configuration within the dash. The resulting displacement of electron charge in each dash forms an electric dipole that is replicated between dashes along the array. The array then spontaneously collapses into an antiferroelectric polarization arrangement.

The most severe drawback to ground state computing systems is the possibility that the system will get "stuck" in metastable states, ones different from the ground state that do not decay quickly. The presence of metastable states can severely hamper the computing performance of the array since it may not reliably collapse into the ground state in a timely manner. As a result, the output cells may sense an improper metastable state response to the input condition thus rendering an incorrect computational result. Of course, with time, the array would ultimately relax into the proper ground state, but that might require an inordinate length of time, not particularly attractive for high speed computing applications.

In addition to the metastable state problem, there exist other limitations to quantum dot computing schemes. One of the key problems encountered in the coupled quantum dot approach is that one electron must be sequentially transferred from one dot to another. Trapping of the transferred electron in a dot other than the desired final dot would result in a serious failure. The devices must be switched slowly to avoid trapping, which necessarily slows the transfer of information within the computer. There are indications that the net transfer rate between devices is orders of magnitude slower than in present day CMOS devices. A second key limitation to quantum dot computing architectures is reproducibility. In order to effect charge transfer by tunneling between different dots under similar conditions, the dots must be made uniformly. In most embodiments of the quantum dot architecture, the barrier widths are defined by lithography, which cannot generally provide the monolayer resolution necessary to precisely control the magnitude of the potential. Therefore, it is questionable that such devices can be utilized in high density integrated circuits. The use of quantum dashes instead of dots with the concomitant avoidance of tunnel transfer eliminates this second limitation. However, the quantum dash polarization array is highly sensitive to temperature. It must be operated at liquid He temperatures, ~ 4.2 K, to avoid thermal smearing of the polarization state. In addition, the quantum dash array is still sensitive to the device geometry, i.e., the spacing and shape of each dash greatly determines the coupling.

There are other approaches that avoid some of the limitations of quantum dot cellular automata. In particular, the usage of the electron's spin presents an alternative paradigm that is more stable than that of quantum dots. Spin based systems have several advantages over quantum dot cellular automata making them attractive alternatives. Discussion of spin based computational methods is beyond the scope of this book and the reader is referred to the book by Brennan and Brown (2002) for an introductory discussion. There are other approaches that go beyond Si CMOS technology but these concepts require some knowledge of quantum mechanics and as such will not be treated here. These approaches are single electron transistors, quantum computing, and resonant tunneling structures. We only mention these here referring the reader to Brennan and Brown (2002) for a full discussion.

Example Problem 8.2 Electron Coulomb energy within a four dot cell

Let us consider the Coulomb potential energy of two electrons within the four quantum dots forming a cell. What we would like to show is that the energy is lower when the electrons occupy diagonally opposite dots than when they occupy adjacent dots. For simplicity let us assume that the cell is square of length a. The Coulomb potential energy for two charges is

$$U = \frac{1}{4\pi\kappa} \frac{q_1 q_2}{r_{12}}$$

where q_1 and q_2 are the charges of each of the two electrons and r_{12} is the spatial separation of the charges. Since both charges are simply single electrons, $q_1 = q_2 = q$. The spatial separation of the two charges if they reside in adjacent dots is simply equal to a. If the electrons occupy dots diagonally apart, the spatial separation, d, is simply

$$d = \sqrt{2}a$$

Thus the potential energy difference between the two arrangements is

$$\text{adjacent dots: } U = \frac{1}{4\pi\kappa} \frac{q^2}{a}$$

$$\text{diagonally separated dots: } U = \frac{1}{4\pi\kappa} \frac{q^2}{\sqrt{2}a}$$

Clearly, the potential energy of the configuration is lower when the two electrons are in dots diagonally separated than when they are in adjacent dots. The potential energy is a factor of the square root of 2 lower in the diagonally separated case. Consequently, the diagonal arrangement constitutes a lower energy system than the adjacent dot arrangement. Thus the minimum energy configuration of the cell occurs when the two electrons occupy dots are arranged diagonally to one another.

Problems

8.1 For two electrons in a four quantum-dot cell determine the ratio of the potential energies of the diagonal and adjacent configurations by treating the electrons as point particles. Assume that the separation between adjacent dots is R and that the four-dot cell is square.

8.2 Consider an eight-node system that is initially configured in a family-tree-like pattern. If the nodes are to be rearranged into a fat-tree architecture such that every node has at least three connections how many more connections does the fat-tree architecture scheme have than the family-tree architecture? Assume that the family-tree architecture plan is a simple circle.

8.3 Assume that the depletion conditions for a carbon nanotube are the same as for the semiconductor region forming a back gate on a nanotube FET. Given that the semiconductor layer has a doping concentration of 10^{17} cm^{-3} and is n-type, determine the reverse bias on a metal gate if the full thickness of the gate and tube is 0.1 mm. Assume that the metal work function is 1.0 V, the affinity is 0.65 V, the energy gap is 1.12 eV, the intrinsic concentration is 10^{10} cm^{-3}, and the relative dielectric constant is 11.8. Assume that the gate is formed by a Schottky contact.

Telecommunications systems – an overview

In this chapter we examine the basics of optical and mobile telecommunications systems that impact semiconductor devices. Our aim is to determine the principal characteristics of lightwave and mobile telecommunications systems that influence device selection. Lightwave communications systems are based on optical fibers and several device components are needed to support lightwave transmission, encoding, amplification, detection, and decoding. The device components used within lightwave systems fall into two general categories: optoelectronic and optical. In mobile telecommunications systems the device types of greatest importance are high frequency, high power transistors. Here we will briefly outline how some of the system requirements influence device choice and dictate their performance.

9.1 Fiber transmission

There are several important advantages to fiber optic communication systems. These can be summarized as:

(1) Smaller diameter, lighter weight, and increased flexibility.
(2) Relatively low cost compared to copper cables. Fiber optic cables are relatively inexpensive due to the low cost of the materials employed.
(3) Good isolation and cross-talk immunity.
(4) Low transmission loss and dispersion.
(5) High security in transmission. There is little signal "spilling" from the fiber if properly shielded in contrast to that for copper systems.
(6) Tremendous capacity. As we will see below, the capacity of existing fiber optic lines as measured by bandwidth is measured in terahertz.

For the above reasons, most of the long distance communication within the USA and many parts of the world is conducted using fiber optics.

A typical optical fiber is sketched in Fig. 9.1. The fiber consists of three different regions: the core, cladding, and encapsulation layers. The signal propagates mainly through the core region of the fiber. The signal is confined principally within the core region by total internal reflection. The index of refraction of the core region is higher than that of the cladding layer. As a result, the lightwave is totally internally reflected as it propagates through the fiber ensuring confinement. Most fibers are made of glass and plastics. The key properties of importance in selecting a material for a fiber are:

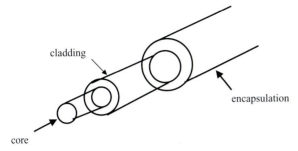

cladding

encapsulation

core

Figure 9.1 Sketch of an optical fiber showing the core, cladding, and encapsulation layers.

(1) slight index of refraction differences that can be exploited to define the core and cladding layers;
(2) flexibility so that the fiber can be bent to go around corners;
(3) ability to be drawn into long fibers.

Usually rather than using only a single fiber as shown in Fig. 9.1, several fibers are stranded together around a central member and encapsulated to form a fiber cable.

Silica glass is the most commonly used material for optical fibers. Other materials that are beginning to be used contain fluorine. Typically, a small concentration of impurities referred to as dopants is added to the glass to alter its refractive index. These dopants also affect the attentuation, or power loss, of the light in the fiber. Lowest attentuation in modern silica glass fibers occurs for specific operating wavelengths. There are three principal transmission bands that are utilized. Early systems operated at wavelengths between 800 and 900 nm. Much lower attentuation can be achieved by operating at longer wavelengths. There are two low loss transmission windows that are centered around 1310 nm and 1550 nm. Most long distance communications systems currently operate using the 1310 nm band, which is often referred to as the medium wave band. Lower attentuation on modern fiber can be achieved by operating within the 1550 nm range, between 1510 and 1600 nm. For silica glass fiber this band, called the long-wave band, has the lowest loss window. However, low cost emitters and detectors that can efficiently operate at these wavelengths are presently unavailable. The capacity of the medium band is about 30 terabits per second. Current technology limits electronic systems to about 10 Gbps, giving a much greater capacity than is presently being used. Hence, the information handling capacity of silica optical fibers is enormous, enabling, at least in principle, enormous growth.

Attenuation generally varies linearly with distance. Thus the longer a fiber optic cable is, the more the signal will be attentuated at the end. The amount by which the signal is attenuated due to cable attentuation is given by the product of the attenuation per kilometer at the signal wavelength and the distance. Clearly, there is a limit to how far a signal can be sent in a fiber optic cable without reamplification. The maximum distance that a signal can be sent depends upon the total attenuation of the system and the initial optical power supplied. The difference between the optical power of the transmitter and the receiver sensitivity is called the "link budget." The link budget is the total amount of power available to compensate for fiber attentuation loss as well as all other losses encountered between the transmitter and receiver. In order to successfully transmit the signal, the system power requirements have to operate within the link budget.

Accounting for the optical power losses during transmission is somewhat compli- cated. The power requirements for some system components vary with the frequency of transmission. The problem is particularly acute for detectors. To preserve the signal to noise ratio of the detector, the power at the receiver must increase with modulation frequency. The signal to noise ratio of the detector depends strongly upon the number of photons detected during a pulse. If the pulse is shortened fewer photons will be present unless the optical power is increased. Therefore, if the optical power remains fixed, and the modulation frequency is increased, fewer photons will be captured dur- ing a pulse leading to a reduction in the signal to noise ratio of the detector. To avoid this reduction, a greater signal power must be transmitted or the sensitivity of the detector must be improved. Laser output power is essentially invariant with modula- tion frequency and thus does not significantly impact the frequency dependent power loss.

There exist several issues that affect the overall system performance that can be overcome to some extent through additional power expenditure. These are: system noise, dispersion, and extinction ratio. The signal to noise ratio, SNR, is defined as the ratio of the signal power to the noise power. In some systems, particularly those that utilize erbium doped fiber amplifiers (EDFAs), to compensate for added noise increased optical power must be supplied. Dispersion is the spreading of the signal pulse, i.e., a short pulse becomes longer in duration. Dispersion arises from several sources. The index of refraction varies with wavelength. The sources used in fiber optic systems emit over a small but nonnegligible range of wavelengths. Therefore, each wavelength will travel at a different speed through the fiber leading to a spreading in the pulse. An additional source of dispersion occurs in multi-mode fibers, those in which more than one propagation mode is excited. The distance traveled by each mode is different since each mode essentially takes a different path as it moves through the fiber leading to dispersion. Finally, the shape and profile of the fiber core can produce dispersion, which is typically referred to as waveguide dispersion. Dispersion acts to increase the noise of the signal, which can be compensated by increasing the transmission power. If the dispersion is particularly pronounced, it is possible for adjacent pulses to blend or merge into one another. This is called intersymbol interference and generally results in the failure to distinguish between adjacent bits in the data stream. The extinction ratio is related to the ratio of the power utilized to encode a "1" bit and a "0" bit. Zero bits can be represented using a nonzero power level rather than a complete absence of power thus narrowing the power difference between a "1" and a "0." The power level of the "0" bit becomes the noise floor of the "1" bit.

The transmitter and receiver selection for a lightwave communications system depends critically upon the desired signal wavelength. Both the transmitter and detec- tor need to be spectrally matched to the transmission wavelength, which is chosen to lie within either the medium or long wavelength bands. In addition, the transmitter selection depends upon wavelength stability, spectral width, output power, and modulation rate. The detector must have sufficient sensitivity and bandwidth. The detector requirements are dependent on signal strength, noise background, and data rate.

There are three different types of optical fibers. These are:

(1) step-index multi-mode;
(2) graded-index multi-mode;
(3) single-mode.

As the name implies in multi-mode fibers, several different paths or modes are available for light propagation. Multiple paths or modes are made available by designing the fiber to have a sufficiently large core diameter to accommodate the different modes. Though the fiber can accommodate multiple paths, light traversing each path travels a different distance. As a result, the light arrives at different times at each location leading to dispersion of the signal pulse. The problem is severe in multi-mode step-index fibers. In multi-mode step-index fibers, the index of refraction changes abruptly from the core to the cladding regions. The graded-index multi-mode fibers overcome this limitation by gradually grading the index of refraction within the core from its center to the edge. The index of refraction is highest near the center of the core and gradually decreases outwards. As the reader may recall, the speed of light decreases with increasing index of refraction. The light propagates at a slower speed near the center of the core, where the index of refraction is higher, than near the cladding. The light traverses a shorter path if it is mainly confined within the center of the fiber. Thus the shorter paths through the center of the fiber have a larger index of refraction and a slower speed than those that traverse closer to the cladding layers. A judicious choice of grading can thus ensure that the light propagating in the different modes all arrives at the same location at the same time, thus mitigating dispersion. The last possibility is the single-mode fiber. In single-mode fibers, the width of the core is sufficiently small that only one mode of the light can propagate and thus dispersion due to multiple paths is avoided altogether.

9.2 **Amplifiers and repeaters**

As discussed in the previous section, there is a limit to how far a signal can propagate due to attenuation. Therefore, to provide long distance propagation the signal must occasionally be recovered, amplified, and retransmitted. This process can be performed in a number of ways. In many of the earlier lightwave systems, the regeneration process was performed by converting the signal into an electrical form, reshaping it electronically, and reconverting it back into light. This procedure was accomplished using a repeater. The advantage of a repeater is that propagation noise and any distortion can be removed. A repeater functions by detecting the signal, extracting its digital content, and then rebuilding and retransmitting the signal. Thus the new signal no longer contains any distortion or noise from propagation and is precisely the same as the original transmitted signal. Repeaters are thus placed along the network to ensure that the signal never attenuates to an unusable level. The problem with the use of repeaters is that an all optical repeater does not exist but some electrooptic conversion must occur. Unfortunately, a new repeater is needed if the transmission code or speed of the signal is changed. Thus in long distance networks, it becomes very expensive to replace the repeaters every time a higher transmission bandwidth is implemented.

Amplifiers are an alternative to repeaters for signal regeneration. In early lightwave systems, amplifiers were avoided because they amplify the signal plus any accompanying distortion and noise. The amplified signal is not a simple clean pulse identical to the original transmitted pulse but one that contains noise and distortion. Therefore, repeated amplification can lead to a highly distorted and noisy signal at the output. For this reason, the early systems avoided usage of amplifiers and relied on repeaters instead. However, in modern digital lightwave systems there is little noise and amplifiers can be utilized. The invention of all optical amplifiers then resulted in the transition from repeaters to amplifiers in fiber optic networks. In the next chapter, we discuss the processes in two important optical amplifiers, EDFAs, and semiconductor optical amplifiers, SOAs.

Many fiber optic systems utilize wavelength-division multiplexing, WDM. In a WDM system, many single-channel transmissions are superimposed onto a single fiber. Each optical channel is independent of the other channels and may have a different encoding scheme and protocols. As the number of channels utilized in a WDM system increases, it is important that the amplifiers provide broad-band amplification. In other words, the gain must be the same for all of the channels implying gain flatness over at least 20 nm of bandwidth. This is one of the key performance figures of merit of an optical amplifier.

Most fiber optic systems presently use EDFAs or SOAs for all optical amplifiers. The basic process in either device is that an incident optical signal is amplified after traversing the optical amplifier. In either case, single-pass stimulated emission is used to amplify the light signal. As we will see below, the manner by which stimulated emission gain is utilized varies between EDFAs and SOAs. It is useful to first discuss how they compare and why both types of amplifiers are used.

EDFAs operate in saturation and have a very slow response time, on the order of microseconds. Since the response time is very much longer than the bit-period used, the amplifier is sensitive to the average signal power. In other words, the amplifier gain is determined by the signal's average power. As a result, an EDFA is immune to distortion due to intrachannel patterning since the gain does not depend upon the pulse sequence. One problem with EDFAs is that their bandwidth is significantly less than that of the fiber. This is not a major problem, since the full bandwidth of the fiber is rarely used. A more significant problem occurs due to the fact that the gain is not constant (flat) over the spectral range of the fiber. Gain "flattening" is difficult to achieve with an EDFA for bandwidths in excess of 33 nm or more. As a result, the gain varies with bandwidth under these conditions, which is undesirable in a WDM system. The gain flattening problem can be somewhat mitigated by using grating and filter technology but this can increase the cost of the system. Nevertheless, most long distance commercial fiber optic systems utilize EDFAs, particularly for undersea and long haul terrestrial applications.

SOAs offer a relatively low cost alternative to EDFAs. An SOA is simply a semiconductor laser with antireflection coatings used for the facets. Therefore, the device is simply a nonresonant single pass laser. Such a device is in principle relatively inexpensive to manufacture. SOAs are less expensive than EDFAs since EDFAs are pumped using a semiconductor laser. Therefore, the cost of an EDFA includes the cost of at least one semiconductor laser, which is essentially the entire cost of an SOA. There are some problems with using SOAs in a communications system. The response time of an SOA is comparable to the bit-period in a Gbps system. The response time is too short to make the amplifier sensitive to the average signal power but too long for it to be sensitive to a single bit. As a result, cross-talk between channels can occur. The processes in both EDFAs and SOAs will be presented in Chapter 10 following the discussion of stimulated emission and laser operation.

Lightwave communications systems utilize either light emitting diodes (LEDs) or semiconductor lasers as emitters. Even if all optical components are used for amplifiers within the fiber, to initially send the signal through the light pipe an optical emitter must be used. Though LEDs are significantly less costly to manufacture than semiconductor lasers, their intensity is far less than that of a laser. As a result, light emitted from an LED does not have the propagation range of that emitted from a laser. For this reason, most systems utilize semiconductor lasers for emitters. As we will see in the next chapter, there are several different laser types.

In addition to emitters, lightwave communications systems also require optoelectronic detectors. It is the function of a detector to convert an input optical signal into an electronic one to form the output that can then interface with existing electronic systems such as telephones, fax machines, computers, etc. As we will see in Chapter 10, it is imperative that a detector be spectrally matched to the input optical signal, operate at very low noise, and often provide gain. Additionally, a detector must have a large bandwidth so it can operate at very high speeds in order to keep up with high speed data transmission.

9.3 **Mobile cellular telecommunications systems**

Mobile telecommunications has undergone a tremendous growth since the 1980s. In 1985 the number of mobile subscribers in the USA totaled about 200 000. In 1999 that number had grown to about 76.3 million users. The explosion in mobile telecommunications customers has resulted in increased demand for capacity as well as bandwidth. Many mobile subscribers are interested in having cell phone operation that can be used for more than simple voice transmission. Specifically, there is a growing market for cellular traffic that delivers pictures, internet connections, data streams, etc. to mobile customers. These demands place important requirements on the cellular system and its constituent components.

A cellular system divides the service area to be covered into a network of contiguous cells. Ideally, each cell is hexagonal in shape such that they completely cover the geographic area of interest. Each cell contains a base station which connects through different radio channels to each user within the cell. The base stations themselves are connected usually by microwave links to a mobile switching center. Thus a cellular system consists of three parts:

(1) mobile units, which contain a control unit, transceiver, and antenna;
(2) cell site or base station;
(3) mobile telephone switching office (MTSO). The MTSO controls wireless telephone call processing and coordination. The MTSO also provides a cellular switch. The cellular switch switches calls to connect mobile users to one another and to connect subscribers to the nationwide network.

A cellular system must operate at a high bandwidth in order to accommodate the large number of users on the network. The choice of frequency of operation is strongly influenced by atmospheric attenuation. The presence of water vapor and oxygen in the atmosphere leads to attenuation due to absorption of microwave radiation. For certain frequencies atmospheric attenuation peaks due to molecular resonance absorption of the oxygen and water vapor molecules. The particular frequencies that suffer the greatest microwave absorption occur at 22.2, 60, 120, and 183 GHz. Conversely, there are certain frequencies at which the absorption is at a minimum. These occur at 35, 94, and 140 GHz. Transmission at these frequencies has minimal attenuation.

The major challenges that next generation wireless systems face are propagation loss, shadowing, and multi-path fading. Propagation loss arises from reflection and scattering. The loss is proportional to the square of the frequency and the distance to the nth power, where n is 2 for free space, 4.35 for rural open space and 3–4.3 for urban/suburban space. Objects, such as autos, buildings, etc. that temporarily block the base station and the receiver cause shadowing. Multi-path fading arises when signals are received after having taken multiple routes. Multi-path fading can cause distortion, which can lead to a significant increase in the bit-error rate.

Other issues that impact wireless communications systems are cell splitting, voice quality, service quality, and hand-off. One of the most important advantages of cellular networks is that the system capacity can be increased by splitting large cells into small cells as the traffic in an area increases. Cells of different size can be utilized to reflect differences in population or user density. Voice quality is somewhat subjective. Basically, the system must provide excellent, clearly recognizable speech for all users. Some noise can be tolerated as long as the speech is still clear and little miscommunication occurs. Service quality can be measured by three different issues. These are coverage, grade of service, and number of dropped calls. Network subscribers demand broad coverage. Current systems enable most users to telephone nationwide from their cell phones. However, there are still regions where cell phone usage is restricted due to either their remoteness or geographical terrain. Users require service on demand. Blocked calls due to overtaxing of the network should be avoided. The traffic each cell contains of course varies. At times, such as during the rush hour, the cellular traffic exceeds the capacity of the network and the user cannot place or take a call. To counteract this problem each cell must contain a sufficient number of channels to ensure that communications can proceed normally. Though dropped calls are less of a problem in modern systems, they can still occur when a user moves between cells or passes through tunnels, etc. To ensure that calls are not dropped mobile systems are designed to efficiently handle hand-offs from one cell to another. A hand-off allows a mobile user to cross cellular boundaries without interruption in service. Base stations control hand-off. They monitor the signal level to adjust which base station is most suited to receive the transmission. When a received signal drops below a preset value the base station hands the call off to a different base station. The new base station is selected by choosing the station that receives the highest power. This new station then continues to handle the call until the subscriber has moved out of range into a different cell. To minimize the number of base stations the coverage of each station should be as large as possible and have the greatest possible capacity. These features require high frequency, high power operation. As we will see below, the requirement of high frequency, high power operation places constraints on the selection of system hardware.

The uneven distribution of users in time and space places severe constraints on a cellular system. One cell may have many users at a given time stretching its capacity while a nearby cell is under utilized. Due to cost considerations it is unwise to add or subdivide cells continuously, unless the traffic density regularly exceeds a cell's capacity. One approach is to design systems that enable shared usage of bandwidth. There exist multiple access techniques that enable users to share the same frequency band. These access techniques depend to some extent upon standards and can vary between geographic regions, such as North America, Europe, and Japan. Three important approaches are as follows. Frequency division multiple access, FDMA, provides for each user an assigned specific frequency channel within a geographical region. Time division multiple access, TDMA, enables multiple users to use a common frequency band. Each user's signal is transmitted on a rotating basis during assigned time slots. TDMA functions much like a time sharing system in that the system cycles through the users sending a portion of each signal with a given periodicity. Provided that the cycling occurs rapidly, the user is unaware of any real time delays in transmission. A third approach is CDMA, code division multiple access. CDMA allows many users to transmit simultaneously in the same frequency band. User signals are distinguished by assigning each user a unique signature sequence. The particular advantages of CDMA are:

(1) resistance to interference;
(2) better immunity to multi-path fading;
(3) increased capacity;
(4) increased security.

Some of the problems associated with CDMA are the need to provide power balancing between signals so that all received signals arrive with the same power. System failure can result if one channel overwhelms an adjacent signal carrying channel.

Synchronization is necessary. Both of these requirements place constraints upon the choice of microelectronic components.

9.4 Device types for cellular systems

The primary driver of the communications industry in the foreseeable future is the ever-growing demand for the rapid, efficient, and accurate transfer of digital information. One of the major issues that confronts the telecommunications/computing industry is the determination of the hardware requirements and specifications for future fiber optic and wireless systems that will ensure high data rate communication at low bit error rates. Feature size is one of the most important parameters that dictates high speed device performance. In addition, the storage of massive amounts of data for rapid retrieval, transmission, and processing requires very small memory devices. Therefore, in both digital and analog electronics the major drive is for continued miniaturization since it provides faster device operation and denser integrated circuits for memory and processing applications.

In a wireless system there are many different functionalities that semiconductor devices must have depending upon the application. A device design that works well for one application may be totally unsuitable for another. For example, power amplifiers require transistors that can deliver high power at high frequencies while switching platforms use transistors that have low "on-resistance" and small "off-capacitance." Optimization of all of the different features needed by transistors for wireless systems cannot be accomplished with a single design. Therefore, different device designs are chosen and employed depending upon the system application. Nevertheless, one can specify a list of attributes that are typically important in device development for wireless systems. These are (see Chang *et al.* (2002)):

(1) maximum power gain bandwidth;
(2) minimum noise figure;
(3) maximum power added efficiency;
(4) low thermal resistance;
(5) high operating temperature and reliability;
(6) low on-resistance, high off-resistance;
(7) high linearity;
(8) low power dissipation;
(9) low leakage current in cutoff;
(10) low $1/f$ noise;
(11) multi-functionality;
(12) low power operation, single power source needed;
(13) semi-insulating substrate;
(14) mature technology;
(15) low cost.

There are numerous device types that can be and are used in wireless systems that have many of these properties and are optimized for some. As we will see in

Chapter 11, MODFETs made from the GaAs and InP materials systems provide very high frequency operation at a very low noise figure making them attractive components for power amplifiers. Additionally, these devices have a high power output and a high power added efficiency. These features make GaAs and InP MODFETs attractive for use in handsets and base station electronics. However, the relative maturity and lower cost of GaAs devices make them more suitable for use in handsets and wireless communications applications in general than InP based structures.

One of the principal operating requirements of the power amplifiers used in wireless systems is linearity. Why though is linearity of the power amplifier so important? Nonlinearity in the output of the power amplifier leads to power leakage out of the signal channel into adjacent channels. Adjacent channel leakage is caused by third order intermodulation distortion, which is intermodulation distortion between the fundamental and second harmonic signals. Thus to avoid leakage of the signal power into other channels it is important that the power amplifier operate linearly.

An additional system constraint results from the current packaging of integrated circuits into modules for RF applications. A mixture of Si circuits and GaAs circuits is common for front-end components. Clearly a single material and circuit technology will lead to lighter and more compact circuits and, therefore, systems. However, each specific material and device technology offers some specific advantages and disadvantages. SiGe technology offers the advantages of heterojunction design to Si technology, but cannot address the RF limitations of the conducting Si substrate. The conducting substrate introduces significant signal loss as the frequency is increased and therefore limits Si as a MMIC (monolithic microwave integrated circuit).

10

Optoelectronic devices – emitters, light amplifiers, and detectors

The primary optoelectronic devices of importance in lightwave communications systems are emitters, amplifiers, modulators, and detectors. Emitters are the front-end components of a lightwave communications system. The signal is input into the fiber using emitters. The most important emitters are light emitting diodes (LEDs) and lasers. As we will see, most communications systems use lasers due to their much higher power and relatively large modulation bandwidth as compared with LEDs. In addition to front-end emitters, modern lightwave communications systems utilize optical amplifiers for long distance communications. The natural attenuation of lightwave signals propagating through a fiber optic cable can be compensated by the use of amplifiers placed at periodic spatial intervals. In this way, very long distance fiber transmission lines can be made for transcontinental and transoceanic communications. The most attractive amplifiers in long distance lightwave networks are all optical devices since these structures are less costly and typically less noisy than their optoelectronic alternatives. In this chapter, we discuss the two most important semiconductor emitters, LEDs and lasers. The operating principles of these devices as well as various device types will be presented. Since the basic physics utilized in lasers, i.e., stimulated emission, is common to optical amplifiers, the chapter includes a discussion of the two most important optical amplifiers used in lightwave communications systems. These are EDFAs and SOAs. Finally, the chapter concludes with a discussion of photodetectors.

10.1 LEDs

The LED has become one of the most ubiquitous compound semiconductor devices. It is commonly employed in numerous applications including clocks, appliances, calculators, lighting, and signs. LEDs make up the largest share of commercial optical semiconductor products. The development of high brightness LEDs has sparked interest in developing LEDs for future commercial lighting applications. Given further improvement in LED performance as measured by the light output in lumens per input watt called the luminous performance, LEDs can potentially become the most efficient artificial light source available. Such a development could have a profound impact on world electrical energy consumption.

An LED is a very simple device; essentially it is just a forward biased p–n junction diode. Under forward bias both electrons and holes are injected into the depletion region. Since the excess carrier concentrations within the depletion region exceed the equilibrium concentrations under forward bias, there will be a net carrier recombination

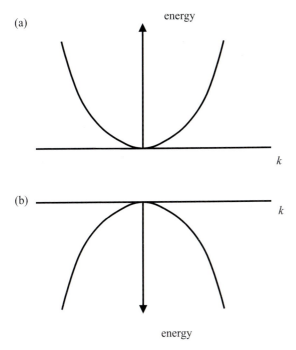

(a)

energy

k

(b)

k

energy

Figure 10.1 Sketch of the energy vs. k relationship for (a) the conduction band and (b) the valence band. Notice that both curves are parabolic.

rate. The device is designed such that the most probable recombination mechanism is radiative recombination. Therefore, the forward bias of an LED leads to minority carrier injection followed by radiative recombination with the concomitant emission of light.

How can the LED be designed such that radiative recombination dominates in the device? To understand the answer to this question we must first revisit the band structure of the material. As discussed in Chapter 1, if an electron is completely free its energy can be expressed as

$$E = \frac{\hbar^2 k^2}{2m} \tag{10.1}$$

In Chapter 1 we found that an electron has a different mass in a crystal than it has in free space: the mass in the crystal is called the effective mass, m^*. The energy of an electron in a crystal is basically the same as that of (10.1) except that the mass used is the effective mass, m^*:

$$T = \frac{\hbar^2 k^2}{2m_e^*} \tag{10.2}$$

The energy vs. k-vector relationship for the conduction band is shown in Fig. 10.1(a). A similar expression holds for holes:

$$T = \frac{\hbar^2 k^2}{2m_h^*} \tag{10.3}$$

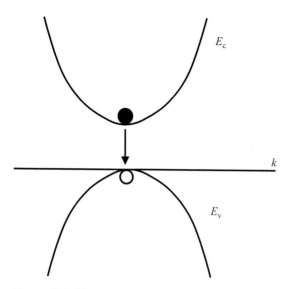

Figure 10.2 Sketch of a direct gap semiconductor showing an electron–hole recombination event. Notice that the electron and hole have minimum energy at the same value of k. Thus the electron and hole have the same momentum, both close to zero.

Hole energy though increases downwards as shown in Fig. 10.1(b). In both cases shown in Fig. 10.1, the conduction and valence bands, the parabolic energy vs. k-vector relationship, simply called $E(k)$, is centered about $k = 0$. As discussed in Chapter 2, such a material is called a direct gap semiconductor. In Chapter 2, we learned that GaAs is a direct gap semiconductor. Si is an indirect gap semiconductor, one in which the conduction band minimum and valence band minimum hole energy occur at different locations in k-space.

Whether a material is direct or indirect greatly influences its radiative generation and recombination rate. In a direct gap material an electron–hole pair at minimum energy has essentially the same value of k as shown in Fig. 10.2. From the discussion above, we recognize that the momentum and k-vector are simply related: $p = \hbar k$. Therefore, the electron and hole in a direct material have the same momentum when near the conduction and valence band edges. In an indirect material the situation is different as shown in Fig. 10.3. As can be seen from the figure, in an indirect material the electron and hole have different values of k or equivalently momentum at minimum energy. This has an important implication in radiative transitions.

Both energy and momentum must be conserved in any transition. Therefore, in a radiative recombination event, the energy arising from the electron–hole pair annihilation is released as a photon. The momentum of the system must also be conserved, i.e., the initial momentum of the electron–hole pair must be taken up by the photon. Let us consider an indirect transition in Si for example and examine the energy and momentum conservation in this system. As can be seen from Fig. 10.3 the electron in the conduction band in an indirect semiconductor has an initial momentum equal to π/a, where a is the lattice constant. If we take Si as an example of an indirect

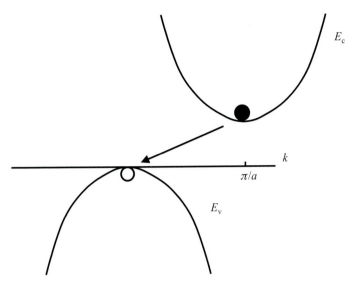

Figure 10.3 Sketch of an electron–hole recombination event in an indirect semiconductor. Notice that the magnitude of k is different for the electron and the hole implying a different momentum for the particles. The electron in the conduction band has a k-vector magnitude of π/a.

semiconductor, its lattice constant is 0.543 nm. The value of the k-vector for the electron is then $\sim 5.78 \times 10^7$ cm^{-1}. In order for momentum to be conserved following a recombination event for the system shown in Fig. 10.3, the photon would have to remove momentum equal to $\hbar k$, which in this case is 3.8×10^{-8} eV s/cm. Energy must also be conserved following the recombination event. For Si the energy gap is 1.12 eV and this amount of energy must be removed by the emitted photon. The wavelength of a photon with energy 1.12 eV is

$$\lambda = \frac{hc}{E} = \frac{1.2 \times 10^{-4} \text{ eV cm}}{1.12 \text{ eV}} = 1.1 \times 10^{-4} \text{ cm} \tag{10.4}$$

The corresponding momentum of this photon is given as $p = h/\lambda$, which for this case is 3.76×10^{-11} eV s/cm which is about three orders of magnitude too small to satisfy momentum conservation. Therefore, we see that a simple radiative recombination event cannot readily occur in an indirect gap semiconductor since both the momentum and energy cannot be conserved. For a direct gap semiconductor a radiative recombination event can occur since both momentum and energy can be conserved. Therefore, we find an important result: to make an efficient semiconductor light emitter it is necessary to use a direct gap semiconductor material. Many of the compound semiconductors are direct gap materials. For example, GaAs and InP are both direct gap semiconductors and can be used for LEDs.

The choice of materials for an LED is dictated primarily by the spectral requirements of the intended application, the nature of the material itself, and its technological maturity. The most commonly used semiconductors for LEDs are GaAs and

GaP. Though GaP is indirect, impurities are added to aid radiative recombination thus ensuring highly efficient operation. GaP LEDs emit light in the green portion of the spectrum. Other colors can be achieved using the ternary compound, $GaAs_{1-x}P_x$, and the quaternary AlGaInP. $GaAs_{1-x}P_x$ changes from direct to indirect as the proportion of phosphorus increases. Red emission can be achieved using direct bandgap emission with a 40% phosphorus composition. Orange and yellow can be obtained by again inserting impurities within the indirect compositions of GaAsP. Blue LEDs are made using the ternary compound $In_{1-x}Ga_xN$. Therefore, the full visible spectral range can be covered using LEDs, making them not only attractive for full color displays but potential candidates for white light illumination. The fact that white light can be achieved with LEDs makes them an attractive replacement for commercial lighting. LEDs have a higher efficiency than incandescent lighting and can ultimately approach or possibly exceed fluorescence in efficiency. LEDs have additional advantages over both incandescent and fluorescent lighting. Specifically, LEDs have a very long lifetime, are highly durable, and reliable. These properties make LEDs attractive in many applications, even if they cost more, since they rarely, if ever, have to be replaced. Additionally, it is possible that LED white light sources can provide significant energy savings compared with conventional lighting. Such a development would have a large impact on world energy consumption.

An LED must be constructed such that the light emitted by the radiative recombination events can escape the structure. LEDs can be designed as either surface or edge emitters, as shown in Fig. 10.4. Surface emitting LEDs can be tailored such that the bottom edge reflects light back toward the top surface to enhance the output intensity. Edge emitters in contrast emit light out of the side of the device. The main advantage of edge emitter LEDs is that the emitted radiation is relatively direct.

The three major issues that influence the commercial success of an LED are: (1) luminous performance, (2) cost, and (3) reliability. The luminous performance is related to the optical yield of the device. The optical yield is a measure of how efficient the LED is at converting input electrical current into output photons. The luminous performance is given as the ratio of the total output flux in lumens to the input power in watts. The distinction between luminous performance and optical yield arises from the use of lumens. The lumen is the unit for measuring photometric flux. Photometry is the science of visible light and its perception by the human eye. The measure of photometric flux, the lumen, describes the psycho-physical effect of radiation on the human eye, in other words how we see light. Therefore, a lumen is a measure of the light flux that is detected by the human eye. The luminous performance is given by the product of the eye sensitivity function, called the CIE curve, and the radiometric power efficiency, where the radiometric power is the total radiated power and is measured in watts. The radiometric power efficiency is defined as the ratio of the output power to the input power. With this definition the radiometric power efficiency, P_E, is given as $P_E = [(photons/s) \times (energy/photon)] / [(current) \times (applied voltage)]$:

$$P_E = \eta_{ext}\frac{E}{V_a} \tag{10.5}$$

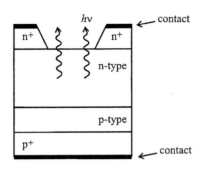

(a)

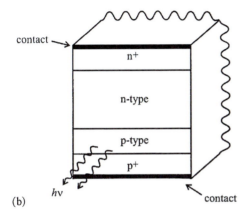

(b)

Figure 10.4 Device structures for (a) surface emitting and (b) edge emitting LEDs (Brennan, 1999).

where η_{ext} is the external quantum efficiency (defined as the ratio of the number of photons emitted from the device per incident injected electron), E the energy of the emitted photon, and V_a the applied voltage. The external quantum efficiency is readily related to the internal quantum efficiency, η_{int}, defined as the number of photons produced for every injected electron, as

$$\eta_{ext} = \eta_{int}C_{ex} \tag{10.6}$$

where C_{ex} is the extraction efficiency. The extraction efficiency is the fraction of generated photons that escape from the device package and emerge from the chip.

The factors that most influence the internal quantum efficiency of a homojunction LED are: (1) the quality and purity of the material and substrate, and (2) the existence of a direct energy gap spectrally matched to the desired emission wavelength. The presence of traps and impurities reduces the internal quantum efficiency since they produce channels by which nonradiative recombination events can occur. The primary nonradiative events that compete with the desired radiative transitions are Shockley–Read–Hall (SRH) recombination, thermal band to band recombination, and Auger recombination (see Chapter 2). Auger recombination is most important in degenerate or nearly degenerate materials. As mentioned above, the radiative efficiency is highest

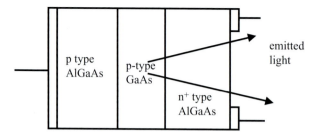

Figure 10.5 Sketch of a heterojunction LED showing light emission out of the right hand edge of the device. Notice that the light is generated within the GaAs but propagates through the AlGaAs towards the external environment. As a result, the wavelength of the generated light is longer than the cutoff wavelength within the AlGaAs.

in direct gap semiconductors. Though the introduction of impurities in some indirect semiconductors, such as GaAsP and GaP, has made these materials useful emitters, their efficiency is still relatively low. Their main advantage stems from their relatively low cost due primarily to the fact that these are familiar materials, and extensive infrastructure has been developed for their manufacture.

The extraction efficiency is the fraction of generated light that actually escapes from the device into the external environment. There are several issues that determine the extraction efficiency. These are: (1) the absorption coefficient of the material at the emission wavelength, (2) the radiation geometry of the LED, and (3) the substrate. Though direct band gap semiconductors have high radiative efficiencies, their corresponding absorption coefficients at the emitted wavelength are also high leading to photon reabsorption. In other words, a photon can be emitted following a radiative recombination event in one part of the device but be reabsorbed before it can reach the surface. One obvious solution is to put the junction close to the surface to avoid reabsorption. However, the presence of surface states, producing nonradiative recombination channels, can result in a reduction in the radiative efficiency of the device. Typically, a heterojunction is utilized instead to alleviate the reabsorption problem. A commonly used heterojunction is that of GaAs and AlGaAs. AlGaAs has an energy gap larger than GaAs. The heterojunction LED is designed such that the recombination occurs in the narrow gap semiconductor, GaAs in this case, as shown in Fig. 10.5. The light is subsequently emitted through an AlGaAs window into the external environment. Since the band gap of the AlGaAs is greater than that of the GaAs, most of the light will transit through the AlGaAs without being absorbed.

The geometry of the LED package must be carefully designed. The index of refraction of the semiconductor is significantly larger than that of the surrounding air. Thus, the generated light upon reaching the surface can be refracted. Depending upon the angle, some of the light will be totally internally reflected from the surface thereby reducing the amount of escaping optical power. The critical angle for total internal reflection is given as

$$\theta_c = \sin^{-1}\left(\frac{n_{air}}{n_s}\right) \qquad (10.7)$$

where n_s is the index of refraction of the semiconductor. For example, the index of refraction of the chip is about 3.4. Assuming that the second medium is air with an index of refraction of 1.0, the resulting critical angle, θ_c, is 17°. Only that light within a cone of semiangle 17° will leave the device.

The extraction efficiency can be improved by choosing a transparent substrate. In such a device the light that is internally reflected but not reabsorbed will eventually escape after being randomly scattered by rough spots on the chip surface. Therefore, the extraction efficiency can be quite high, ~ 0.3 in most structures.

New LED structures are being pursued using organic semiconductor materials. These devices are referred to as OLEDs, for organic light emitting diodes. An OLED consists of several layers of organic material sandwiched between two metallic electrodes, one of which is transparent. Given that the layers of organic material contain a variety of organic substances, multi-color light is emitted during their operation. As such, OLEDs can naturally produce white light with a suitable mix of organic molecules. There are two general types of OLED devices. These are devices made with small organic molecules and devices made with organic polymers. In either case, the organic material is highly disordered and thus can be readily deposited at very low manufacturing cost.

OLED operation can be understood as follows. The two sandwiching electrodes form a cathode, an electron emitter, and an anode, an electron absorber. The organic materials have extended molecular states much like those discussed in Chapter 8. Again LUMO and HOMO states exist within the molecular layers. The work function of the cathode is chosen such that it is close in value to the LUMO states in the organics. Similarly, the work function of the anode is chosen such that it is close in magnitude to the HOMO states in the organics. The situation is shown in Fig. 10.6(a). In equilibrium the Fermi level aligns throughout the system. Thus the Fermi levels of the cathode and anode metallic contacts must align as shown in Fig. 10.6(b). The fact that the Fermi levels align produces a built-in potential as shown in the figure. As in the case of a p–n junction diode, the built-in potential acts to block carrier injection from either contact. Application of a bias across the metallic contacts produces the band bending shown in Fig. 10.6(c). In this case, a positive voltage is applied to the anode with respect to the cathode thus lowering the electron energies near the anode and raising them near the cathode. The forward bias produces carrier injection via tunneling. Electrons tunnel into the LUMO states from the cathode while holes tunnel into the HOMO states from the anode. The injected electrons and holes are attracted by the positive and negative potentials respectively as they make their way through the organic layers. The mobility of both carriers is very small. Consequently, the electrons and holes are likely to recombine before they reach the collecting contacts. Most recombination events are radiative and thus the OLED structure emits light.

The main advantage of OLEDs over standard LEDs is their cost. Since OLEDs are noncrystalline materials, they can be readily deposited in a relatively low cost manner. However, OLEDs have some limitations. One problem faced by OLEDs is lifetime. The organic materials used in OLEDs are very sensitive to water and oxygen. Thus they must be deposited on glass substrates and covered with a second sheet of

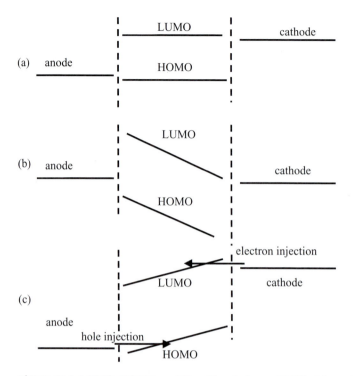

Figure 10.6 LUMO, HOMO, and Fermi levels for an OLED: (a) apart, (b) equilibrium band diagram, (c) forward bias condition.

glass. Unfortunately, the usage of glass makes the cost of manufacturing higher. Future developments based on new materials may solve this problem.

10.2 **Stimulated emission**

There are two general mechanisms by which optical radiation is emitted. These are spontaneous and stimulated emission. As the name implies, in spontaneous emission a radiative emission event occurs naturally without external cause. An example of a spontaneous emission event is the emission of a photon following a natural recombination of an electron from the conduction band with a hole in the valence band of a semiconductor. The recombination energy of the event is carried off by the emitted photon. In general, a spontaneous emission event corresponds to the emission of radiation following a spontaneous decay of an electron from a high energy state into a lower energy state. Most radiative emission events are spontaneous. Light emitted from the sun, incandescent bulbs, LEDs, etc. arises from spontaneous emission.

Alternatively, light can be emitted as a result of an external cause leading to stimulated emission. Stimulated emission is induced by the action of an external electromagnetic field. The external field perturbs an electron in a high energy state inducing it to decay into a lower energy state with the subsequent emission of a photon. An absorption event is simply the inverse of a stimulated emission event, an electron

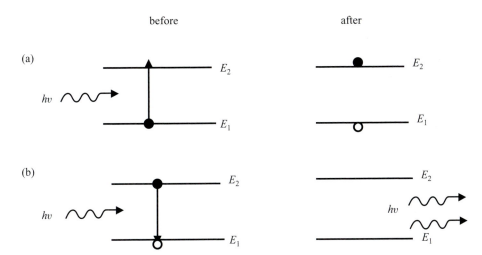

Figure 10.7 Sketch of (a) absorption and (b) stimulated emission processes. Notice that in both cases there is an input photon. In the case of absorption, an electron initially within the valence band absorbs the input photon and makes a transition into the conduction band producing an electron–hole pair. In the case of stimulated emission, an incident photon causes an initial electron within the conduction band to radiatively recombine with a hole initially in the valence band. An additional photon is produced along with the initial photon.

within a low energy state absorbs a photon and makes a transition into a high energy state An incident photon is necessary in order to induce the transition.

The absorption and stimulated emission processes within a semiconductor are diagrammatically sketched in Fig. 10.7. As can be seen from the figure, absorption occurs when an electron initially within the valence band absorbs a photon and makes an upward transition in energy into the conduction band. The absorption rate, r_{12}, the rate from level 1 to level 2, can be determined as follows. The absorption rate obviously depends upon the probability that the low energy state, state 1 in Fig. 10.7, is initially occupied and that the corresponding high energy state, state 2 in Fig. 10.7, is initially empty. Let f_1 represent the probability that level 1 is initially occupied. Then $(1 - f_2)$ must be the probability that level 2 is initially empty. The absorption rate also depends of course on the presence of input light. If no light is present, obviously the system cannot make a radiative absorption transition. Thus r_{12} depends directly upon the input light intensity, $P(E_{21})$, where E_{21} is the energy difference between the 1 and 2 energy levels, and can be written as

$$r_{12} \sim P(E_{21}) f_1 (1 - f_2) \tag{10.8}$$

Calling the constant of proportionality B_{12}, the absorption rate becomes

$$r_{12} = B_{12} P(E_{21}) f_1 (1 - f_2) \tag{10.9}$$

The stimulated emission rate, $r_{21}(\text{stim})$, also depends upon the input light intensity. As mentioned above, a stimulated emission event occurs only in the presence of an input light signal. The process is shown in Fig. 10.7(b). As can be seen from the figure, stimulated emission occurs when an electron initially within the conduction band recombines with a hole initially within the valence band through the action of an input photon. Two photons are then emitted, the initial one and another produced from the energy of recombination of the electron–hole pair. The rate, $r_{21}(\text{stim})$, depends upon the probability that the high energy state is initially occupied, f_2, the low energy state is initially empty, $(1 - f_1)$, and the incident light intensity, $P(E_{21})$ as

$$r_{21}(\text{stim}) \sim f_2(1 - f_1)P(E_{21}) \tag{10.10}$$

Defining the constant of proportionality as B_{21}, the stimulated emission rate becomes

$$r_{21}(\text{stim}) = B_{21}f_2(1 - f_1)P(E_{21}) \tag{10.11}$$

The spontaneous emission rate does not depend upon the input light. In a spontaneous emission event, light is emitted following a spontaneous transition of a high energy electron into a low energy state. As in the case for stimulated emission, a photon is emitted with energy equal to the difference between the two energy levels. The spontaneous emission rate, $r_{21}(\text{spont})$, like the stimulated emission rate, depends upon the probability that the high energy state is initially occupied and the low energy state is initially empty. Calling the constant of proportionality A_{21}, the spontaneous emission rate is given as

$$r_{21}(\text{spont}) = A_{21}f_2(1 - f_1) \tag{10.12}$$

In equilibrium the net transition rate must be zero. Therefore, the absorption rate must be equal to the sum of the spontaneous and stimulated emission rates which is given as

$$r_{12} = r_{21}(\text{stim}) + r_{21}(\text{spont})$$
$$B_{12}f_1(1 - f_2)P(E_{21}) = B_{21}f_2(1 - f_1)P(E_{21}) + A_{21}f_2(1 - f_1) \tag{10.13}$$

Solving (10.13) for $P(E_{21})$ obtains

$$P(E_{21}) = \frac{A_{21}}{B_{12}\dfrac{f_1(1 - f_2)}{f_2(1 - f_1)} - B_{21}} \tag{10.14}$$

Since the system is assumed to be in equilibrium, f_1 and f_2 are equilibrium distribution functions. Their form is given by (1.8) as,

$$f_1 = \frac{1}{1 + e^{\frac{(E_1 - E_f)}{kT}}} \qquad f_2 = \frac{1}{1 + e^{\frac{(E_2 - E_f)}{kT}}} \tag{10.15}$$

Using (10.15), (10.14) can be simplified to

$$P(E_{21}) = \frac{A_{21}}{B_{12}e^{\frac{E_{21}}{kT}} - B_{21}} \tag{10.16}$$

The optical power can be expressed in another manner (Brennan, 1999, Chapter 5) as

$$P(E_{21}) = \frac{8\pi \bar{n}^3 E_{21}^2}{(hc)^3} \frac{1}{e^{\frac{E_{21}}{kT}} - 1} \tag{10.17}$$

where n is the index of refraction, h is Planck's constant, and c is the speed of light. Equating (10.16) and (10.17) obtains

$$\frac{A_{21}}{B_{12}e^{\frac{E_{21}}{kT}} - B_{21}} = \frac{8\pi \bar{n}^3 E_{21}^2}{(hc)^3} \frac{1}{e^{\frac{E_{21}}{kT}} - 1} \tag{10.18}$$

Both the temperature dependent and temperature independent terms in (10.18) must be equal. Equating the temperature dependent terms in (10.18) the following relationship between A_{21} and B_{12} is obtained:

$$A_{21} = \frac{8\pi \bar{n}^3 E_{21}^2}{(hc)^3} B_{12} \tag{10.19}$$

Relating the temperature independent terms yields

$$A_{21} = \frac{8\pi \bar{n}^3 E_{21}^2}{(hc)^3} B_{21} \tag{10.20}$$

Combining (10.19) and (10.20), the B coefficients are related as

$$B_{12} = B_{21} \tag{10.21}$$

A and B are called the Einstein A and B coefficients. Notice that (10.20) relates A to B.

Let us now consider a nonequilibrium condition in which there is either a net emission or a net absorption rate. For simplicity let us neglect the spontaneous emission rate compared with the stimulated emission rate. With this assumption, the system will exhibit a net absorption rate if the absorption rate exceeds the stimulated emission rate. Conversely, the system will exhibit a net stimulated emission rate if the stimulated emission rate exceeds the absorption rate. Let us examine the conditions for either net absorption or net stimulated emission. It is useful to compare the absorption process with stimulated emission. The two mechanisms are diagrammatically shown in Fig. 10.8. To have a net absorption rate, the electron concentration within the valence band must exceed that in the conduction band. Conversely, to have a net stimulated emission rate the electron concentration within the conduction band must exceed that in the valence band. Under usual conditions, the electron concentration within the valence band exceeds that in the conduction band. Therefore, to have a net stimulated emission rate, the populations must be inverted. This condition is called population inversion.

The population inversion condition for a semiconductor can be derived as follows. Neglecting the spontaneous emission rate, a system will exhibit a net stimulated emission rate if $r_{21}(\text{stim})$ is greater than r_{12}. Using (10.9) and (10.11), the condition for the system to have a net stimulated emission rate is

$$B_{21} f_2 (1 - f_1) P(E_{21}) > B_{12} f_1 (1 - f_2) P(E_{21}) \tag{10.22}$$

(a)

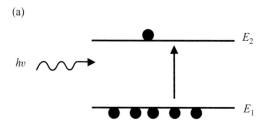

(b)

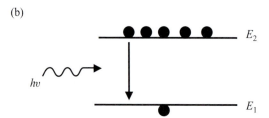

Figure 10.8 Diagrammatic sketch of the conditions for (a) net absorption and (b) net stimulated emission. Notice that for net absorption the electron concentration within the valence band must be greater than that in the conduction band. For net stimulated emission the opposite condition must be met, the electron concentration within the conduction band must exceed that in the valence band. Such a condition requires population inversion.

Recognizing that $B_{12} = B_{21}$ (10.22) simplifies to

$$f_2(1 - f_1) > f_1(1 - f_2) \tag{10.23}$$

which becomes

$$f_2 > f_1 \tag{10.24}$$

The distribution functions f_1 and f_2 in this case are nonequilibrium distributions. Generally, the nonequilibrium distributions are difficult to obtain. However, one can approximate their form using the concept of quasi-Fermi levels. If the system is weakly coupled, then each separate subsystem, in this case the electrons and holes, can be treated as being separately in equilibrium even though the full system is in nonequilibrium. Using this approximation, the nonequilibrium distribution functions resemble the equilibrium distribution functions but with the Fermi level replaced by the quasi-Fermi level as

$$f_1(E_1) = \frac{1}{1 + e^{\frac{(E_1 - q\varphi_1)}{kT}}} \qquad f_2(E_2) = \frac{1}{1 + e^{\frac{(E_2 - q\varphi_2)}{kT}}} \tag{10.25}$$

where φ_1 is the quasi-Fermi level for the lower energy level, the valence band, and φ_2 is the quasi-Fermi level for the higher energy level, the conduction band. The physical meaning of the quasi-Fermi levels can be understood as follows. In Chapter 2 we examined the operation of p–n junctions. In equilibrium it was found that the Fermi level was flat everywhere in the device as shown in Fig. 10.9(a). If a forward bias is applied to the junction, the potential energy barrier separating the n and p sides of the

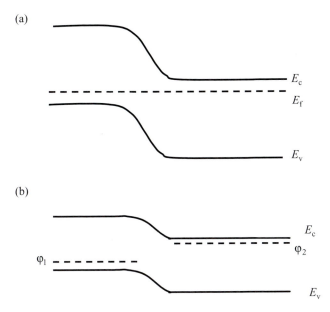

Figure 10.9 Sketch of a p–n junction diode in (a) equilibrium with a flat Fermi level and (b) under forward bias showing the two quasi-Fermi levels formed for the electrons, φ_2 and holes, φ_1.

junction is reduced. The Fermi level is no longer flat since the device is no longer in equilibrium. Instead, the Fermi level breaks into two parts, the electron quasi-Fermi level and the hole quasi-Fermi level. It is reasonable to assume that the electrons on the n side of the junction and the holes on the p side of the junction are separately in equilibrium even though the overall structure is clearly not in equilibrium. Under this assumption, the quasi-Fermi levels can be defined for either side of the junction. The resulting band diagram for the diode under forward bias including the quasi-Fermi levels is shown in Fig. 10.9b. The difference between the quasi-Fermi levels is equal to the applied external bias.

Substituting in for f_1 and f_2 in (10.24) the forms given by (10.25) yields

$$\frac{1}{1+e^{\frac{(E_2-q\varphi_2)}{kT}}} > \frac{1}{1+e^{\frac{(E_1-q\varphi_1)}{kT}}} \tag{10.26}$$

Simplifying, (10.26) becomes

$$q(\varphi_2 - \varphi_1) > (E_2 - E_1) \tag{10.27}$$

Equation (10.27) is the condition for net stimulated emission in a semiconductor. In words, (10.27) implies that in order for a semiconductor to have a net stimulated emission rate, the difference in the quasi-Fermi levels must be greater than the difference between the conduction and valence bands or the band gap energy.

From inspection of Fig. 10.9(b) it is clear for this particular device that the difference in the quasi-Fermi levels is less than the energy band gap. From (10.27), the diode

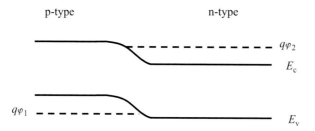

p-type   n-type

$q\varphi_2$

E_c

$q\varphi_1$

E_v

Figure 10.10 Sketch of a forward biased degenerately doped p–n junction diode. Notice that the difference in the quasi-Fermi levels is greater than the energy gap. Therefore, the condition for net stimulated emission is satisfied.

cannot exhibit a net stimulated emission rate. In order that the difference in the quasi-Fermi levels be greater than the band gap, it is necessary that the quasi-Fermi level for the electrons lies above the conduction band while the quasi-Fermi level for the holes lies below the valence band. Such a condition can be achieved by degenerately doping the n- and p-type semiconductors forming the p–n junction, as discussed in Section 1.2. If the p and n sides of the junction are degenerately doped, then under forward bias the difference in the quasi-Fermi levels within the junction region exceeds the band gap as shown in Fig. 10.10. As a result, the concentration of electrons in the conduction band exceeds the electron concentration in the valence band within the junction producing population inversion, the condition for net stimulated emission. In the next section we will see how a degenerately doped p–n junction diode can be made into a semiconductor laser.

Before we end this section it is useful to provide expressions for the nonequilibrium carrier concentrations using the quasi-Fermi levels. The nonequilibrium electron and hole carrier concentrations can be expressed as

$$n = n_\text{i} e^{\frac{(F_\text{n} - E_\text{i})}{kT}} \qquad p = n_\text{i} e^{\frac{(E_\text{i} - F_\text{p})}{kT}} \qquad (10.28)$$

10.3 **Laser operation**

Minority carrier injection occurs within a forward biased p–n junction diode leading to a net recombination rate. If the radiative recombination channel is greater than either thermal or Auger recombination, then the device will provide a net emission of radiation. As discussed above, this is the basis of the operation of an LED. In an LED the emitted light results from spontaneous emission. However, in a laser the light is emitted through stimulated emission. As discussed in Section 10.2, a semiconductor exhibits a net stimulated emission rate when the difference in the quasi-Fermi levels exceeds the band gap. This condition is called population inversion and can be realized using a forward biased, degenerately doped p–n junction diode. Although in order for a system to lase there must be a net stimulated emission rate, this by itself is insufficient to produce lasing. In order for a system to lase, the optical gain must exceed all optical losses such that the optical field grows with distance and the field must be resonantly amplified.

Resonant amplification can be achieved by confining the emitted light in a resonant cavity. The simplest such design is formed using two reflective surfaces bounding the active region in which stimulated emission occurs, often referred to as the gain medium. The overall structure forms a waveguide such that the light is confined to bounce back and forth through the gain medium between the two reflective surfaces. In a gas laser, the reflective surfaces are simply mirrors. In a semiconductor laser, cleaving the ends of the waveguide forms the reflective surfaces. By confining the radiation to a resonant cavity essentially only one wavelength oscillates and thus is resonantly amplified. As a result, the light output is nearly monochromatic. This is often referred to as the oscillation condition. The spacing of the reflective surfaces influences which wavelength of light is resonantly enhanced. Constructive interference of the light occurs when the light is coherently reflected back and forth within the cavity. Thus only an integral number of half-wavelengths can be supported by the cavity. This leads to the following condition on the cavity length, L:

$$L = \frac{m\lambda}{2} \tag{10.29}$$

where m is an integer.

The resonance condition can be developed as follows. The optical field, E, propagating in the cavity can be expressed as

$$E = E_i e^{-\Gamma x} \tag{10.30}$$

where x is the distance along the cavity, E_i is the amplitude and the propagation constant, Γ, is a complex number. The complex propagation constant can be written in terms of its real and imaginary parts as

$$\Gamma = j\beta k_0 - \alpha \tag{10.31}$$

The imaginary part of Γ leads to a propagating wave. Generally, the real part of the propagation constant for most media is negative and as a result the wave attenuates with distance traveled. This is the usual situation since most media are lossy. For example, an optical signal propagating through an optical fiber attenuates with distance traveled. However, in a laser or more generally a system with optical gain, the real part has a positive sign and as a result the amplitude of the wave increases in magnitude as a function of distance traveled. Coupled with the oscillation condition, a nearly monochromatic amplified light beam is produced.

Generally, the optical field within the resonant cavity of a laser experiences multiple reflections between the two reflective surfaces passing numerous times through the gain medium. Each time the field passes through the gain medium it is amplified by the addition of photons produced through stimulated emission events. The additional photons produced during each pass from stimulated emission events add coherently to the optical field. A fraction of the light is transmitted through the second reflective surface, E_t, providing the output beam. Assuming that the reflective surfaces have transmission coefficients t_1 and t_2 and reflection coefficients r_1 and r_2, the transmitted field, E_t, can be written as (Brennan, 1999, Chapter 13)

$$E_t = \frac{E_i t_1 t_2 e^{-\Gamma L}}{1 - r_1 r_2 e^{-2\Gamma L}} \tag{10.32}$$

When the denominator vanishes in (10.32), the system is in resonance. Hence the oscillation or resonance condition becomes

$$1 = r_1 r_2 e^{-2\Gamma L} \tag{10.33}$$

Substituting for Γ the expression given by (10.31), the resonance condition becomes

$$1 = r_1 r_2 e^{-2j\beta k_0 L} e^{2\alpha L} \tag{10.34}$$

The phase term, $e^{-2j\beta kL}$, does not change the amplitude of the field since it has unity magnitude. Therefore, the amplitude of the field depends upon the product of the reflectivities and the real exponential term:

$$1 = r_1 r_2 e^{2\alpha L} \tag{10.35}$$

The parameter α can have either a negative or positive sign. As discussed above, if α is negative then the field attenuates with distance and the propagation medium must be lossy. Alternatively, if α is positive then the field grows with distance and the propagation medium provides gain. In a laser α has both positive and negative real components and can be written as

$$\alpha = g - \alpha_i \tag{10.36}$$

where g represents the gain and α_i the losses within the propagation medium. Inspection of (10.36) shows that α is positive if g is greater than α_i and conversely negative if g is less than α_i. Hence, for the medium to deliver a net gain the sum of all of the losses, α_i, must be less than the gain, g. Under this condition, α is positive and the amplitude grows with distance traveled and the system will lase. If, on the other hand, the sum of all of the losses exceeds the gain then α is negative and the amplitude decays with distance and the system will not lase.

In general, there are significant losses within a semiconductor laser. The effect of the loss mechanisms on the gain can be seen by substituting into (10.35) the expression for α given by (10.36) to obtain

$$g = \alpha_i + \frac{1}{2L} \ln\left(\frac{1}{r_1 r_2}\right) \tag{10.37}$$

The α_i represents various loss factors. Among these are free-carrier absorption in the active and cladding layers, scattering loss, and coupling loss. The active layer is defined as the region in which the population inversion exists, i.e., the gain medium. The cladding layers are defined as the surrounding semiconductor layers. Within the cladding layers no gain occurs. Free-carrier absorption arises, as the name implies, from absorption of photons by free carriers within the conduction or valence bands.

Upon absorption of a photon, the free carriers attain relatively high energies in either the conduction or valence bands. These high-energy carriers can relax through a non-radiative channel resulting in the net loss of a photon from the gain medium. Scattering loss is due to the scattering of radiation out of the optical waveguide by either nonplanar heterojunction interfaces or imperfections in the dielectric layers. Coupling losses occur when the optical fields spread beyond the wide band-gap confining layers. The last factor in (10.37) represents the mirror loss term, which includes output transmission.

The threshold condition for lasing can be understood from (10.37). When the gain is equal to the sum of the loss terms, including the mirror loss, the system can begin to lase. The corresponding gain is the threshold gain. To achieve threshold gain there must be a minimum current density injected into the active region of the laser. This minimum current density is called the threshold current density, J_{th}. The threshold current density can be determined by summing the current densities corresponding to the loss terms and to the output signal. In steady-state the net input current must be balanced by the net recombination rate. The continuity equation is given then as

$$\frac{dn}{dt} = \frac{J}{qd} - \frac{n}{\tau} = 0 \tag{10.38}$$

where τ is the recombination time, d the width of the active region and n the electron concentration within the active region. The threshold current density, J_{th}, is given then as

$$J_{th} = \frac{qdn_{th}}{\tau} \tag{10.39}$$

where n_{th} is the threshold electron concentration. Therefore, a minimum current density and corresponding minimum electron concentration must be electrically injected by the action of the forward bias in order to achieve threshold and subsequently lasing.

Example Problem 10.1 Calculation of the gain in a semiconductor laser

The reflection coefficients for a semiconductor laser are equal and have the value 0.5. The cavity length is 1.0 μm. Assume that there are no additional loss terms and calculate the gain of the laser.

From (10.36) the gain, g, is equal to α when the internal losses are zero. The value of α can be obtained from (10.35). Equivalently when there are no internal losses, the gain is equal to α and (10.37) can also be used. Thus the value of α can be found from

$$\alpha = g = \frac{1}{2L} \ln\left(\frac{1}{r_1 r_2}\right)$$

Substituting in, g is given as

$$g = \frac{1}{2(1.0 \times 10^{-4}\text{ cm})} \ln\left(\frac{1}{(0.5)^2}\right) = 6.93 \times 10^3 \text{ cm}^{-1}$$

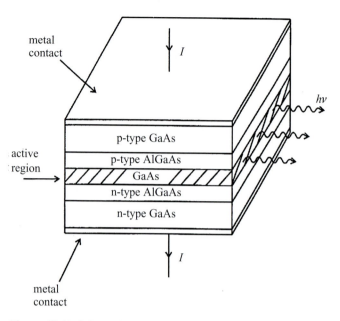

Figure 10.11 Schematic drawing of a p–n double heterojunction laser. The active region consists of the hatched GaAs layer sandwiched between the p-type and n-type A1GaAs layers as shown in the diagram.

10.4 **Types of semiconductor lasers**

Simple homojunction semiconductor lasers typically have relatively high threshold current densities since both carrier and optical confinement within the active region of the device are minimal. Carriers injected by the forward bias into the active region may simply pass through without recombining or diffuse outward and again not contribute to the lasing process. In addition, the optical fields leak into the surrounding inactive layers bordering the gain medium. Both of these effects conspire to increase the threshold current density for lasing. With increased threshold current density the laser efficiency is reduced.

Heterostructures provide a means of confining both the optical fields and carriers within the active region reducing the threshold current density. As we discussed earlier, a heterostructure is formed by combining two dissimilar semiconductor materials. In many instances it is desired that the lattice constants of the two materials be lattice matched as is the case for GaAs and AlGaAs. A typical GaAs–AlGaAs heterostructure laser is sketched in Fig. 10.11. The device shown in Fig. 10.11 is called a double heterostructure laser since it contains two AlGaAs–GaAs interfaces. The energy band gap of the AlGaAs layers is greater than that of the GaAs layer. As a result there is an energy gap discontinuity between the two materials. GaAs and AlGaAs are said to form a Type I heterojunction. A Type I heterostructure is defined as a structure in which the energy gap discontinuity is equal to the sum of the conduction and valence band edge discontinuities.

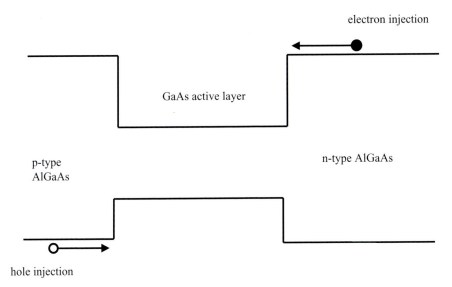

electron injection

GaAs active layer

n-type AlGaAs

p-type
AlGaAs

hole injection

Figure 10.12 Conduction and valence band energy diagrams for the AlGaAs–GaAs double heterostructure within the double heterostructure laser sketched in Fig. 10.11. Holes are injected from the p-type AlGaAs and electrons from the n-type AlGaAs into the active GaAs layer as shown. The presence of the conduction and valence band energy discontinuities results in capturing the carriers within the GaAs layer leading to radiative recombination.

Double heterostructure lasers like that shown in Fig. 10.11 with a narrow band gap material surrounded on either side by a wide band gap material have a substantially lower threshold current density than homojunction lasers. The presence of the conduction and valence band energy discontinuities in the two heterostructures acts to greatly confine the injected carriers within the narrow gap GaAs layer. As a result, the injected electrons and holes are trapped within the GaAs layer as shown in Fig. 10.12, which ultimately can result in a radiative recombination event with the subsequent emission of a photon. The double heterostructure also acts to confine the optical fields within the GaAs layer due to the different indices of refraction of the constituent semiconductors. As can be seen from Fig. 10.11, current is injected perpendicular to the heterostructures. Notice that the structure consists of p- and n-type GaAs and AlGaAs to form a diode. The diode is operated in forward bias so that holes are injected from the p side and electrons from the n side. Following radiative recombination of an injected electron and hole a photon of energy roughly equal to the energy band gap of GaAs is emitted.

It is important to note that the photon is emitted by stimulated emission. Light bounces back and forth by being reflected at the boundaries of the device. The reflecting surfaces are produced by cleaving the crystal such that the cleaved surface provides for nearly complete reflection. Stimulated emission within the GaAs layer is triggered by the reflected light. The emitted photon has the same phase as that of the stimulating light and thus the light adds coherently. The fact that the phase is the same between the emitted and stimulating photons can be understood as follows. In an absorption event, some of the incident light is absorbed and some is transmitted. The amplitude of the transmitted light is less than that of the incident light since some of the incident light is absorbed. The transmitted light is of course at the same frequency, in the same direction, and has the same polarization as the incident light. Stimulated emission is the inverse to absorption. In a stimulated emission event, the transmitted light amplitude is greater than the incident light amplitude implying a net emission of radiation. Again the transmitted and incident radiation have the same frequency and polarization, and propagate in the same direction. As a result, the phase of the transmitted light (which includes emitted photons as well as the incident photons) is the same as the incident light.

In a double heterostructure laser the emitted light has an energy close to that of the energy band gap of the narrow gap, confining semiconductor. The energy of the emitted photon thus depends upon the intrinsic properties of the active medium. Quantum well lasers offer an alternative to double heterostructures in that the energy of the emitted photon can be made substantially higher than that of the band gap. In addition, by a judicious choice of the well width, the energy of the emitted photon can be adjusted thus enabling lasing at various wavelengths using the same materials system. A single quantum well laser is similar in design to a double heterostructure laser except that the narrow gap, active region semiconductor layer is made very small in width such that spatial quantization effects occur. As mentioned in Section 5.2, when the dimensions of a confining region (in this case a simple well) are comparable to that of the electron de Broglie wavelength, quantum mechanical effects arise (see Brennan (1999) for a full discussion of spatial quantization). When a system behaves quantum mechanically, only certain discrete energy states are available. These states are called the allowed energy levels of the system. For example, there are only certain allowed electronic energy states in an atom. Upon the absorption or emission of radiation, an electron jumps from one quantum state to another. The electron can only occupy the allowed energy states and thus cannot be found at any energy other than that of an allowed quantum state. A similar situation occurs for spatial quantization. Again, only certain discrete energy states within the well are allowed and the electron or hole can only be found in one of these states. Spatial quantization results in the production of energy levels above the conduction band minimum in a quantum well. An example is shown in Fig. 10.13. Notice that within both the conduction and valence bands quantum levels are introduced. The minimum electron and hole energies within the well are both above the conduction and valence band minima respectively. If the electron and hole recombine as shown in Fig. 10.13, the energy of the emitted photon is greater than that of the energy gap. The energy of the quantum states is a strong function of the well width: the narrower the well, the greater the quantum state energy. Thus in a very narrow well width device, the energy of the emitted photon is greater than that of a wider well width structure. By adjusting the well width the energy of the emitted photon can be tuned.

The primary limitation to single quantum well lasers is that the cross-section for carrier capture in the well is relatively small. In order to obtain a radiative recombination event it is necessary for the injected electrons and holes to be captured within the

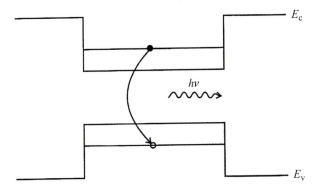

Figure 10.13 Energy band structure of a single quantum well laser showing a confined quantum state transition. Notice that an electron in a confined quantum state in the conduction band well recombines with a hole in a confined quantum state in the valence band well. The recombination leads to the emission of a photon as shown (Brennan, 1999).

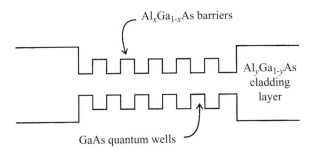

Figure 10.14 Energy band diagram for a multi-quantum well laser structure.

quantum well. Generally, the carriers must undergo an inelastic scattering event in order to lose sufficient energy that they can become confined to the well. If no scattering events occur within the well width, the injected carriers will traverse the well without being captured. Thus the injected carriers have a reasonable likelihood of passing directly over the well without becoming trapped and do not contribute to the optical output. As a result, the threshold current density of the device can be relatively high.

To improve the capture probability of the injected carriers within the quantum well, alternative device structures have been developed. One approach is to utilize multiple quantum wells instead of a single well. A representative multiple quantum well structure is shown in Fig. 10.14. As can be seen from the figure the single quantum well within the active region is replaced by a series of quantum wells. As for the single quantum well device, carriers are injected from either side of the structure. Given that there are more quantum wells present than in the single quantum well device, there is a higher probability that an injected carrier will suffer an inelastic scattering event and be captured within a well before it can completely traverse the device. One drawback to the multiple quantum well device is that the injected electrons and holes enter the active region (multiple well region) from opposite ends. Therefore, the electrons

(a)

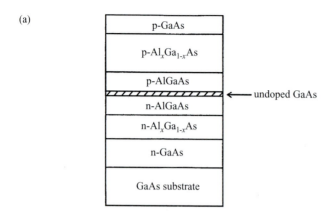

(b)

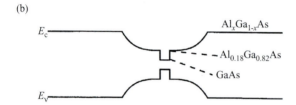

Figure 10.15 (a) Layers and (b) energy band structure for a GRINSCH laser.

and holes are not necessarily captured within the same well which acts to reduce the radiative recombination rate. For this reason, multiple quantum well devices do not have the lowest threshold current density of semiconductor lasers.

The device structure shown in Fig. 10.15 has the lowest threshold current density of existing semiconductor lasers. It is called a GRINSCH laser. The acronym stands for graded index separate confinement heterostructure. As can be seen from Fig. 10.15 a single quantum well is embedded within a graded funnellike region. As is shown in the figure the graded region surrounding the quantum well provides a means by which the injected carriers can be funneled into the well. The band gap grading creates an electric field that directs the electrons and holes into the same quantum well where they can radiatively recombine. As a result, fewer electrons and holes can traverse the quantum well without becoming trapped. The threshold current density of the GRINSCH laser is significantly lower than that of either the single or multiple quantum well devices. The improvement stems from the fact that the carriers thermalize more efficiently into the well owing to the funnellike region surrounding the well.

We conclude this section with a discussion of vertical cavity surface emitting lasers, VCSELs. In these structures the light is emitted normal to the surface as shown in Fig. 10.16. The primary advantages of VCSELs is that single mode operation can be achieved due to the short cavity length of the device, and these devices are better suited than edge emitting lasers to forming a two-dimensional array. Single mode operation is possible with VCSELs since the mode spacing is inversely proportional to the cavity length. Thus the smaller the cavity length the greater the mode spacing which leads

light output

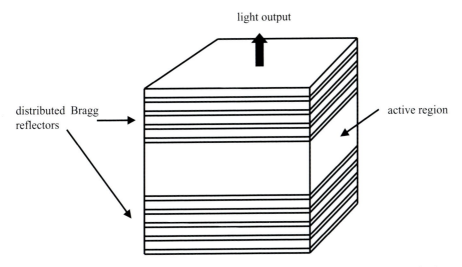

distributed Bragg reflectors

active region

Figure 10.16 Sketch of a VCSEL. The light is output from the top surface as shown in the diagram. The alternating layers comprise the distributed Bragg reflectors. The center region is the active region of the laser.

then to single mode operation. The mode spacing in a laser can be obtained using (10.29) as follows. The output light wavelength, λ_o, is given as

$$\lambda_o = \lambda n' \tag{10.40}$$

where n' is the index of refraction within the laser and λ is the wavelength of the light in the laser. The mode index, m, can be described using (10.29) and (10.40) as

$$m = \frac{2Ln'}{\lambda_o} \tag{10.41}$$

The mode spacing with wavelength is given as $dm/d\lambda_o$ which becomes

$$\frac{dm}{d\lambda_o} = -\frac{2Ln'}{\lambda_o^2} + \frac{2L}{\lambda_o}\frac{dn'}{d\lambda_o} \tag{10.42}$$

Solving (10.42) for $d\lambda_o$ obtains

$$-d\lambda_o = \frac{\frac{\lambda_o^2}{2Ln'}dm}{\left(1 - \frac{\lambda_o}{n'}\frac{dn'}{d\lambda_o}\right)} \tag{10.43}$$

The value of dm is 1 for adjacent modes. Inspection of (10.43) shows that the wavelength separation between adjacent modes is inversely proportional to the cavity length in agreement with the statement above. Thus in order to obtain a large separation between adjacent modes and have only a single mode excited in the laser, the cavity length needs to be short.

One of the major disadvantages of VCSELs is that since the cavity length is short, the round trip gain of the device is low. Consequently, the threshold current density of a VCSEL is significantly higher than that of an edge emitting laser. The mirrors of the VCSEL are made using distributed Bragg reflectors, or DBRs. DBR mirrors are formed using quarter wavelength plates alternating between GaAs and AlGaAs as shown in the figure. The DBR mirrors provide excellent reflection such that the quality factor of the resonator, Q, is high. The quality factor Q is defined as the ratio of the energy stored in the system at resonance to the energy lost in a cycle of oscillation. A high value of Q implies that the energy stored in the cavity is much greater than the energy loss per cycle. Another advantage of DBR mirrors is that they can be readily grown using epitaxial growth techniques. For edge emission lasers, the mirrors are typically made by cleaving the crystal. This is not a practical solution for a surface emitting laser and DBR mirrors are used instead.

Example Problem 10.2 Estimate of the emitted photon energy from a single quantum well laser

Using an infinite square well approximation, estimate the energy of a photon emitted from a single quantum well laser.

The energy levels for an infinite square well are given by Brennan (1999, (2.30)) as

$$E_n = \frac{n^2 \pi^2 \hbar^2}{2ma^2}$$

where n is an integer with value 1, 2, 3, ..., m is the electron mass within the well, a is the well width and $\hbar$ is the reduced Planck constant. Assume that the electron mass is 0.067 times the free space mass of an electron, the hole mass is 0.62 times the free space mass, the length of the well is 10.0 nm. Physical constants are: m_0 (free space electron mass) $= 0.511 \times 10^6$ eV/C^2; $\hbar = 6.58 \times 10^{-16}$ eV s; $c = 3.0 \times 10^{10}$ cm/s. The energy gap of GaAs is 1.42 eV. Assume that only the $n = 1$ states for both the electron and hole participate in the transition.

Inspection of Fig. 10.13 shows that the photon energy emitted from the well is equal to the sum of the confined state energies of the electron and hole and the band gap energy. Let us determine the confined state energies for the $n = 1$ states for the electron and hole. The confined state energy is given as

$$E_n = \frac{n^2 \pi^2 \hbar^2}{2ma^2} = \frac{\pi^2 \hbar^2 c^2}{2(0.067)(0.511 \times 10^6)(10.0 \times 10^{-7})^2} = \frac{3.76 \times 10^{-3} \text{ eV}}{m}$$

where the m appearing in the last term above is the relative mass. For the electron the energy is

$$E = \frac{3.76 \times 10^{-3} \text{ eV}}{0.067} = 0.056 \text{ eV}$$

For the hole the energy is

$$E = \frac{3.76 \times 10^{-3} \text{ eV}}{0.62} = 0.0061 \text{ eV}$$

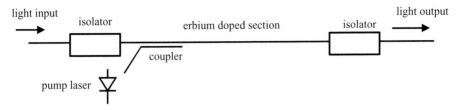

Figure 10.17 Sketch of an EDFA system. A short section of fiber optic cable doped with erbium is used to provide amplification. The erbium doped fiber is pumped using a semiconductor pump laser as shown in the diagram.

The photon energy is given then as the sum of the confined state energies and the band gap as

$$h\nu = 1.42\,\text{eV} + 0.056\,\text{eV} + 0.0061\,\text{eV} = 1.48\,\text{eV}$$

10.5 **EDFAs**

The basic scheme of an erbium-doped fiber amplifier (EDFA) is shown in Fig. 10.17. The essential components are a laser pump, a wavelength selective coupler, an erbium (Er) doped fiber, and optical isolators. As can be seen from the diagram a short section of fiber optic cable doped in its core with erbium is connected to the cable through isolators. The isolators at the beginning and end of the EDFA ensure that the optical signals pass in only one direction and prevent the pump light from propagating back or forward in the main fiber. The operation of the EDFA can be understood as follows. The processes in an EDFA are similar to those of a single pass laser. As in a laser, light amplification is achieved through stimulated emission of radiation. In an EDFA the stimulated emission is produced using an excited state of the Er^{3+} ions which are intentionally added to the fiber as dopants. A simplified energy level scheme for the Er^{3+} ions is shown in Fig. 10.18. The system is first pumped using a semiconductor laser, called the pump laser, to achieve population inversion. The laser light excites electrons within the Er atoms to the level marked as $^4I_{11/2}$ in Fig. 10.18 using photons of energy equal to the energy difference between the $^4I_{11/2}$ and $^4I_{15/2}$ levels. The lifetime of electrons within the $^4I_{11/2}$ level is relatively short and they decay nonradiatively into the $^4I_{13/2}$ level. The $^4I_{13/2}$ level is metastable implying that the electrons can remain there for a relatively long time without decaying. As a result the population of the $^4I_{13/2}$ level is greater than that of the $^4I_{15/2}$ level and the system is said to have a population inversion. If the excited electrons within the $^4I_{13/2}$ level are exposed to a radiation field with energy equal to the transition energy between $^4I_{13/2}$ and $^4I_{15/2}$, then stimulated emission can occur. The optical signal has an energy equal to this transition energy and thus stimulates radiative emission. The photons produced by the stimulated emission events add coherently to the optical signal thereby producing amplification. There are several states within each level thus generating many possible transitions from one state to the next giving rise to an amplified spectrum of wavelengths. It is precisely in this way that

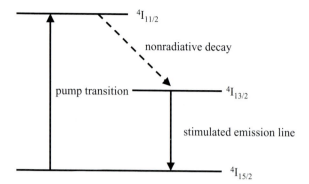

Figure 10.18 Sketch of simplified description of the energy levels and transitions within an Er^{3+} ion used in an EDFA. The wavelength of the stimulated emission line is between 1520 and 1570 nm, spectrally matched to that of the input optical signal. Each line represents a series of substates around one level. Due to the relatively large number of substates in each level, this results in a relatively wide range of emitted energies.

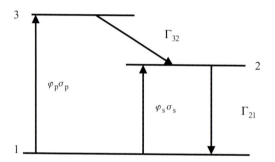

Figure 10.19 Sketch of a three-level system used to model the workings of an EDFA. The notation is as described in the text. Note that stimulated emission occurs between the 2 and 1 lines. The populations in levels 3, 2, 1 are N_3, N_2, and N_1 respectively.

gain can be delivered to wavelengths within the 20–30 nm bandwidth range mentioned above.

We can obtain a simplified picture of the workings of an EDFA using a three-level system in one dimension. Let us represent the transitions using the notation shown in Fig. 10.19. Let us assume that the laser pump has intensity flux φ_p, the incident signal flux intensity is φ_s, and the cross-sections for 1–3 and 1–2 transitions are σ_p and σ_s respectively. The transition probability from 3 to 2 is given as Γ_{32} and is assumed to be mainly nonradiative. The transition probability from 2 to 1 is mainly radiative and is represented as Γ_{21}. Let N_3, N_2, and N_1 represent the populations of states 3, 2, and 1 respectively. The total population, N, is then simply equal to the sum of the populations in each level:

$$N = N_1 + N_2 + N_3 \tag{10.44}$$

The rate equations for the populations in each level follow immediately from Fig. 10.19 as,

$$\frac{dN_3}{dt} = -\Gamma_{32}N_3 + (N_1 - N_3)\varphi_p\sigma_p$$

$$\frac{dN_2}{dt} = -\Gamma_{21}N_2 + \Gamma_{32}N_3 + (N_1 - N_2)\varphi_s\sigma_s \qquad (10.45)$$

$$\frac{dN_1}{dt} = \Gamma_{21}N_2 - (N_1 - N_3)\varphi_p\sigma_p - (N_1 - N_2)\varphi_s\sigma_s$$

In order to obtain gain there must be a population inversion between levels 2 and 1. This means that the population in level 2 must be greater than that in level 1, $N_2 > N_1$. In steady-state the time derivatives are, of course, equal to zero. The population in the third level, N_3, can then be expressed in terms of the population in level 1, N_1, as

$$\Gamma_{32}N_3 = (N_1 - N_3)\varphi_p\sigma_p$$

$$N_3 = \frac{N_1}{1 + \Gamma_{32}/\varphi_p\sigma_p} \qquad (10.46)$$

If we assume that Γ_{32} is large, then the expression for N_3 given by (10.46) can be approximated as

$$N_3 \sim \frac{\varphi_p\sigma_p}{\Gamma_{32}}N_1 \qquad (10.47)$$

Substituting into the second of equations (10.45) under steady-state conditions the expression for N_3 given by (10.47), N_2 can be written as

$$N_2 = \frac{\varphi_s\sigma_s + \varphi_p\sigma_p}{\Gamma_{21} + \varphi_s\sigma_s}N_1 \qquad (10.48)$$

At threshold, the population of level 2 must be equal to that of level 1. Therefore, equating N_2 and N_1 in (10.48) an expression for the threshold pump intensity flux, φ_{th}, is given as,

$$\varphi_{th} = \frac{\Gamma_{21}}{\sigma_p} \qquad (10.49)$$

φ_{th} is the minimum pump photon flux necessary in order to achieve population inversion between the first and second levels. For pump photon fluxes equal to or greater than φ_{th}, gain can be achieved and the signal can be boosted.

The pump energy intensity is given by the product of φ_p and the energy of a photon, $h\nu_p$. Thus the pump threshold energy intensity, I_{th}, is

$$I_{th} = \frac{h\nu_p\Gamma_{21}}{\sigma_p} \qquad (10.50)$$

Inspection of (10.50) shows that as the cross-section for pump photon absorption, σ_p, increases then I_{th} decreases, implying that fewer pump photons are required in order to achieve threshold. This of course follows from the obvious observation that the greater the chance that a pump photon will be absorbed, the fewer incident pump photons are needed to achieve population inversion. Additionally, the smaller Γ_{21} is, the fewer pump photons are needed per unit time to achieve threshold. Physically a small value of Γ_{21} implies that level 2 is metastable, transitions from 2 to 1 do not occur rapidly resulting in a buildup of population in level 2. The transition rate Γ_{21} is simply equal to the inverse lifetime of level 2, which is called τ_p. Therefore, the two major criteria that determine the threshold pump intensity are the absorption cross-section for the pump photons and the lifetime of level 2. Thus if a small threshold pump intensity is desired it is important to have a system with a large absorption cross-section between levels 1 and 3 and a long lifetime of level 2.

In addition to the threshold pump intensity the other major figure of merit of an EDFA is the gain. The gain describes how much the signal is boosted when passing through the Er doped region of the fiber. The signal intensity, $I_s(z)$, increases with distance traveled (assuming a one-dimensional system oriented along the z direction) in the presence of gain as

$$I_s(z) = I_s(0)e^{\alpha_p z} \tag{10.51}$$

where α_p is the gain coefficient. An approximate expression for the gain coefficient can be obtained in the limit where the pump intensity is very strong such that all of the N Er ions are inverted. Under this condition the gain coefficient can be approximated as

$$\alpha_p = \sigma_s N \tag{10.52}$$

where we recall that σ_s is the signal transition cross-section. Equation (10.52) implies that all of the ions contribute to the gain coefficient under high pumping conditions. However, in practice this is not correct, since some of the excited ions decay via spontaneous emission and thus do not contribute to the amplification of the signal through stimulated emission.

10.6 **SOAs**

Though EDFAs are of great importance to optical fiber networks, they suffer from two major limitations. As discussed in Section 10.5, EDFAs do not generally provide flattened gain over a broad band. Additionally, EDFAs are somewhat expensive since they require a pump laser, couplers, isolators, and a section of Er doped fiber. Semiconductor optical amplifiers (SOAs) offer an attractive alternative to EDFAs. The primary advantages of SOAs are their cost and compactness. An SOA is simply a semiconductor laser with antireflection coated facets at both ends of the structure. Therefore, the cost is significantly less than that of an EDFA since only a modification of the laser component of the EDFA is needed. Additionally, the gain bandwidth of an SOA is parabolic in shape. Consequently, gain equalization can be more readily achieved with an SOA than an EDFA. The other major advantages of SOAs are that they can be made to operate through a suitable choice of materials and designed to operate over a considerable range of wavelengths, they can be integrated with other devices, and they have a low power consumption.

A sketch of the basic structure of an SOA is shown in Fig. 10.20. The basic processes in an SOA can be understood as follows. As opposed to a laser, each facet has an

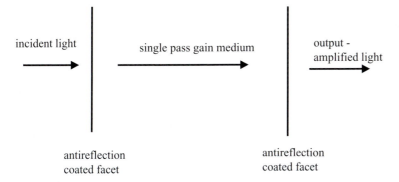

Figure 10.20 Diagrammatic sketch of an SOA, showing the light progression and gain medium. Since both facets have antireflection coatings, the light experiences a single pass during which it is amplified by stimulated emission.

antireflection coating to ensure that the signal is not reflected at either end of the device. Such a structure is often referred to as a traveling wave amplifier, TWA. The incident signal light passes through the first facet and enters the gain medium as shown in Fig. 10.20. The gain medium is the active region of a semiconductor laser. In a similar manner as for any semiconductor laser the system is pumped electrically using a forward biased degenerately doped p–n junction diode. The external pumping creates a population inversion. As was discussed in Section 10.2, the population inversion enables the amplification of light through stimulated emission. The energy gap of the SOA is spectrally matched to the signal energy. The incident signal light can thus induce stimulated emission events within the gain medium resulting in light amplification. In this way, the signal light is amplified. The light makes only a single pass through the gain medium and is transmitted through the second facet. No significant reflection occurs at the second facet since it also has an antireflection coating. The SOA uses only a single pass in order to avoid gain ripple, which is modulation of the gain spectrum arising from residual resonances.

Despite performance improvements, SOAs still suffer from several problems. The conversion of a standard semiconductor laser structure to an SOA through the deposition of antireflection coatings on the facets has two major problems. These are polarization sensitivity of the gain and difficulty in achieving low facet reflectivity to avoid reflections. Antireflection coatings are generally optimized for a particular wavelength and thus have a low reflectivity only over a relatively limited range of wavelengths. In contrast, in an EDFA the index of refraction of the Er doped section of fiber is matched to the transmission fiber thus reducing reflections. Noise and nonlinearities also present problems for SOAs. Noise is introduced in any amplifier. Noise arises in SOAs from spontaneous emission within the gain medium. Though spontaneous emission is not as strong as stimulated emission, within the gain medium some of the carriers recombine through spontaneous emission events. These events lead to random fluctuations or noise in the output signal. The operating characteristics of SOAs can be highly nonlinear. Nonlinearities in the response of an SOA arise from the fact that the gain and refractive index within the gain medium are highly concentration dependent.

In addition, the short lifetime within an SOA can lead to gain modulation, which can result in cross-talk between channels. Cross-talk is a particularly important problem since it can lead to interference between different channels resulting in signal corruption. Cross-talk in EDFAs is avoided due to the very long lifetime of the metastable state used for stimulated emission.

Many of the characteristics of an SOA make for an unfavorable comparison with EDFAs. However, due to the fact that SOAs are compact and inexpensive it is not quite clear whether they will ever fully compete with EDFAs and replace them in practical systems.

10.7 p–i–n photodetectors

In this section we examine photodetectors, specifically those made using a p–i–n diode. In the next section we will discuss a related device that is most often used in lightwave communications systems, the avalanche photodiode. However, we will begin our discussion using the simpler p–i–n photodiode.

The fundamental purpose of any photonic detector is to convert an input photonic signal into an electrical signal. The application in which the detector is used greatly affects the performance criteria of the detector. For example, a detector can be used for imaging or in lightwave telecommunications. In imaging applications the detector must provide a high degree of spatial resolution, gray-scale resolution (defined as the ability to distinguish different shades on a totally white to totally black contrast scale), operate at both high and low level illumination, and be amenable to two-dimensional geometries. In this chapter, we will be most interested in photodetectors that are used in lightwave communications systems. These devices have different requirements than photodetectors used for imaging. Specifically, in modern lightwave communications systems, semiconductor detectors are used to capture the output optical signals from optical fibers. The detector must be designed to ensure signal recognition, avoid inter-symbol interference, have a low bit-error rate and provide for very high speed operation. The speed of response of a detector used in a lightwave communications system is critical to the detector performance. In contrast, speed is not nearly as important in imaging applications. Therefore, a detector optimized for lightwave communications is not necessarily optimized for imaging applications and vice versa.

One of the key parameters that defines the usefulness of different materials for photonic detection is the cutoff wavelength. Most semiconductor photonic detectors utilize band to band transitions to detect radiation. The minimum photon energy that a semiconductor can detect is given by the difference between the conduction band and valence bands or equivalently the band gap energy. There exists then a cutoff energy below which photons cannot be detected by the semiconductor. It is useful to define a cutoff wavelength, λ_c, that represents the maximum wavelength of light that can be detected. For example, GaAs has an energy band gap of 1.42 eV and a corresponding cutoff wavelength given by

$$\lambda_c = \frac{hc}{E_g}$$

(10.53)

which for GaAs computes to 8.43×10^{-5} cm. Radiation with a wavelength greater than the cutoff wavelength cannot be detected by the GaAs and will pass through it instead. In other words, radiation longer than the cutoff wavelength of a material is transparent in that material.

The detector types with the highest speed of response and thus applicable to light-wave communications systems are p–i–n photodiodes, avalanche photodiodes, and metal–semiconductor–metal (MSM) detectors. There are other detectors of importance such as charge coupled devices, CCDs, photoconductors, and phototransistors (for a full discussion of these topics see Brennan (1999)), but these devices are typically not employed in lightwave communications systems. Therefore, we will omit any further discussion of these detectors referring the interested reader to Brennan (1999, Chapter 12).

A diode can be used as a photodetector when operated in reverse bias. As discussed in Chapter 3 the reverse saturation current within an ideal diode arises from generation–recombination events within a diffusion length of the depletion region on either side of the junction. As the generation rate increases the reverse saturation current increases. A more accurate formulation of the reverse saturation current accounts for generation–recombination events within the depletion region of the diode itself. Generation of electron–hole pairs within the depletion region adds to the reverse current. The magnitude of the reverse current is thus a measure of the carrier generation rate within the device: the greater the generation rate, the higher the reverse saturation current. Incident light spectrally matched to the band gap of the material forming the diode leads to optical generation of carriers and a subsequent increase in the reverse saturation current. As the incident light intensity increases the generation rate increases and consequently so does the reverse saturation current. Thus the change in the magnitude of the reverse saturation current from dark conditions provides a measure of the input light intensity. This is the underlying principle behind the operation of a reverse biased p–n junction photodetector. The effect of the optical generation rate on the reverse saturation current is shown in Fig. 10.21. Often a p–i–n junction diode is used in place of a p–n junction diode since the addition of an intrinsic region increases the depletion region width and consequently the amount of light absorbed and its concomitant generation rate.

Photodiodes are attractive for lightwave communications systems because they provide high quantum efficiency and very high bandwidth. The quantum efficiency is defined as the number of electron–hole pairs generated per incident photon. At 100% efficiency, for every incident photon, one electron–hole pair is produced. Typically, the quantum efficiency is less than 100% in a device with no gain implying that not all incident photons are successfully converted into electron–hole pairs. The bandwidth is inversely proportional to the temporal response of the diode. The bandwidth depends upon three factors. These are the carrier diffusion to the depletion region edge, drift time within the depletion region, and the capacitance of the depletion region. Each of these entities adds to the time delay of the device and thus reduces the bandwidth. Those carriers photogenerated outside of the depletion region must first diffuse to the depletion region edges after which they are swept out of the device by the action of the reverse bias field. The diffusion process is relatively slow and can take a long

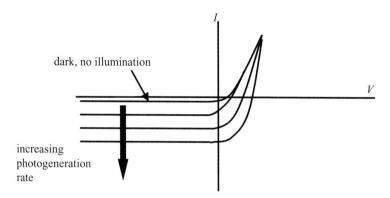

Figure 10.21 Current–voltage characteristic for a junction diode showing the effects of photogeneration on the reverse saturation current. As can clearly be seen from the figure, as the light intensity increases the photogeneration rate increases leading to an increase in the reverse saturation current.

time depending upon the width of the absorbing layer. Since the device is operated in reverse bias, electrons drift toward the n contact while the holes drift toward the p contact. The time delay due to the carrier drift within the depletion region depends upon the width of the depletion region, W, and the relative speeds of the carriers. In order to fully absorb the incident radiation it is necessary to utilize a wide depletion region. As mentioned above this is the reason why a p–i–n junction diode is often used. However, as the width of the depletion region increases the transit time across the device increases thereby reducing the device bandwidth.

Typically, a p–i–n junction diode is illuminated from the p side as shown in Fig. 10.22(a). Recall that the device is reverse biased. The depletion region of the device contains the intrinsic region as shown in the diagram. Within the intrinsic region the electric field strength is constant as shown in Fig. 10.23. Holes generated near the edge of the depletion region on the p side generally recombine at the contact or within the p region. In general, the electrons can be photogenerated within the p^+ region as well as the intrinsic region of the diode. As we will see below, most p–i–n photodiodes are designed such that the vast majority of carriers are photogenerated within the intrinsic region. Nevertheless, for a general photodiode, there are some carriers generated outside of the depletion region. As discussed in Section 10.8, most of the photogenerated carriers within an avalanche photodiode (a special type of p–i–n diode) are produced within the p^+ region. Therefore, it is useful to first consider the general case wherein carriers are photogenerated in both the p^+ and intrinsic regions. The photogenerated electrons produced within the p^+ region diffuse to the edge of the depletion region or intrinsic region in a p–i–n structure, after which they are swept out by the action of the reverse bias field as shown in Fig. 10.22. Once within the depletion/intrinsic region the current flows mainly from drift owing to the relatively high electric field. Further inspection of Fig. 10.22 shows that by illuminating the p side rather than the n side electrons are the carrier species that drifts through the depletion region. Owing to the fact that the electron mobility and the concomitant

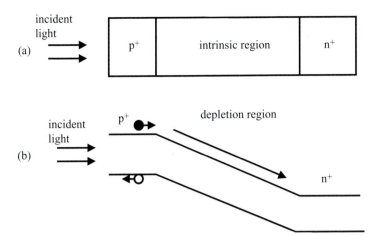

Figure 10.22 Reverse biased photodiode showing: (a) the general device design and location of incident light and (b) the energy band diagram under reverse bias. The electron–hole pairs are assumed to be generated within the p⁺ region as shown in the diagram. The holes recombine in the p⁺ region or the contact while the electrons diffuse to the depletion region after which they are swept out. Given that the electrons have a much higher mobility and consequently drift velocity the prefered illumination is on the p side such that electrons drift through the depletion region rather than holes.

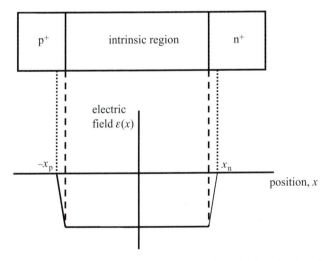

Figure 10.23 Electric field profile of a uniformly doped p–i–n photodiode. Notice that the electric field is constant within the intrinsic region and varies linearly in the depletion regions formed on either side of the junction in the n and p regions as shown in the diagram.

drift velocity are higher for electrons than for holes, it is desirable for high speed operation that the electrons make the relatively long transit through the depletion region.

In most p–i–n photodiodes the p$^+$ layer is relatively thin so that most of the photogenerated carriers are produced within the intrinsic region. This has the beneficial effect of mitigating the slow diffusion process that would occur if the carriers were mainly produced within the highly doped p$^+$ or n$^+$ regions. The photogenerated carriers are then immediately swept out of the intrinsic region by the action of the reverse bias electric field leading to a fast speed of response of the device.

Photodiodes can be operated in basically two different ways. If the applied voltage is such that the electric field within the intrinsic region is less than the critical field (see Section 3.3) then relatively few if any impact ionization events occur in the device. The photodiode has unity gain; there is no carrier multiplication or gain in the structure. However, if the magnitude of the reverse bias is increased to the point at which the field within the intrinsic region becomes equal to or exceeds the critical field then carrier multiplication through impact ionization occurs. The presence of carrier multiplication results in an exponential increase in the reverse current flowing through the diode and the diode has gain. Photodiodes operated at sufficiently high reverse bias that carrier multiplication occurs are called avalanche photodiodes. The carrier multiplication provides a current gain mechanism that amplifies the input photocurrent. Avalanche photodiodes are the topic of the next section.

The quantum efficiency of a p–i–n photodiode can be estimated as follows. Since there is no internal gain within a p–i–n photodiode, the maximum quantum efficiency is 100% and is typically less than 100%. The quantum efficiency is determined from the ratio of the power absorbed within the intrinsic and depletion regions to the total incident power onto the device. Reflection at the top surface of the device reduces the amount of absorbed power by a factor R. The amount of transmitted power is then $(1 - R)P_0$, where P_0 is the input radiative power. The light passes through the p$^+$ region in which some is absorbed prior to reaching the intrinsic region of the diode. Let the p$^+$ region have width W_p. If the absorption coefficient is assumed to be the same for each of the three regions of the diode, p$^+$, n$^+$ and i, then the amount of power incident onto the intrinsic region is

$$P(W_p) = P_0(1 - R)e^{-\alpha W_p} \tag{10.54}$$

where α is the absorption coefficient and R the reflection coefficient at the top surface of the photodiode. Let the width of the intrinsic region be W_i, then P_i, the amount of power absorbed within the intrinsic region of the diode, is

$$P_i = P_0(1 - R)e^{-\alpha W_p}(1 - e^{-\alpha W_i}) \tag{10.55}$$

The quantum efficiency is then given from the ratio of P_i to P_0 as

$$\eta = (1 - R)e^{-\alpha W_p}(1 - e^{-\alpha W_i}) \tag{10.56}$$

Inspection of (10.56) indicates that the quantum efficiency can be optimized by reducing the width of the p$^+$ region, reducing the reflectivity, and by increasing the width of the intrinsic region. In (10.56) losses due to recombination within the intrinsic region are not included. This is generally an acceptable approximation since the carriers spend relatively little time within the intrinsic region owing to the high drift field. Additionally, the concentration of free carriers within the intrinsic region is relatively small so that electron–hole pair recombination is generally small.

There is a limit to how large the intrinsic region of a p–i–n photodetector can be which is determined by the speed of response or bandwidth of the diode. The frequency response of a p–i–n detector is limited by transit time effects or circuit parameters. The most important issue impacting the carrier transit time within a p–i–n diode is the width of the intrinsic region. As the region increases in width, the total distance traveled by the photogenerated carriers increases thus reducing the device bandwidth. However, if the intrinsic region is made too small, the quantum efficiency is reduced as shown by (10.56). Selecting the optimal intrinsic region width is one of the complications associated with p–i–n photodiode design.

The speed of response of a p–i–n photodiode can be estimated as follows. As mentioned above, a wide intrinsic region improves the quantum efficiency of the photodiode and also reduces the junction capacitance. However, a wide intrinsic region also increases the carrier transit time through the device. Typically, a p–i–n photodiode is designed such that the transit time through the intrinsic region is equal to one-half of the modulation period of the optical signal. Therefore, the ideal modulation frequency has a period of twice the transit time within the intrinsic region of the diode,

$$f = \frac{1}{2\tau_t} \tag{10.57}$$

where τ_t is the transit time within the intrinsic region.

10.8 **Avalanche photodiodes**

One of the most important detectors used in lightwave communications systems is the avalanche photodiode, or APD. The front-end gain provided by an APD is important since it boosts the signal above the noise floor of the following electronics used in the system. There are several different APD device concepts. Some of the more important of these device concepts are the conventional APD, separate absorption multiplication (SAM) APD, thin multiplication region APDs, and the superlattice/multiquantum well APD. Within each of these major categories of APDs there are several different device types. A discussion of all of the different types of APDs is beyond the scope of this book. The interested reader is referred to Brennan and Haralson (2000) and Brennan (1999, Chapter 12). Here we will confine ourselves to discussing only the conventional APD structure and its basic operation.

A conventional APD device is simply a p–i–n photodiode except that the applied bias is sufficiently large that impact ionization occurs. As in the p–i–n photodiode the intrinsic region has a uniform constant electric field within it. Depending upon the applied reverse bias, the field can be made as high as or higher than the critical field. Once the critical field is reached carrier multiplication through impact ionization

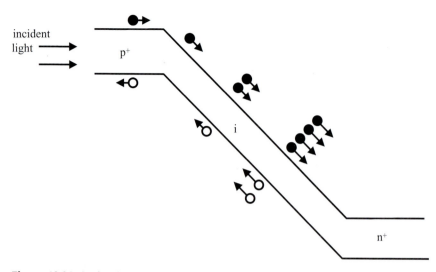

Figure 10.24 Avalanche multiplication process in an APD. The solid circles are electrons and the open circles are holes. A single electron–hole pair is photogenerated within the p⁺ region. The photogenerated electron diffuses to the edge of the depletion region where the electric field acts to sweep the electron through the intrinsic region as shown. If the magnitude of the reverse bias voltage is sufficiently high, impact ionization can occur. As can be seen from the figure, impact ionization is an avalanching process. One electron gives rise to an additional electron–hole pair, two electrons create two electron–hole pairs, etc.

begins to occur. The impact ionization process is an avalanching process in that the production of one electron–hole pair can lead to the generation of additional pairs. The multiplication scheme of an APD is shown in Fig. 10.24. As can be seen from the figure, incident light photogenerates an electron–hole pair within the p⁺ region. The hole is immediately swept out and recombines, while the electron diffuses to the edge of the depletion region. Once within the depletion region the electron is swept through the intrinsic region gaining energy from the electric field. The electron can attain the threshold energy for impact ionization and subsequently cause an ionization event.

Following an impact ionization event there is a new electron–hole pair generated along with the initial electron as can be seen from Fig. 10.24. The initiating electron is called the primary and the resulting impact ionization event produces secondary carriers. The primary and secondary electrons continue their flight through the intrinsic region towards the n⁺ region of the diode. Impact ionization events involving both carriers can occur. If both electrons impact ionize then two additional electron–hole pairs are created, yielding a total of four electrons and two holes as sketched in Fig. 10.24. The secondary holes generated in the process drift in the opposite direction to the electrons within the intrinsic region. Assuming that the secondary holes do not ionize, they ultimately drift to the p⁺ region after which they recombine. If only one carrier species is injected and ionizes the condition is called single carrier injection, single carrier multiplication or SCISCM. APDs operated under SCISCM conditions have the

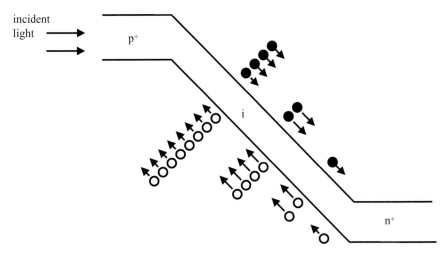

Figure 10.25 APD device showing secondary hole multiplication. It is assumed that a single hole produced near the end of the intrinsic region (n side of the APD) will ionize during its flight towards the p region of the diode. A single hole is thus shown to produce first one additional electron–hole pair. The two holes, the primary and the secondary hole, then multiply to produce two additional electron–hole pairs. Finally, the four holes then multiply to produce four additional electron–hole pairs resulting in a total of eight holes and four total electrons. The generated electrons can also multiply during their flight through the intrinsic region leading to additional multiplication.

lowest noise and highest bandwidth. Conversely, if an APD operates such that both the electrons and holes impact ionize then its bandwidth is significantly less and the device is much noisier. This can be understood as follows.

Impact ionization can be initiated by a high energy electron or hole. In some materials the electron and hole impact ionization rates are comparable. Thus if a hole is generated within the intrinsic region following an electron initiated impact ionization event, while drifting towards the p^+ contact the hole can become sufficiently energetic that it too can ionize as shown in Fig. 10.25. In Fig. 10.25 a single hole near the n^+ region is assumed to exist. This hole is the product of an earlier electron initiated impact ionization event. As the hole moves through the depletion region it experiences the same electric field strength as the electron does. Consequently, the hole can also gain high energy and impact ionize. When the hole impact ionizes an additional electron–hole pair is produced as shown in Fig. 10.25. After one impact ionization event, there are two holes, the primary and secondary, and a secondary electron. The two holes continue their flight and possibly impact ionize producing four holes and two electrons. Thus when the hole ionization rate is comparable to the electron ionization rate many secondary carriers are produced. In fact if the field is sufficiently high that both the injected electrons and secondary holes born from electron initiated impact ionization events always impact ionize during their drift through the depletion region, the carriers continuously regenerate themselves and the process never terminates. This is called avalanche breakdown and the device becomes unstable. An APD should always be operated below the avalanche breakdown condition.

Let us examine the avalanche breakdown condition in more detail. As mentioned above when the device enters avalanche breakdown, the avalanching process continuously regenerates itself and the device never shuts off. Alternatively, the fastest device response of an APD occurs when only the injected carrier species ionizes. In other words, highest bandwidth operation occurs when only one carrier species impact ionizes and that carrier species is the only one injected to initiate the impact ionization process. In addition to optimizing the bandwidth, SCISCM conditions result in the lowest noise performance of the APD as well. The noise of the device is worst when the impact ionization rates are comparable.

Noise arises from fluctuations from the average behavior. A system is said to be noiseless if it undergoes no fluctuations from its average behavior. For a multiplicative device, such as an APD, as well as the usual noise sources in a semiconductor device, there is an additional noise that arises from fluctuations in the number of collected carriers. The variation in the number of carriers arises from randomness in the multiplication rate; each initial photogenerated carrier gives rise to a random number of secondary electron–hole pairs. The additional randomness due to the multiplicative gain process leads to additional noise, which is called the excess noise. The excess noise is a function of the mean multiplication rate or gain of the APD, M, as $F(M)$.

Since the multiplication process that creates gain produces an additional noise component, one might wonder, what is the value of the gain in an APD? The gain produced within the APD from carrier multiplication serves to boost the signal over the noise floor of the following electronics (for a full discussion see Brennan (1999, Chapter 12)). If the gain can be produced at a relatively low excess noise then the overall SNR of the system can be improved. If, on the other hand, the gain mechanism is very noisy then the overall SNR may not be substantially improved.

Let us now consider the noise figure of a conventional APD. If it is assumed that only electrons are injected and initiate the impact ionization process, then the excess noise factor, $F(M)$, is

$$F(M) = kM + \left(2 - \frac{1}{M}\right)(1 - k) \tag{10.58}$$

where k is the ratio of the hole to electron impact ionization rate coefficients and M is the mean gain. A simple expression for the average gain is obtained when the electron and hole ionization coefficients are the same (see Brennan (1999, Chapter 12)). Under these conditions the gain becomes

$$M = \frac{1}{1 - \alpha W} \tag{10.59}$$

where α is the electron impact ionization rate coefficient and W is the width of the depletion region. The hole impact ionization rate coefficient is often represented as

β. Notice that the average gain, M, approaches infinity if the product of α and W is 1. When this occurs the device is in avalanche breakdown and as mentioned above is unstable. When the device is in avalanche breakdown the excess noise factor also diverges and thus the device is very noisy. Alternatively, when the device operates in SCISCM conditions, the gain never diverges. Under SCISCM conditions the gain becomes (Brennan (1999, Section 12.5))

$$M_{\mathrm{n}} = e^{\alpha W} \tag{10.60}$$

and avalanche breakdown never occurs. The device remains stable for all finite gain. Under these conditions the factor k in (10.58) becomes zero. Thus the excess noise factor reduces to

$$F(M) = \left(2 - \frac{1}{M}\right) \tag{10.61}$$

which approaches 2 as the gain increases. The device is far less noisy when only one carrier species ionizes than when both ionize.

APDs play important roles in long distance, high bit-rate fiber optic telecommunications networks. The lowest loss windows for most fiber optic systems occur at wavelengths of 1.3 and 1.55 μm. The materials spectrally matched to these windows are mainly compound semiconductors such as InP, GaInAs, and InGaAsP. Unfortunately, most compound semiconductor materials have nearly equal electron and hole impact ionization rate coefficients. Thus conventional APDs made from these materials would have low bandwidth and high excess noise. For example, bulk GaAs and InP materials, both compound semiconductors, have nearly equal electron and hole impact ionization rate coefficients. Therefore, the use of GaAs, InP, or another similar material in conventional APD structures provides unsatisfactory device performance. An important exception is Si. In bulk Si the electron impact ionization rate is significantly higher than the hole ionization rate, by a factor of nearly 20 at some field strengths. For this reason, Si APDs are attractive devices. However, Si is an indirect gap semiconductor and, as such, it has a much lower absorption coefficient than GaAs or InP. Additionally, Si cannot be used for most mid-to-long wavelength infrared detectors due to its band gap energy of 1.1 eV.

Ideally, one would like to find a material that has a large difference in the impact ionization rates ratio and has a direct band gap. Such a material would exhibit a high quantum efficiency and could be used in an APD to provide ultra-low noise and high bandwidth detection. Unfortunately, no material has been identified to date that has these characteristics and also matches the spectrum used for lightwave communications. Alternatively, work has been done to develop structures in which the ionization rates ratio can be altered. These devices utilize multi-quantum well structures, graded layers, etc. to change the ionization rates ratio. A full discussion of these different APD designs is beyond the level of this book. The interested reader is referred to the book by Brennan (1999) or the review article by Brennan and Haralson (2000) for a full discussion.

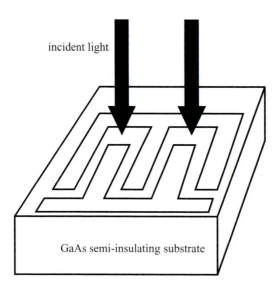

incident light

GaAs semi-insulating substrate

Figure 10.26 Sketch of a MSM photodetector. The light is incident from the top of the device as shown. Two sets of interdigitated fingers form the contacts. The contacts are Schottky barriers.

The last photodetector that we will consider in this chapter is the MSM photodetector. The MSM photodetector is of interest to lightwave communications systems due to its relative ease of fabrication. A representative sketch of an MSM detector is shown in Fig. 10.26. Light is incident upon the top of the device as shown in the diagram. The Schottky barrier contacts form two sets of interdigitated fingers on the top surface of the device. One set is typically grounded while the other set of fingers is reverse biased. Electron–hole pairs are photogenerated within the bulk region of the device. The application of a bias to the metallic fingers creates an electric field within the underlying semiconductor that acts to sweep the photogenerated carriers out of the device. If a sufficiently large reverse bias is applied the semiconductor layer between the two contacts as well as the region into the device becomes depleted. As the voltage is increased the semiconductor layer becomes fully depleted between the contacts. The bias under which this occurs is called the reach-through voltage, V_{RT}. At the reach-through voltage the current flow is dominated by drift rather than the much slower diffusion process. How fast the carriers are collected and how many of them actually survive to be collected at the contacts within a particular time determines the speed and responsivity, respectively, of the detector. Carriers generated deep within the material must traverse a greater distance before they are collected at the contacts compared with those generated near the surface. Depending upon the magnitude of the electric field within the semiconductor, the time needed to collect those carriers generated deep within the device can vary drastically. If the applied bias is low, such that the semiconductor region beneath the contacts is not fully depleted, the speed of the response of the device may be too slow, since the carriers must first diffuse into the depletion region before drifting to the collecting contacts.

Example Problem 10.3 Breakdown field of a GaAs APD

Consider a simple conventional APD made from GaAs. Assume that the device is symmetric and consists of p^+ and n^+ regions that are uniformly doped at 6.42×10^{18} cm^{-3}. Assume that the width of the p^+ and n^+ regions is 5.0 nm and there is an intrinsic region sandwiched between them of length 10.0 nm. If the p^+ and n^+ regions are fully depleted calculate the field within the intrinsic region. The relative dielectric constant in GaAs is 12.9, $\kappa_0 = 8.85 \times 10^{-14}$ F/cm, and $q = 1.6 \times 10^{-19}$ C.

Start with (3.19). On the p side of the junction the Poisson equation includes only ionized acceptors. All of the acceptors within the p region are ionized. Therefore, the Poisson equation gives

$$\frac{d\varepsilon}{dx} = -\frac{qN_a}{\kappa}$$

The field is thus found by integrating over the full width of the p^+ region. The maximum field is

$$\varepsilon_{max} = -\frac{qN_a d}{\kappa}$$

where d is the width of the p^+ region. Substituting in the values for each variable, the maximum field becomes

$$\varepsilon_{max} = -\frac{(1.6 \times 10^{-19}\,\text{C})(6.42 \times 10^{18}\,\text{cm}^{-3})(50.0 \times 10^{-8}\,\text{cm})}{(12.9)(8.85 \times 10^{-14}\,\text{F/cm})} = 450\,\text{kV/cm}$$

Inspection of Fig. 10.23 shows that the field within the intrinsic region is equal to the maximum electric field. Thus the field in the intrinsic region is 450 kV/cm.

Example Problem 10.4 Noise and gain performance of a Si APD

Calculate the excess noise factor for a Si p–n junction APD if the ratio of the electron to hole ionization rates is 20, the electron ionization coefficient is 1.0×10^4 cm^{-1} at the maximum electric field within the junction of 250 kV/cm, the n- and p-type doping concentrations are 2×10^{17} cm^{-3}, and the intrinsic concentration of Si is 10^{10} cm^{-3}. Assume that the mean gain can be calculated by assuming only single carrier multiplication.

Since the doping concentrations on the n and p sides are the same, then the depletion region width, W, is equal to twice the depletion region width on the n or p sides of the junction. The depletion region width can be calculated from the expression for the maximum electric field given by (3.23). The depletion region width is determined from

$$\varepsilon_{max} = -\frac{qN_a x_p}{\kappa}$$

which leads to

$$W = \frac{2\kappa\varepsilon_{max}}{qN_a} = 1.64 \times 10^{-5}\,\text{cm}$$

Assuming that the mean gain can be determined by assuming only single carrier multiplication, the mean gain is given as

$$M = e^{\alpha W}$$

where α is the electron impact ionization rate coefficient. At 250 kV/cm the value of α is 1.0×10^4 cm^{-1}. Substituting in for α and W, the mean gain is calculated to be

$$M = e^{\alpha W} = e^{10^4(1.64 \times 10^{-5})} = 1.18$$

The excess noise factor, $F(M)$, is given then as

$$F(M) = kM + \left(2 - \frac{1}{M}\right)(1 - k)$$

Substituting $1/20$ for k, and 1.18 for M, the excess noise factor is calculated to be 1.154.

Example Problem 10.5 Calculation of the excess noise factor and gain for a GaAs APD

Determine the excess noise factor and gain for a GaAS APD device with a maximum electric field of 400 kV/cm. The electron and hole ionization rate coefficients are essentially equal. At a field of 400 kV/cm the electron ionization rate coefficient is 2.0×10^4 cm^{-1}. The relative dielectric constant for GaAs is 12.9. Determine the excess noise factor for the APD. Assume that the doping concentration is 2.0×10^{17} cm^{-3}.

The depletion region width is determined from

$$\varepsilon_{\text{max}} = -\frac{q N_a x_p}{\kappa}$$

which leads to

$$W = \frac{2\kappa \varepsilon_{\text{max}}}{q N_a} = 2.85 \times 10^{-5} \text{ cm}$$

The mean multiplication rate, M, is

$$M = \frac{1}{1 - \alpha W} = 2.33$$

The excess noise factor is given then from

$$F(M) = kM + \left(2 - \frac{1}{M}\right)(1 - k)$$

Assume that $k = 1$, the value of $F(M)$ is then just equal to M or 2.33.

Notice that the GaAs device is much noisier than the Si device examined in Example Problem 10.4.

Problems

10.1 Determine the mirror reflectivity of a semiconductor laser if the system is at threshold, the mirrors have identical reflectivities, the cavity length is 10 μm, the collective losses are $10.0 \, \text{cm}^{-1}$, and the threshold gain is $10^3 \, \text{cm}^{-1}$. Assume that the optical confinement factor is unity.

10.2 Determine the threshold gain of a semiconductor laser with mirror reflectivities of 0.25 and 0.30, a cavity length of 20 μm, and collective losses of $20 \, \text{cm}^{-1}$.

10.3 Determine the length of a semiconductor laser cavity if its collective losses are $50 \, \text{cm}^{-1}$ and the collective losses equal that of the mirror loss. Assume that the two mirrors have identical reflectivities of 0.32.

10.4 The electron and hole masses in a GaAs semiconductor are 0.067 and 0.62 times the free space mass respectively. The energy band gap of GaAs is 1.42 eV. Determine the width of the well in a single quantum well laser if the emitted photon from a recombination event within the well has an energy of 1.52 eV. Assume that the transition occurs between the first confined energy states in the conduction and valence band wells and that the energy levels in the well can be calculated using the infinite square well approximation.

10.5 A p^+–n diode is used as an APD. What must the impact ionization rate coefficient, α, be equal to such that the device is in avalanche breakdown, i.e., its gain has diverged? The donor doping concentration is $1.0 \times 10^{16} \, \text{cm}^{-3}$ and the breakdown voltage is 50 V. Assume that the electron and hole impact ionization coefficients are the same and neglect the built-in voltage of the junction. The relative dielectric constant is 11.8.

10.6 Determine the gain of an APD if only one carrier species ionizes and the impact ionization coefficient is $1.0 \times 10^4 \, \text{cm}^{-1}$. Assume that the diode is p^+–n with a donor concentration of $5.0 \times 10^{16} \, \text{cm}^{-3}$, the relative dielectric constant is 13.85, and the breakdown voltage is 30 V. Neglect the built-in voltage of the diode.

10.7 If the gain of a p^+–n GaAs APD is 5.0, compute the breakdown voltage of the diode if the ionization coefficients of both carrier species are equal. The ionization coefficient at the applied field within the APD is $1.0 \times 10^4 \, \text{cm}^{-1}$, and the donor doping concentration is $1.0 \times 10^{16} \, \text{cm}^{-3}$. The relative dielectric constant for GaAs is 12.9. Neglect the built-in voltage of the diode.

10.8 Compute the excess noise factor for the GaAs APD described in Problem 10.7.

10.9 Consider a Si APD. The electron ionization coefficient is much larger than the hole coefficient such that single carrier multiplication can be assumed. Compute the excess noise factor for the APD if the electron impact ionization coefficient is $1.0 \times 10^4 \, \text{cm}^{-1}$ and the junction is p^+–n with a $10^{16} \, \text{cm}^{-3}$ donor doping concentration. The breakdown voltage of the APD is 40 V.

10.10 Calculate the minimum carrier concentration for band–band stimulated emission transitions in GaAs if it is assumed that the electron and hole carrier concentrations are equal. Assume that the intrinsic level lies at midgap, where the energy gap of GaAs is 1.42 eV. The intrinsic concentration of GaAs is about 10^6 cm^{-3}.

10.11 Compare the gains of two different semiconductor lasers that have identical cavity lengths and losses if one of the lasers has an identical reflectivity of 0.9 for both mirrors while the other laser has mirror reflectivities of 1.0 and 0.8. Neglect the loss term in both instances.

10.12 Calculate the maximum time needed for a photogenerated electron to be collected in a p$^+$–n junction photodiode if all of the carriers are produced within the depletion region of the diode. Assume that the carrier is collected as soon as it arrives at the p$^+$ region. Assume the following information: Si photodiode, $n_i = 10^{10}$ cm^{-3}, $N_d = 10^{15}$ cm^{-3}, $N_a = 10^{18}$ cm^{-3}, relative dielectric constant $= 11.8$, applied bias $= 3$ V. Assume that the average velocity of an electron is 10^7 cm/s.

11

Transistors for high frequency, high power amplifiers for wireless systems

Solid state devices and their associated circuits used in wireless systems operate from the ultra-high frequency (UHF) to millimeter-wave frequencies. Each system requires various RF functionalities such as switching, amplification, mixing, filtering, sampling, dividing, and combining. Some of these functions are produced using a mixture of active and passive components. The active components are typically transistors, primarily FETs and BJTs.

The power amplifier is the critical portion of the RF front end in a wireless telecommunications system. The two most prevalent applications of power amplifiers in wireless systems are in cellular telephones and base stations. Power amplifier performance depends to some extent on some or all of the following: power, gain, efficiency, linearity, reliability, and thermal management issues. Any one of these specifications can be traded off against any other one. Additionally, none of these issues is solely dependent upon the device itself. For example, reliability and thermal management depend upon packaging and bonding in addition to their dependence on the circuit elements. The performance attributes are also coupled. The higher the device output power the more important thermal management becomes. Thus optimization for one parameter often impacts another device parameter that in turn may require different optimization. In the first part of this chapter we will discuss some of the performance metrics and examine how they vary with device type and design when used in power amplifiers. In addition to providing high power, the frequency of operation of a power transistor used in wireless systems must be high. Therefore, transistors must be optimized to deliver high output power at high frequency. This constraint influences the design, and material choice in device selection. Additionally, the parasitics of the devices must be reduced. The balance of this chapter will discuss two different device types, MODFETs and heterostructure bipolar transistors, HBTs, that are the major device structures used for high frequency, high power operation.

11.1 Transistor figures of merit for wireless systems

The basic performance requirements of devices used in wireless systems depend in part upon the application. For example, the requirements for power amplifiers for wireless systems are primarily high gain, high linearity, high power added efficiency (PAE), high reliability, high output power, small size, and low cost. Alternatively, in satellite communications, radio astronomy, and electronic warfare applications, most transistor power amplifiers must be optimized for low noise operation. Therefore, the transistors

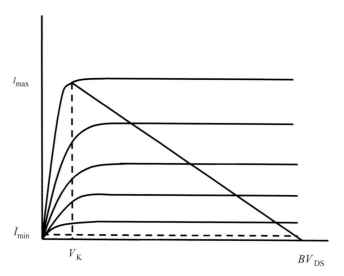

Figure 11.1 Current–voltage characteristic of a representative FET showing the load line for maximum power delivery operating in Class A. BV_{DS} is the drain to source breakdown voltage, V_K is the knee voltage, and I_{max} and I_{min} are the maximum and minimum currents respectively.

used in the power amplifiers for these various applications must be optimized for different qualities. The wide usage of power amplifiers for handsets and base stations makes both power consumption and cost key issues in their design. The RF portion of a microwave chip consumes about 25% of the power. Of this 25%, 40% of the power is consumed by the power amplifier. Standards imposed by the industry and government on the operating characteristics of a wireless network also influence the optimization of the constituent devices forming power amplifiers, etc. For example, digital standards require that the power amplifiers have high linearity. As we will see below, the requirement of high linearity favors the selection of BJTs over FETs in most situations. Personal communications systems require high efficiency and linear power amplifiers for both hand-held units and base stations. Presently, power amplifiers for base stations require frequency operation between 0.8 and 2 GHz, power levels ranging from 1 to 100 W, and an operating efficiency of 15–40%. In addition, these power amplifiers need to be highly linear, reliable, and low cost.

The first questions that the reader might raise are, how are the key figures of merit that are used to discriminate between candidate devices defined and what does optimization of each figure of merit require? Let us consider some of the more important figures of merit. The ac output power of the transistor is of course one of its most important attributes. Let us use a FET as an example for finding the output power. Consider the current–voltage characteristic shown in Fig. 11.1. The maximum ac output power is given by

$$P_0 = \tfrac{1}{8}(I_{max} - I_{min})(BV_{DS} - V_K) \tag{11.1}$$

where BV_{DS} is the drain to source breakdown voltage, V_K is the knee voltage defined as the voltage at which the current becomes saturated, and I_{max} and I_{min} are the maximum and minimum currents respectively. Notice that the output power increases with increasing breakdown voltage. As discussed in Section 3.3 breakdown can occur from impact ionization or tunneling. In most transistor designs, breakdown is due to impact ionization either near the drain in a FET or in the reverse biased collector–base junction in a BJT. The breakdown voltage in turn depends upon the energy gap of the semiconductor. Thus the wider the energy band gap the greater the breakdown voltage and the higher the output power of the device. This is one of the most important reasons why wide band gap semiconductors are emerging as the materials of choice for high power transistor amplifiers. We will discuss wide band gap semiconductors in a later section of this chapter.

The PAE is defined as

$$\text{PAE} = \frac{P_{\text{o}} - P_{\text{i}}}{P_{\text{dc}}} \tag{11.2}$$

where P_{dc} is the dc power from the power supply and P_{i} is the ac input power. If the knee voltage and the minimum current can be neglected, the dc power becomes

$$P_{\text{dc}} = \tfrac{1}{4} B V_{\text{DS}} I_{\text{max}} \tag{11.3}$$

and the output power is

$$P_{\text{o}} = \tfrac{1}{8}(I_{\text{max}})(B V_{\text{DS}}) \tag{11.4}$$

The input power can be written in terms of the gain and the output power as

$$P_{\text{i}} = \frac{P_{\text{o}}}{G} \tag{11.5}$$

where G is the gain. The gain is defined as the ratio of the power delivered to the load to the power available from the source to the amplifier. Substituting (11.3), (11.4), and (11.5) into (11.2), the PAE becomes

$$\text{PAE} = \frac{\left(1 - \dfrac{1}{G}\right)}{2} \tag{11.6}$$

Inspection of (11.6) shows that as the gain increases the PAE approaches $\frac{1}{2}$. At unity gain, the PAE approaches zero. Thus the PAE lies somewhere between 0 and $\frac{1}{2}$. Note that the gain cannot be less than 1 so the PAE remains positive.

As mentioned above, digital applications require that the amplifiers and their constituent transistors exhibit high linearity. An amplifier is said to be linear when the output power increases linearly with the input power. Wireless systems employing CDMA (those that transmit multiple users over the same frequency band) require precise control of the output amplifier power. This is because of what is termed the near–far problem. The received signal levels at a base station from multiple users are very different depending upon the distance between the signal source and the base station. Clearly, if one user is much closer to the base station than another user then the received power in that user's channel will be much greater than that in the other user's channel. This can lead to significant interference in adjacent channels. The need

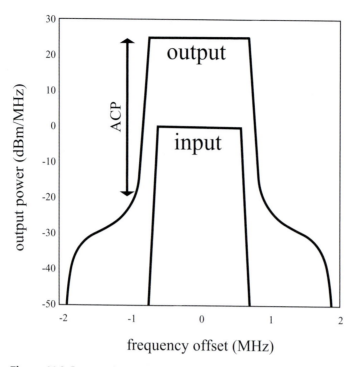

Figure 11.2 Input and output waveforms showing the requirement for adjacent channel power (ACP). Ideally, little leakage out of the main channel is desired, which can be achieved using an amplifier with high linearity (Brennan, 1999).

for high linearity can be understood then as follows. The original signal does not have any signal power outside of the main channel. However, due to amplifier non-linearities there is a significant leakage of signal power into adjacent channels. This leakage can lead to significant power in adjacent channels resulting in signal distortion as seen in Fig. 11.2. The power leakage into an adjacent channel is due to third order intermodulation distortion. Third order intermodulation distortion is defined as the intermodulation distortion between the fundamental and second harmonic signals. If the amplifying transistor is highly linear, then the second harmonic signal is less and hence the intermodulation distortion is lowered. Therefore, it is important to have highly linear transistors forming the power amplifier to ensure that the power leakage into adjacent channels is a minimum.

The maximum operating frequency of the transistor is an important figure of merit. There are two different frequencies that are often used to characterize the frequency response of the transistor. These are the cutoff frequency, f_t, and f_{max}. The cutoff frequency is defined as the frequency at which the device can no longer amplify the input signal. The cutoff frequency can be determined by calculating the frequency corresponding to unity gain with the output short circuited as was done in Section 6.4. The cutoff frequency in general for a MESFET was calculated in Section 6.4 to be

$$f_t = \frac{g_m}{2\pi C_G} \tag{11.7}$$

The cutoff frequency for a MESFET/JFET can be reexpressed in terms of the device parameters as (see Brennan and Brown (2002))

$$f_t \leq \frac{q\mu N_d a^2}{4\pi\kappa L^2} \tag{11.8}$$

where μ is the mobility, N_d the donor doping concentration, L the channel length, and a the width of the channel for a MESFET and half-width of the channel for a JFET in equilibrium. Notice that the cutoff frequency increases with increasing mobility and decreasing gate length. Thus to have a high frequency device it is desirable that the gate length be very small. This is the reason why short-channel devices are used for high frequency operation. Additionally, the fastest devices are those made with high mobility material.

The maximum frequency of oscillation, f_{max}, is defined in Section 6.4 as the frequency at which the unilateral power gain of the transistor rolls off to unity. The value of f_{max} is given by (6.65) as

$$f_{max} = \sqrt{\frac{f_t}{8\pi R_G C_{GD}}} \tag{11.9}$$

where R_G is the gate resistance and C_{GD} is the gate to drain capacitance. Inspection of (11.9) shows that f_{max} depends strongly on the parasitics.

In many instances, such as satellite communications, radio astronomy, and electronic warfare, the noise properties of the amplifier are highly important. The noise figure of a two-port network is defined as the ratio of the available SNR at the input to the SNR at the output. This is given as

$$F = \frac{S_i/N_i}{S_o/N_o} \tag{11.10}$$

where S_i, N_i, S_o, and N_o are the signal and noise power at the input and output respectively. Equation (11.10) can be rewritten in terms of the gain, G, and the total noise, N_a, at an input temperature of 290 K as

$$F = 1 + \frac{N_a}{GkTB} \tag{11.11}$$

where T is the temperature, k Boltzmann's constant, and B the bandwidth.

The minimum noise figure of the device is most often determined using a semi-empirical relationship known as the Fukui equation. The minimum noise figure is

$$F_{min} = 1 + K\frac{f}{f_t}\sqrt{g_m(R_G + R_S)} \tag{11.12}$$

where K is a fitting factor that relates to the material properties of the device. R_G and R_S are the gate and source resistances respectively. Inspection of the equation for F_{min} and (11.11) indicates that an optimized FET for noise performance must exhibit a high cutoff frequency, high associated gain and f_{max}, and low parasitic resistances.

From the above discussion it is clear that to optimize a device for frequency performance as measured by the cutoff frequency and f_{max}, it is necessary to maximize the transconductance and minimize the parasitics. The transconductance depends in turn upon the mobility, doping concentration, channel width (MESFET), and channel length. As a result, the cutoff frequency is optimized for a high mobility material with a short-channel length. The maximum frequency of oscillation is optimized by a high cutoff frequency with low parasitics. If the device is to be further optimized for low noise performance then in addition to having a high cutoff frequency and low parasitics it should have a high gain. If the device is optimized for output power, it must have a relatively high breakdown voltage. In this way, maximum power can be provided by the transistor amplifier. Further if the transistor is to be optimized such that there is little "spillover" from one channel to an adjacent channel, then it must be optimized for linearity.

The selection of a transistor type and design is predicated on the performance attributes of each transistor type. Once a transistor type has been selected it must be designed such that the key operating principles necessary for the particular application are optimized in accordance with the above discussion. There are several different device types that are commonly used in wireless systems. These are Si BJTs, Si BiCMOS (hybrid of bipolar and CMOS), GaAs FETs, GaAs HBTs, GaAs PHEMTs (pseudomorphic high electron mobility transistors, or equivalently pseudomorphic MODFETs), InP based HEMTs and PHEMTs, and wide band gap semiconductor devices such as GaN and SiC FETs. HEMTs and MODFETs are precisely the same device. There are two different names since there were multiple independent inventors. HEMT stands for high electron mobility transistor while MODFET is an acronym for modulation doped field effect transistor.

As mentioned above the frequency or speed of operation of a device depends upon the active device dimensions (for a FET the channel length and for a BJT the base width), the material parameters such as mobility, saturation velocity, etc., and the parasitics. Of the device types listed above, Si is inherently the slowest material. The electron mobility in Si is much less than in GaAs or InP based compounds. Though Si devices that have very small active dimensions can be readily made, their high frequency performance is poorer than that of the compound semiconductors due principally to the mobility difference. Therefore, Si based structures are unsuitable for transistors optimized for frequency response. Nevertheless, due to their low cost, Si based electronics is used in some low noise, low power amplifiers. Additionally, device structures such as laterally diffused MOS, LDMOS, are designed such that the electric field strength is reduced thereby increasing the breakdown voltage of the device. In an LDMOS structure, a laterally diffused n region is formed between the gate and the n^+ drain. As a result, the voltage drop near the drain is spread out spatially thus reducing the magnitude of the electric field strength and thereby raising the breakdown voltage. As a result an LDMOS device can operate at a higher voltage and thus produce higher output power. For high speed operation the base material is usually chosen to be a compound semiconductor with a concomitant high mobility.

As discussed in Section 5.2, the device type that has the highest mobility is the MODFET. The material and device structure of choice for high speed operation have evolved over time from GaAs/AlGaAs MODFETs to GaInAs/AlInAs pseudomorphic MODFETs to GaInAs/AlInAs lattice matched or pseudomorphic MODFETs made on an InP substrate. Though the latter material has the highest mobility and thus the highest operating frequency its technological immaturity has made it less attractive than either GaAs/AlGaAs or GaInAs/AlGaAs MODFETs. Improvements in the material quality of InP based compounds (those either incorporating InP or grown on InP substrates) have resulted in their application to high frequency, high power transistors.

In actual system applications the transistors are used in various manners, called classes. There are several classes of power amplifiers that describe the operating conditions of the device. The classification scheme is based on the amount of time that the output transistors are conducting or in the on state. The selection of the type of class of the amplifier depends upon the desired parameter or parameters that are to be optimized.

The most important class for wireless systems is Class AB. Class AB is somewhat intermediate between Class A and Class B operation. In Class A amplifiers the output transistor is biased such that it is conducting for the entire cycle of the input signal. The quiescent current is selected near the center of the device current such that for the full current swing the device is conducting. The output signal from a Class A amplifier is an exact replica of the input signal. Thus the Class A amplifier has a high linearity. However, the maximum efficiency, defined as the ratio of the signal load power to the supply power, of a Class A amplifier is only 25%. This efficiency is relatively low as is the PAE. For applications that require a high output signal power Class A amplifiers are not commonly used since their efficiency is too poor.

In Class B amplifiers the output transistor is conducting for only one-half of each input sinusoidal signal. To attain this behavior the quiescent current is set at the cutoff of the device current and as such only one-half of the input sinusoid leads to conductance. Class B amplifiers supply optimal gain, output power, and PAE. Unfortunately, Class B amplifiers are not very linear. In Class B amplifiers linearity is sacrificed for the output power and PAE.

For wireless systems the PAE and linearity should be simultaneously optimized. In Class A and Class B amplifiers there is a tradeoff in PAE and output power vs. linearity. Class B optimizes PAE and output power but at the expense of the linearity. Conversely, Class A amplifiers have excellent linearity but at the expense of poor PAE and output power. Thus some intermediate configuration between Class A and Class B operation would give the best overall performance in terms of PAE, power, and linearity. For this reason, the most commonly used class of single ended power amplifiers for wireless systems is Class AB. In Class AB amplifiers the quiescent current is small but nonzero. As a result, Class AB amplifiers conduct for slightly more than one-half of the input cycle, and thus operate somewhere in between Class A and Class B amplifiers. Cellular systems tend to use Class AB amplifiers since they have better PAE than Class A but better linearity than Class B amplifiers. Thus Class AB amplifiers provide an optimized tradeoff in PAE and linearity.

11.2 **Heterostructures**

As discussed in Section 5.2, heterostructures are formed by placing two dissimilar semiconductor materials into contact. Typically, one semiconductor is grown epitaxially (grown on top of a thin layer) onto another material. The most commonly used

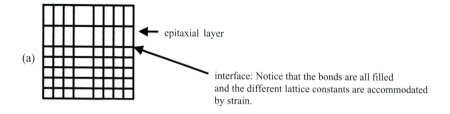

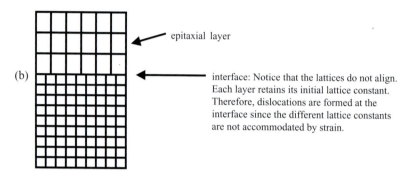

Figure 11.3 Sketch of (a) a thin epitaxial layer strained to accommodate the different lattice constant of the underlying semiconductor layer and (b) a thicker epitaxial layer that has relaxed. In (b) the epitaxial layer is thicker than the critical thickness and dislocations appear at the interface.

heterostructure is that formed between GaAs and AlGaAs. The GaAs/AlGaAs heterostructure is particularly important since these two materials are lattice matched. As mentioned in Section 5.2, two materials are said to be lattice matched if their lattice constants are identical. Under this condition, the two lattices, one from GaAs and the other from AlGaAs, are neatly accommodated at the interface, and there are few dangling bonds and imperfections.

In many instances heterostructures are formed between two semiconductors with different lattice constants. As was discussed in Chapter 8 for the specific case of Si and SiGe, two possible outcomes can occur depending upon the thickness of the top epitaxial semiconductor layer. If the layer is sufficiently thin, then the top layer will adopt the lattice constant of the bottom semiconductor layer. Thus the top layer becomes strained as shown in Fig. 11.3(a). If the top layer is too thick, then the layer relaxes and the system is no longer lattice matched. As can be seen from Fig. 11.3(b), dislocations appear at the interface. These dislocations can act as scattering centers thus reducing the mobility of the carriers at the heterointerface. In order that the lattice constant differences be accommodated at the interface, the thickness of the top layer must be less than the critical thickness. Typically, the critical thickness is very small and of course it varies with material. The thin top layer, if grown thinner than the critical thickness, accommodates the lattice mismatch through either tensile or compressive strain. The order in which the layers are grown determines whether the layers are in compressive or tensile strain. The situation for Si and SiGe is shown in Fig. 8.4.

A similar situation arises between any two nonlattice-matched semiconductors grown together to form a heterostructure. A key example is the GaN–AlGaN heterostructure system. Again, the lattice constants of the two constituent semiconductors, GaN and AlGaN, are not the same. Therefore, the heterostructure can be strained if the top layer is grown thinner than the critical thickness, or relaxed otherwise. The situation is essentially identical to that of Fig. 8.4 but in this case, the constituent materials are GaN and AlGaN. The lattice constant of GaN is larger than that of AlGaN. Therefore, if the top layer is grown with AlGaN, then the top layer accommodates the lattice mismatch through tensile strain. Conversely, if the top layer is grown with GaN then the top layer accommodates the lattice mismatch through compressive strain. The thin strained layers within the heterostructure are commonly said to be pseudomorphic. Thus pseudomorphic layers accommodate lattice mismatch through strain. As we will see below, another materials system that is typically grown pseudomorphically is InGaAs on GaAs.

In any heterostructure there is necessarily a difference in the energy gap between the two materials. For instance in GaAs and $Al_{0.32}Ga_{0.68}As$ the energy gaps are 1.42 eV and 1.82 eV respectively. At the interface the difference in the energy gaps must be accommodated by discontinuities in the conduction and valence band edges. There are three general ways in which the energy gap difference can be accommodated at the heterointerface. These three structures are called Type I, Type II, and Type III heterostructures. They are sketched in Fig. 11.4. As can be seen from Fig. 11.4, in a Type I heterostructure the energy gap difference is accommodated by the sum of the conduction and valence band edge discontinuities. In other words, the energy gap difference, ΔE_g is given as

$$\Delta E_g = \Delta E_c + \Delta E_v \tag{11.13}$$

GaAs and AlGaAs form a Type I heterostructure or, equivalently, heterojunction. Typically, most heterostructures are Type I. The other two types are less important and will not be further discussed here.

Carrier flow in a heterostructure can proceed either laterally, along the heterostructure direction, or transversely, perpendicular to the heterointerface. In a MODFET the carrier transport is along the heterostructure direction while in a heterostructure bipolar transistor the carrier transport is perpendicular to the heterojunction. Let us first consider lateral transport in a heterostructure using the GaAs–AlGaAs system as an example. Let us further assume that the AlGaAs is doped n-type while the GaAs is unintentionally doped. Therefore, the system is modulation doped. As discussed in Section 5.2, a modulation doped heterostructure is used to supply free electrons to the GaAs layer without the introduction of dopant atoms in the GaAs. The free electrons come from ionized donor atoms within the AlGaAs. To equilibrate the Fermi level, electrons are transferred from the AlGaAs into the GaAs which results in a strong band bending in the GaAs layer as shown in Fig. 11.5. A triangularlike potential well is formed on the GaAs side of the heterointerface. On one side, the potential well is

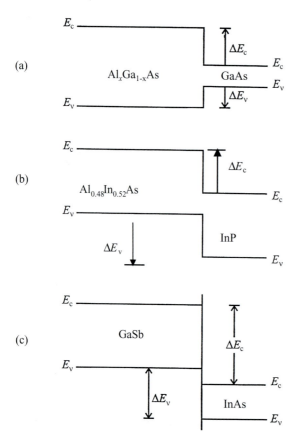

Figure 11.4 Three types of heterostructures and representative materials that form each one: (a) Type I; (b) Type II; (c) Type III (Brennan, 1999).

bounded by the conduction band discontinuity and on the other side it is bounded by the band bending within the GaAs. If the potential well is narrow, less than the de Broglie wavelength, then spatial quantization effects occur. As discussed in Chapter 5, when the energy is quantized, only certain discrete energy levels are allowed. The quantization occurs in only one direction, that being the z direction as shown in Fig. 11.5. The electrons are not quantized in the other directions. The resulting electronic structure forms a two-dimensional electron gas, since two degrees of freedom are retained (see Section 5.2). Each level is called a subband. In the diagram sketched in Fig. 11.5, there are two subbands.

If an electric field is applied in the lateral direction, electrons will flow within the subbands parallel to the heterointerface as shown in Fig. 11.5. The electrons can transfer between subbands as a consequence of scattering events. In addition, as the electrons gain energy from the field, they can reach energies that enable them to scatter out of the subbands to states within the bulk GaAs away from the interface. Once the electrons leave the potential well, they behave as ordinary three-dimensional free electrons; they experience no spatial quantization effects. If the energy gained by

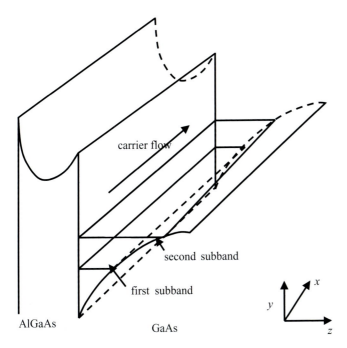

carrier flow

second subband

first subband

AlGaAs

GaAs

x

y

z

Figure 11.5 Sketch of the band bending at a heterointerface and the resulting formation of two energy subbands. The carrier flow is parallel to the heterointerface as shown in the figure.

the electrons is greater than that of the conduction band edge discontinuity, they can also scatter into the AlGaAs. Thus the electrons can return to the AlGaAs layer due to the action of the field. This is called real space transfer.

Carrier flow can also occur perpendicular to the heterointerface. In this case, carriers encounter either a potential barrier or potential step as they travel from one material to the other as shown in Fig. 11.6. When the carriers are incident from the narrow gap to the wide gap semiconductor, they must have sufficient kinetic energy in order to overcome the potential barrier. This is the usual classical condition. If the carrier has insufficient kinetic energy, then it cannot generally transfer into the wide gap semiconductor and is reflected at the interface instead. However, if the potential barrier at the heterointerface is very thin, corresponding to a very thin layer of the wide band gap semiconductor, then the incident electron can possibly tunnel through the potential barrier and reemerge on the other side of the barrier. For tunneling to be possible, the barrier width should be comparable with or less than a de Broglie wavelength. The second condition, shown in Fig. 11.6(b), corresponds to the case where the electron is incident from the wide band gap semiconductor into the narrow gap semiconductor. In this case, the electron gains a kinetic energy boost from the potential discontinuity formed at the heterointerface. Consequently, the electron energy is suddenly increased as it crosses the interface. The increase in the kinetic energy of the electron can potentially increase its velocity. This property is used in HBTs to reduce the base transit time of the injected electrons and thus increase the speed of the device. We will discuss HBTs in more detail in Section 11.4.

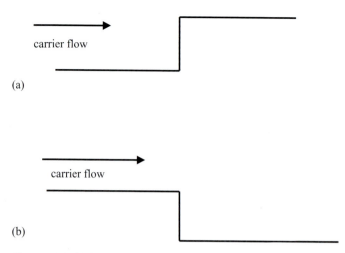

Figure 11.6 (a) Carrier flow in a heterostructure from the narrow gap semiconductor towards the wider gap semiconductor. In this case, the carriers encounter a potential barrier at the interface that arises from the band edge discontinuity. The electrons can overcome the barrier and enter the wide gap material provided they have sufficient kinetic energy. (b) Carrier flow in a heterostructure from the wide gap semiconductor towards the narrow gap semiconductor. In this case, the carriers gain energy from crossing the potential step.

11.3 **MODFET devices**

As discussed in Section 5.2, modulation doping provides a means of increasing the free carrier concentration without introducing donor atoms into the channel. Devices that exploit modulation doping in their operation are called modulation doped field effect transistors, MODFETs, or equivalently high electron mobility transistors, HEMTs. The particular advantage of MODFET devices is that through modulation doping the inherently high mobility of the constituent narrow gap semiconductor material is retained while the carrier concentration is increased. The principle of modulation doping gives rise to the excellent current carrying capability at high frequency of operation of MODFET structures. Comparison of the performance of GaAs based MODFETs to MESFETs shows that the MODFET devices operate at higher frequency, higher gain, and higher PAE. For this reason, MODFETs are particularly attractive devices for high power, high frequency operation. Additionally, MODFETs generally operate at low noise. In this section, we examine the different types of MODFET structures and their suitability to power amplifier operation.

The first MODFETs were made using the GaAs–AlGaAs materials system. This was because GaAs and AlGaAs are lattice matched over the full range of compositions of the ternary and because of the technological maturity of the GaAs based materials system. GaAs is used as the channel layer, while the AlGaAs layer is doped n-type. The Fermi level becomes equilibrated by the transfer of free electrons to the GaAs layer from the doped AlGaAs layer. Thus the free carrier concentration is increased without the introduction of ionized impurities within the GaAs channel layer. As a result the relatively high mobility of intrinsic GaAs is retained yet the carrier concentration is

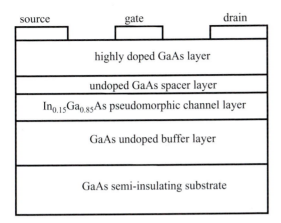

Figure 11.7 Cross-sectional view of a representative GaAs–InGaAs pseudomorphic MODFET. The thickness of the InGaAs layer is sufficiently thin that the lattice mismatch between the GaAs and InGaAs layers is accommodated by strain. For this reason the InGaAs layer is said to be pseudomorphic.

greatly increased making the device suitable for high power, high frequency operation. The mobility of intrinsic, bulk GaAs is about 8500 cm^2/(V s), which is substantially greater than that for bulk, intrinsic Si which is about 1450 cm^2/(V s). The mobility can be determined from the mean time between collisions and the effective mass using (2.13) as

$$\mu = \frac{q\tau}{m_e^*} \tag{11.14}$$

The higher mobility of GaAs compared to Si is mainly due to the much smaller electron effective mass at low applied fields in GaAs than in Si. The electron effective mass in GaAs is 0.067 multiplied by the free space mass, while the averaged relative mass in Si is about 0.32.

One limitation of GaAs based MODFETs is the presence of deep level trap states within the heavily doped AlGaAs at Al compositions in excess of about 23%. These trap states are called DX centers and their presence in MODFET devices can lead to undesirable effects. Partly in order to overcome the limitations posed by DX centers in AlGaAs, a new MODFET device structure was invented (Rosenberg *et al.* 1985) in which the channel layer is formed with In$_{0.15}$Ga$_{0.85}$As and the doped layer is GaAs. In this structure no DX centers are present and the limitations faced by AlGaAs–GaAs MODFETs are overcome. The layers forming the device structure are shown in Fig. 11.7. As can be seen from the figure GaAs is used as the substrate material with a high quality GaAs buffer layer epitaxially grown on top. The InGaAs channel layer is grown next followed by a thin GaAs spacer layer and a top highly doped GaAs layer. Though the GaAs and In$_{0.15}$Ga$_{0.85}$As are not lattice matched, if the InGaAs layer is grown sufficiently thin it will adopt the lattice constant of the underlying GaAs layer. The InGaAs layer is pseudomorphic. To accommodate the difference in the lattice constants the InGaAs layer must necessarily be strained as we discussed in the previous section. If on the other hand the InGaAs layer is relatively thick, then the lattice mismatch cannot be accommodated by strain and the InGaAs layer is no longer pseudomorphic. In this case, the interface between the InGaAs and GaAs layers will contain many dislocations, dangling bond states, etc., which will act to trap the channel electrons and increase the scattering rate. As a result, the performance of the MODFET will be greatly degraded. Thus the InGaAs layer needs to be pseudomorphic and its thickness cannot be greater than the critical thickness of about 20.0 nm.

The addition of In to GaAs to form the ternary compound InGaAs has some advantages and disadvantages. The binary compound InAs has a very small effective mass of 0.023 times the free space effective mass. This is significantly less than that of GaAs, which is 0.067 times the free space mass, and very much less than that of Si. The electron mobility of bulk InAs is very high about 3.3×10^4 cm^2/(V s), significantly higher than that of GaAs, which is 8500 cm^2/(V s). Thus InAs is seemingly a good choice for high frequency devices. Unfortunately, the energy band gap of InAs is very small, about 0.35 eV at room temperature compared with 1.42 eV for GaAs at room temperature. The resulting small energy band gap of InAs results in a relatively low breakdown voltage thus limiting the output power of an InAs based device. Consequently, a tradeoff exists in the composition of InGaAs. A high concentration of In produces a high mobility, high speed device but at the expense of a significant decrease in the breakdown voltage and accordingly a reduction in the output power of the device. On the other hand, a low concentration of In results in a lower mobility material and a concomitant lower frequency of operation. However, the breakdown voltage is higher so the output power of the device is greater. Below we examine different choices of In composition in the InGaAs channel layer.

The frequency performance of a pseudomorphic MODFET (more commonly referred to as a pseudomorphic HEMT or PHEMT) is typically greater than that of a GaAs–AlGaAs MODFET because of the addition of In into the channel layer. The mobility of the InGaAs layer is higher than that of bulk GaAs due to the lower effective mass of the InGaAs. The heterointerface formed between the GaAs and InGaAs layer has a relatively small energy band discontinuity. Thus the well in which the two-dimensional electron gas forms at the interface is relatively shallow. This has the unfortunate effect of the electrons gaining sufficient energy to leave the well even at relatively low bias. Once the electrons leave the well, they can transfer into the GaAs layer beneath the gate and on top of the InGaAs layer thus increasing the gate leakage current. As mentioned in Section 11.2, this is called real space transfer. In addition, the mobility is generally higher within the two-dimensional system than in the bulk so some deterioration in the frequency response of the MODFET occurs when the electrons leave the two-dimensional system. Greater confinement of the electrons can be achieved by making the well deeper using a heterostructure that has a greater conduction band edge discontinuity at the interface. This can be achieved by using two materials with a greater energy band gap difference than that of GaAs and In$_{0.15}$Ga$_{0.85}$As, such as Al$_{0.15}$Ga$_{0.85}$As and In$_{0.15}$Ga$_{0.85}$As. Though this structure contains AlGaAs, no DX centers are present. DX centers are not a significant factor in AlGaAs unless the Al percentage is greater than 23%. Pseudomorphic MODFETs

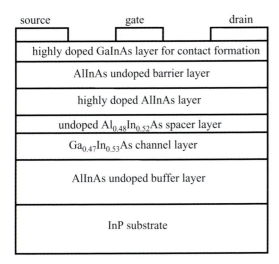

Figure 11.8 Layers used to form a $Ga_{0.47}In_{0.53}As$–$Al_{0.48}In_{0.52}As$ MODFET.

made with $Al_{0.15}Ga_{0.85}As$–$In_{0.15}Ga_{0.85}As$ materials system generally exhibit better performance than those with GaAs–$In_{0.15}Ga_{0.85}As$.

MODFETs with the highest frequency of operation are produced by growing either lattice matched or pseudomorphic AlInAs–GaInAs structures. As mentioned above, by increasing the In percentage within the InGaAs channel layer, the electron mobility and hence the frequency of operation of the MODFET can be increased significantly though at the expense of a lower breakdown voltage. The AlInAs–GaInAs layers are typically grown on an InP substrate as shown in Fig. 11.8. The ternary compositions $Al_{0.48}In_{0.52}As$–$Ga_{0.47}In_{0.53}As$ are lattice matched to each other and to the InP substrate. As a result the system has no internal strain. An $Al_{0.48}In_{0.52}As$ buffer layer is grown on top of the InP substrate followed by the $Ga_{0.47}In_{0.53}As$ channel layer. On top of the $Ga_{0.47}In_{0.53}As$ channel layer is an AlInAs layer consisting of three portions: an undoped spacer layer, a highly-doped n-type layer, and a barrier layer of undoped AlInAs. This is capped with a highly-doped GaInAs layer to form the contacts (Brown *et al.*, 1989). The channel is formed within the high mobility, low effective mass $Ga_{0.47}In_{0.53}As$ layer leading to improved high frequency performance over either the PHEMTs or GaAs based HEMTs described above.

The electron mobility within the $Ga_{0.47}In_{0.53}As$ layer is about 12 000 cm^2/(V s) and has a peak velocity of 3.0×10^7 cm/s (Mishra *et al.*, 1988). In addition there is a much higher conduction band edge discontinuity formed between the GaInAs and AlInAs layers, producing a very deep well. As a result the electrons are well confined within the two-dimensional system formed at the interface for a wide range of applied bias. The deeper well has the additional advantage of enabling a higher electron sheet carrier concentration within the well. Coupled with the higher velocity the increased carrier concentration leads to a higher current density. This results in a greater current drive capability and higher transconductance. Submicron gate length MODFET devices with lattice matched GaInAs–AlInAs have been made. Devices with gate lengths of 0.1 μm have been reported to have a cutoff frequency of 170 GHz (Brown *et al.*, 1989). In addition the noise performance of these devices is lower than that of comparable GaAs based structures.

Further improvement of the operating frequency of GaInAs–AlInAs MODFETs can be achieved by using pseudomorphic layers rather than lattice matched layers. Instead of $Ga_{0.47}In_{0.53}As$, a different ternary composition containing a greater proportion of In, $Ga_{0.4}In_{0.6}As$, which is no longer lattice matched to $Al_{0.48}In_{0.52}As$ or InP, is used. A thin layer of $Ga_{0.4}In_{0.6}As$ forms the channel region. The increased In concentration within the InGaAs layer acts to reduce the effective mass and increase the mobility. Of course, a much higher In composition than the 60% typically used results in a much lower breakdown voltage. The 60% In composition represents a reasonable compromise in frequency performance vs. breakdown voltage. The higher In content of the InGaAs layer has the additional advantage in that the well depth of the two-dimensional system is increased yielding even better electron confinement than that of the PHEMTs, or lattice matched AlInAs–GaInAs MODFETs (Fathimulla *et al.*, 1988). Submicron gate length pseudomorphic $Ga_{0.47}In_{0.53}As$ –$Al_{0.48}In_{0.52}As$ devices have been made (Kuo *et al.*, 1987). Experimental results for a 0.05 μm gate length device have been reported (Nguyen *et al.*, 1992). This structure exhibited a cutoff frequency of 340 GHz making it one of the highest frequency MODFET devices ever reported.

MODFETs provide the highest frequency performance of any transistor device. In addition, ultra-low noise operation has been observed in MODFETs. BJTs offer higher linearity than MODFETs and other FET devices. As discussed in Section 11.1, linearity is important in cellular telephony to minimize intermodulation distortion and thus avoid leakage between adjacent channels. For this reason, BJTs and the related HBTs are of importance in cellular systems. In the next section we present a brief description of HBT behavior.

11.4 **HBTs**

The operation of bipolar transistors was covered in Chapter 4. In that chapter we discussed BJTs that are made of a single semiconductor material. However, one can make BJTs using heterostructures. The most important heterostructure BJT uses a heterojunction emitter. Such devices are called heterostructure bipolar transistors or HBTs. An HBT device consists of a heterostructure emitter–base junction typically produced by a wide band gap emitter in contact with a narrower band gap base region. The heterojunction can be either abrupt, in which the materials composition changes suddenly, or graded, wherein a much more gradual change from one material to another occurs via continuous compositional grading. Here, we will discuss mainly abrupt HBT structures referring the reader to the references, particularly the book by Brennan and Brown (2002) for more information. The advantage of using a heterostructure emitter in a BJT device can be understood as follows.

The common emitter current gain, β_{dc}, is defined as the ratio of the collector to base currents. To have a common emitter current gain of about 100, it is necessary to have the collector current about 100 times larger than the base current. The value of β_{dc} is given for a pnp BJT by (4.46) as

$$\beta_{dc} = \frac{1}{\dfrac{D_E N_{dB} W_B}{D_B N_{aE} L_E} + \dfrac{1}{2}\left(\dfrac{W_B}{L_B}\right)^2} \tag{11.15}$$

If it is assumed that the base width is much smaller than the base diffusion length, then β_{dc} simplifies to the expression given by (4.47):

$$\beta_{dc} \sim \frac{1}{\dfrac{D_E N_{dB} W_B}{D_B N_{aE} L_E}} \tag{11.16}$$

Inspection of (11.16) shows that β_{dc} depends upon the ratios of the base to emitter diffusion constants, emitter diffusion length to base width, and the emitter to base doping concentrations. To first order of approximation, the ratios of the diffusion constants and emitter diffusion length to base width are about unity. Therefore, the factors that most influence the value of β_{dc} that can be engineered are the doping concentrations within the emitter and base regions. If the emitter doping concentration, N_{aE}, is significantly larger than the base doping concentration, N_{dB}, β_{dc} is increased. Thus it is desirable to dope the emitter more highly than the base and to achieve a gain of about 100 the emitter would need to be doped about 100 times more heavily than the base. However, if the base doping is relatively low, the base resistance increases. This has the unfortunate effect of lowering the frequency performance of the device, particularly f_{max}. In addition, a high emitter doping concentration results in an increased emitter–base capacitance further lowering the frequency performance of the device.

As discussed above, in an ordinary BJT, an engineering tradeoff occurs in its design since a high current gain requires a high ratio of emitter to base doping concentrations but results in a reduction in frequency performance. This tradeoff can be clearly seen by examining the frequency dependence of the device. The maximum frequency of operation, f_{max}, is one of the most important figures of merit that characterizes RF performance. As we stated in Section 11.1, f_{max} is defined as the frequency at which the unilateral (condition under which there is no reverse transmission) power gain of the transistor goes to unity. The most common formulation for f_{max} for a BJT is

$$f_{max} = \sqrt{\frac{f_t}{8\pi r_{bb} C_C}} \tag{11.17}$$

where r_{bb} is the base resistance, f_t the cutoff frequency, and C_C the collector–base junction capacitance. Inspection of (11.17) indicates that f_{max} is inversely proportional to the square root of the base resistance. Therefore, as the base resistance increases, which results from lowering the base doping concentration, the maximum frequency of operation is reduced. Clearly, to maintain a high frequency of operation it is important to reduce the base resistance, which in turn requires a high base doping concentration.

The insertion of a wide band gap semiconductor material for the emitter enables the usage of a higher base doping concentration without compromising the dc common

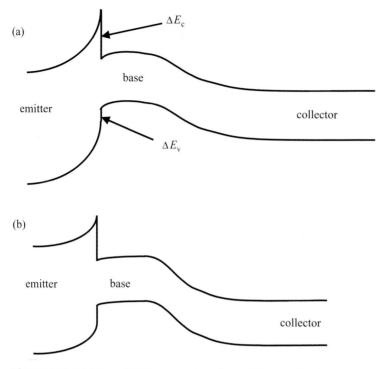

Figure 11.9 (a) Abrupt HBT npn transistor in equilibrium; (b) abrupt HBT npn transistor under active biasing conditions (Brennan, 1999).

emitter current gain. Thus, both high gain and high frequency performance can be simultaneously obtained. The energy band diagram of an HBT in equilibrium and under bias is shown in Fig. 11.9. The heterojunction is assumed to be abrupt and the device type is npn implying that the emitter is doped n-type, the base p-type, and the collector n-type. An npn transistor is used here instead of the pnp discussed in Section 4.1 since npn HBTs are far more commonly employed. This is due to the fact that npn HBTs offer far superior frequency performance than pnp devices due to the much higher velocity of the electrons. Examination of Fig. 11.9 shows that the energy barrier for electron injection from the emitter into the base is significantly less than the corresponding energy barrier for hole injection from the base into the emitter. Notice that the band bending is the same for both the conduction and valence bands within the emitter, but the valence band edge discontinuity adds to the energy barrier for the holes. Thus hole injection from the base into the emitter requires a higher energy to surmount the potential barrier at the emitter–base junction than electron injection from the emitter into the base. Under active biasing conditions, electrons are more readily injected from the emitter into the base than holes from the base into the emitter. The base current due to hole injection is significantly reduced without compromising the emitter current. Therefore, the base doping concentration can be increased without altering the common emitter current gain so high frequency performance can be maintained at high β_{dc}. The processes in HBTs are discussed in detail in the book by Brennan and Brown (2002) and will not be repeated here. Suffice it to say that HBTs offer substantial improvement in performance over BJTs since they enable a higher base doping concentration with a concomitant reduction in the base resistance without sacrificing the dc gain.

HBTs are important devices for power amplifiers. Several different materials systems are used to fabricate HBT structures. The major heterostructures utilized in HBT technology are GaAs–AlGaAs, SiGe, and InP based heterostructures, primarily InGaAs–InP and InGaAs–AlInAs. The SiGe heterostructure is not lattice matched and thus is strained upon growth. A Type I heterostructure can be formed by growing strained SiGe onto a relaxed Si layer. This is commonly used to form the emitter of a SiGe HBT device. As mentioned above, a Type I heterostructure is best suited to improving bipolar transistor behavior since back injection of the holes from the base is reduced while high energy injection of the electrons from the emitter and concomitant base transit time reduction can be achieved. The major advantage of the use of SiGe for HBT devices is that it represents a low cost alternative to compound semiconductor devices. However, SiGe HBT devices have limited frequency performance and thus they are more favored for low frequency, below 2 GHz, operation. At higher frequencies HBTs made from the compound semiconductors are more attractive.

GaAs based HBTs are the most utilized and studied HBT structures. However, InP based HBTs are potentially better candidates for high power at higher frequency operation. The double heterostructure InP HBT, wherein heterostructures are used for both the emitter and collector regions, can potentially deliver better performance than GaAs devices. These devices are referred to as InP DHBT structures. The particular advantages of InP DHBT devices are that the electron saturation velocity and breakdown field in the InP collector are both higher than in an AlGaAs–GaAs HBT structure. Since the breakdown field is higher, the maximum output power of the device should also be higher. The major drawback to the use of DHBTs is the fact that the second heterostructure formed at the collector–base junction can potentially block the current. As we discussed in Section 11.2, in a Type I heterostructure when electrons are incident from a narrow gap semiconductor, in this case the base InGaAs layer, to a wide gap semiconductor layer, in this case the InP collector layer, they experience a potential barrier due to the conduction band edge discontinuity. If the electron energies are insufficient to overcome this barrier then they can be reflected and will accumulate in the base region. To combat this effect, a different base layer is used. Specifically, GaAsSb base layers have been utilized which form a Type II heterojunction with the InP collector (Nguyen and Micovic, 2001). As a result, the conduction band edge of the GaAsSb base lies above that of the InP collector and no pileup of electrons occurs at the collector–base junction (Nguyen and Micovic, 2001).

In power amplifiers, since the current and voltage are both necessarily high, the heat generated by the devices is significant and can substantially raise the operating temperature of the structure. Consequently, most power devices operate at elevated temperatures. The operating temperature greatly limits the behavior of the device since at very high temperatures the semiconductor material can become intrinsic depending upon the magnitude of the energy gap. As described earlier, by intrinsic we mean that the vast majority of free carriers produced within the conduction band are created by directly promoting electrons from the valence band rather than from the dopant atoms. At elevated temperatures, the thermal generation rate can be very high such that the interband generation rate vastly exceeds the free carrier concentration arising from ionized donors. As a result, the effect of doping on the semiconductor is lost. The semiconductors all behave the same way independent of doping. The semiconductor becomes less tolerant of thermal heating the smaller its energy gap. Conversely, wide band gap semiconductors have much greater tolerance of thermal heating effects and can be operated at much higher temperatures.

Thermal management is an important issue for Si, GaAs, and InP based electronics because these materials do not have a particularly large energy gap. Thus the removal of waste heat is critical to maintaining their operation. Waste heat can be removed using what is called heat sinking techniques. In these approaches, the device is packaged such that waste heat is removed quickly from the active region of the device.

The major advantage of HBTs over MODFETs is their linearity. Additionally, HBTs are typically more uniform in device characteristics so they are better suited to millimeter wave integrated circuit fabrication (Nguyen and Micovic, 2001). HBTs can deliver high power at X band, 8–12 GHz frequencies. The total output power and efficiency for a single device are comparable between MODFETs and HBTs up to K band frequencies, ranging from 18 to 27 GHz. Within the K_a band, 27–40 GHz, InP based HBTs are superior to GaAs based HBT devices. However, above ~40 GHz, MODFETs particularly PHEMTs and InP based MODFETs are exclusively used. Therefore, for the highest frequency performance for power amplifiers, PHEMTs and InP based MODFET structures are used. MODFETs also have better low noise performance than HBTs. Ultimately, higher power, higher frequency electronics requires the use of new semiconductor materials that have much wider energy band gaps than either GaAs or InP. In the next section we discuss the potential of the wide band gap semiconductors for high power, high frequency amplifiers.

11.5 **Wide band gap semiconductors**

The wide band gap semiconductors are defined as materials that have energy band gaps roughly 2–3 times that of Si. The most important wide band gap semiconductors are the various polytypes of SiC and the group of materials called the III-nitrides. The III-nitrides are compound semiconductors formed by combining an element from column III of the Periodic Table with nitrogen. The most important such compounds are GaN, AlN, and InN and their related ternary compounds, GaInN, AlGaN, and InAlN. The wide band gap semiconductors, as their name implies, have substantially higher band gap energies and thus have much higher breakdown voltages and can operate at much higher temperatures than do Si or GaAs. In addition, the wide band gap semiconductors typically have a higher saturation drift velocity, relatively small dielectric constant, and in the case of SiC a much higher thermal conductivity than either Si or GaAs. The fact that the relative dielectric constants are smaller than for Si and GaAs implies that the capacitance of the devices is smaller leading to less RC loss. For these reasons the wide band gap semiconductors are attractive for high power, high temperature, high frequency device operation.

The wide band gap semiconductor materials have attracted much attention for possible insertion in military power electronic device applications, such as high power switching devices, high frequency, high power amplifiers for radar and fire control systems, electronic warfare, and multi-functional and RF systems. Some attractive commercial applications of wide band gap electronic devices include power amplifiers for base stations in wireless networks, automotive and aviation electronics, where high temperature operation is needed, and in high voltage switching for power systems and electric vehicles (Trew, 2002).

The relatively low breakdown voltages of GaAs, InP, and Si based electronic devices limit their maximum output power levels. The maximum power for a Class A amplifier is given by an equivalent form of (11.1) as

$$P_{max} = \frac{(V_{BR} - V_{knee})^2}{8R_L} \tag{11.18}$$

where R_L is the load resistance, V_{BR} the breakdown voltage, and V_{knee} the knee voltage defined as the voltage at which the transistor current saturates. Inspection of (11.18) indicates that the output power of a RF transistor can be greatly improved with a high breakdown voltage and a low knee voltage. The breakdown voltage is, in turn, primarily affected by the energy gap of the material. Therefore, improved power performance requires usage of the wide band gap materials.

To date the most studied wide band gap materials for transistor device applications are GaN (Kahn et al., 1996) and SiC (Weitzel et al., 1996). In both of these materials the breakdown electric field strengths are expected to be at least four times larger than in either Si or GaAs (Kolnik et al., 1997; Oguzman et al., 1997). Experimental work has shown that GaN can be successfully grown on a SiC substrate (Sriram et al., 1997) which has a relatively high thermal conductivity. Using this material, an AlGaN–GaN MODFET has been demonstrated to operate at 10 GHz with a maximum output power of 2.66 W/mm, about 4.5 times that of a comparable GaAs X-band FET.

GaN offers two additional advantageous qualities that make it highly attractive for device design. The first is that GaN can form Type I heterojunctions with the ternary compounds AlGaN and InGaN. This enables the use of modulation doping techniques and the adoption of GaN in MODFET devices. In addition, the lattice mismatch between GaN and AlN can be exploited to alter the carrier concentrations near the interface (Kuech et al., 1990; Asbeck et al., 1997). If the AlN or AlGaN layers are made sufficiently thin, the lattice mismatch is accommodated by internal strains rather than by the formation of misfit dislocations. Through the action of the piezoelectric effect, these strains can induce electric fields in III-nitride heterostructures grown with hexagonal symmetry. The strain induced polarization fields in turn can alter the band bending and carrier concentration at the interface. The strain associated with the lattice mismatch of the GaN and AlGaN layers generates a piezoelectric polarization that acts perpendicular to the heterointerface. The piezoelectric polarization of the strained AlGaN layer is much greater than that of AlGaAs/GaAs structures by about a factor

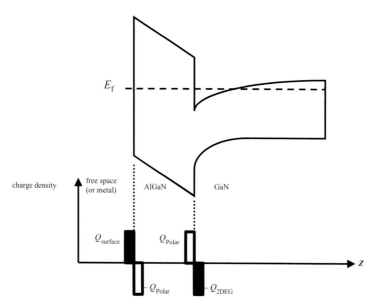

Figure 11.10 Band diagram and schematic charge density resulting from polarization induced sheet charge in a GaN–AlGaN MODFET design (Brennan, 1999).

of 5 (Ambacher *et al.*, 2000) and thus has a significant effect. The resulting action of the piezoelectric fields produces a piezoelectrically induced charge density given by

$$\nabla \circ \vec{P} = \sigma_{pz} \tag{11.19}$$

where $\vec{P}$ is the polarization field and σ_{pz} is the piezoelectrically induced sheet charge density. Equation (11.19) states that the polarization fields within the strained semiconductor layer can be replaced by an equivalent bound sheet charge. Sheet charge, as the name suggests, is charge arrayed on a two-dimensional plane at the heterointerface. The situation is shown in Fig. 11.10. As can be seen from the figure, the action of the strain polarization field within the AlGaN layer has been replaced by two sheet charge distributions on either side of the AlGaN. The layers are grown such that the positive direction of the polarization field occurs at the AlGaN–GaN interface. Thus the bands bend down within the AlGaN as they approach the interface as shown in Fig. 11.10. The action of the polarization field is equivalent to the action of a positive sheet charge induced within the AlGaN at the AlGaN–GaN heterojunction. The potential is lower, and hence the electron energies are higher, at the other end of the AlGaN layer near the metal interface due to the polarization field. Hence, a negative charge sheet is induced at the other end of the AlGaN layer as shown in the figure. The negative charge sheet in the AlGaN layer arising from the polarization field induces a positive charge sheet at the surface of the metal overlaid onto the AlGaN layer. This is the usual situation for a MODFET gate, a top metal gate contact is formed that creates a Schottky barrier with the wide band gap semiconductor material (in this case, AlGaN) which is in turn grown onto a narrower gap semiconductor (in this case GaN). The presence of the positive charge sheet in the AlGaN layer induces an equal but opposite charge sheet in the GaN well. Thus electrons are induced within the GaAs channel by the action of the strain polarization field within the AlGaN. The strain induced polarization charge can greatly increase the free carrier concentration within the channel region of a heterojunction FET beyond that achievable by modulation doping alone. In this way, strain induced polarization provides a new additional engineering degree of freedom by which device structures and their behavior can be altered.

Very high sheet electron concentrations have been obtained in piezoelectrically strained GaN–AlGaN MODFETs. These devices exhibit electron mobilities at room temperature of about 1500 cm^2/(V s) equal to that of bulk, intrinsic GaN. As a result, high output power, provided by the large sheet carrier concentration within the GaN channel, at high frequency operation, due to the reasonably high mobility, has been achieved (Wu *et al.*, 1997). Generally, GaN MODFETs have demonstrated about one order of magnitude higher power density and higher efficiency than comparable Si or GaAs devices. The improved power density implies that for the same output power, a GaN device can be an order of magnitude smaller than a GaAs structure (Mishra *et al.*, 2002).

Though GaN and more generally III-nitride based electronics are highly attractive for power amplifier applications due to the many features discussed above, their usage has been frustrated to date by the lack of high quality, defect free material. The major limitation to the development of III-nitride devices is the lack of a suitable, inexpensive substrate upon which GaN and the other III-nitrides can be grown. Most of the early devices were grown on sapphire substrates which are not lattice matched to GaAs. In fact the relative mismatch of the GaN and sapphire is such that many dislocations and defects are present in the GaN grown on top of the sapphire. As a result, device quality suffered. More recently, improved material quality has been achieved using SiC substrates which provide closer lattice matching than sapphire. Far fewer dislocations are generated and the quality of the GaN films is far superior to that typically grown on sapphire. However, the cost of SiC substrates has been relatively high compared with sapphire thus limiting their use. Future improvements in device quality will undoubtedly be made. It is possible that GaN substrates of sufficient quality can be grown thus eliminating the substrate selection problem. Alternatively, other substrates may prove more suitable to GaN device development.

Because of the materials complexity associated with the III-nitrides there has been considerable interest in developing SiC based electronics. SiC crystallizes in many different forms, called polytypes. The most important polytypes are called 3C-SiC, 4H-SiC, and 6H-SiC. 3C-SiC crystallizes with cubic symmetry, identical lattice constants exist along all three perpendicular directions. Both 4H-SiC and 6H-SiC have hexagonal symmetry: in two dimensions the lattice constants are the same (called the basal plane) and in the perpendicular direction (often referred to as the *z* axis) the lattice constant is different. 4H-SiC and 6H-SiC differ by the number of atoms per unit cell with 6H having more. High quality 4H-SiC and 6H-SiC polytype substrates and epitaxial layers are commercially available today. 3C-SiC has been grown successfully in the form of films on Si(100). However, bulk 3C-SiC is typically polycrystalline. 3C-SiC has a smaller band gap and a lower mobility than 4H-SiC and is generally less attractive. Therefore, most SiC devices are made from either 4H-SiC or 6H-SiC.

SiC has some advantages over the III-nitrides. SiC substrates do exist and this enables high quality epitaxial growth of SiC layers. As mentioned above, one of the major difficulties frustrating the development of the III-nitrides is the lack of a suitable substrate. SiC has the additional feature of a high thermal conductivity. The thermal conductivity of GaN is 1.3 W/(cm K) while for SiC it is 4.0 W/(cm K). Consequently, the thermal conductivity of SiC is roughly three times that of GaN. A high thermal conductivity aids in heat removal and thus enables higher device operating temperatures. The primary drawback of SiC compared with the III-nitrides is that heterostructure formation is far more limited in SiC. To date, no modulation doped structures or piezoelectrically enhanced devices have been made with any of the SiC polytypes. Therefore, it is expected that SiC devices will not be competitive for very high frequency device applications.

In conclusion, the wide band gap semiconductors offer numerous advantages over more conventional Si, GaAs, and InP electronics. Currently, both the III-nitrides and SiC are being developed for high power, high operating temperature electronics. It is expected that these materials will ultimately replace conventional electronics for use in satellite and cellular communications (mainly in base stations applications). Additionally, these materials should become of great importance in military and high temperature environments, such as automotive and aviation electronics.

Problems

11.1 A GaAs MESFET is to be used as a high frequency, power transistor. Given the following information determine the maximum cutoff frequency of the device. The gate length is 0.25 μm, the active layer thickness is 0.1 μm, the doping concentration is 10^{17} cm^{-3}, the dielectric constant is 12.90, and the mobility is 8500 cm^2/(V s).

11.2 A GaN n-channel JFET power transistor has a knee voltage of 1.0 V at a gate bias of -1.1 V. Neglect the built-in voltage of the p$^+$–n junction. Assume that the breakdown voltage is equal to 15 times $V_{D,\text{sat}}$ and that the minimum current can be neglected compared with the maximum current of 700 mA. The device has a doping concentration of 3×10^{17} cm^{-3} and an active layer thickness of 0.1 μm. The dielectric constant of GaN is 9.5, the electron mobility is about 1000 cm^2/(V s), and the channel length is 0.1 μm. Determine
  (a) the maximum output power of the device;
  (b) the cutoff frequency of the device;
  (c) the PAE of the device. Assume that the input power is 20% of the output power.

11.3 An HBT device doped in the base at 10^{16} cm^{-3} has an f_{max} of 20 GHz. If the base doping concentration is increased by a factor of 100 to 10^{18} cm^{-3} and all other parameters are held constant, what is the new value of f_{max}? Assume that the device is an npn device.

11.4 Assume that the breakdown voltage of a GaAs power device is 20 V. If GaN is used in place of GaAs in a power circuit with the same load resistance, determine the relative increase in the maximum power for a Class A amplifier. Assume that the knee voltages are the same for the GaAs and GaN devices and are equal to 0.5 V. Further assume that the breakdown voltage for GaN is, as stated in the text, about 4 times larger than that for GaAs.

References

Ambacher, O., Foutz, B., Smart, J., *et al.* (2000). Two dimensional electron gases induced by spontaneous and piezoelectric polarization in undoped and doped AlGaN/GaN heterostructures. *J. Appl. Phys.*, **87**, 334–44.

Asbeck, P. M., Yu, E. T., Lau, S. S., Sullivan, G. J., Van Hove, J., and Redwing, J. (1997). Piezoelectric charge densities in AlGaN/GaN HFETs. *Electron. Lett.*, **33** 1230–1.

Baccarani, G., Wordeman, M. R., and Dennard, R. H. (1984). Generalized scaling theory and its application to a $\frac{1}{4}$ micrometer MOSFET design. *IEEE Trans. Electron Dev.*, **ED-31**, 452–62.

Bakshi, P., Broido, D. A., and Kempa, K. (1991). Spontaneous polarization of electrons in quantum dashes. *J. Appl. Phys.*, **70**, 5150–2.

Birge, R. R. (1995). Protein based computers. *Sci. Am.*, March, 90–95.

Brennan, K. F. (1999). *The Physics of Semiconductors with Applications to Optoelectronic Devices*. Cambridge: Cambridge University Press.

Brennan, K. F., and Brown A. S. (2002). *Theory of Modern Electronic Semiconductor Devices,* New York: John Wiley and Sons.

Brennan, K. F., and Haralson J., II (2000). Superlattice and multiquantum well avalanche photodetectors: physics, concepts and performance. *Superlattices and Microstructures* **28**, 77–104.

Brews, J. R. (1990). The submicron MOSFET. In *High Speed Semiconductor Devices*, edited by S. M. Sze. New York: John Wiley and Sons.

Brews, J. R., Fichtner, W., Nicollian, E. H., and Sze, S. M. (1980). Generalized guide for MOSFET miniaturization. *IEEE Electron Device. Lett.*, **EDL-1**, 2–4.

Brown, A. S., Mishra, U. K., and Rosenbaum, S. E. (1989). The effect of interface and alloy quality on the DC and RF performance of Ga0.47In0.53As-Al0.48In0.52As HEMTs. *IEEE Trans. Electron Dev.*, **36**, 641–5.

Casady, J. B., and Johnson, R. W. (1996). Status of silicon carbide (SiC) as a wide-bandgap semiconductor for high-temperature applications: A review. *Solid-State Electron.*, **39**, 1409–22.

Chang, K., Bahl, I., and Nair, V. (2002). *RF and Microwave Circuit and Component Design for Wireless Systems*. New York: John Wiley and Sons.

Chatterjee, P. K., Hunter, W. R., Holloway, T. C., and Lin, Y. T. (1980). The impact of scaling laws on the choice of n-channel or p-channel for MOS VLSI. *IEEE Electron Dev. Lett.*, **EDL-1**, 220–3.

Dekker, C. (1999). Carbon nanotubes as molecular quantum wires. *Phys. Today*, May, **1999**, 22–8.

Dennard, R. H., Gaensslen, F. H., Yu, H-N., Rideout, V. L., Bassous, E. and LeBlanc, A. R. (1974). Design of ion-implanted MOSFETs with very small physical dimensions. *IEEE J. Solid-State Circuits*, **SC-9**, 256–67.

Farahmand, M., and Brennan, K. F. (1999). Full band Monte Carlo simulation of zincblende GaN MESFETs including realistic impact ionization rates. *IEEE Trans. Electron Dev.*, **46**, 1319–25.

Fathimulla, A., Heir, H., and Abrahams, J. (1988). High-current, planar–doped pseudomorphic $Ga_{0.4}In_{0.6}As/Al_{0.48}In_{0.52}As$ HEMTs. *Electron. Lett.*, **24**, 717–18.

Garfinkel, S. L. (2000). Biological computing. *Technology Rev.*, May/June, 71–7.

Garzon, M. H., and Deaton, R. J. (1999). Biomolecular computing and programming. *IEEE Trans. Evol. Computation*, **3**, 236–50.

Grotjohn, T., and Hoefflinger, B. (1984). A parametric short-channel MOS transistor model for subthreshold and strong inversion current. *IEEE Trans Electron Dev.*, **ED-31**, 234–46.

Harriott, L. R. (2001). Limits of lithography. *Proc. IEEE*, **89**, 366–74.

Heath, J. R., Kuekes, P. J., Snider, G. S., and Williams, R. S. (1998). A defect tolerant computer architecture: opportunities for nanotechnology. *Science*, **280**, 1716–21.

Ismail, K., Meyerson, B. S., and Wang, P. J. (1991). High electron mobility in modulation-doped Si/SiGe. *Appl. Phys. Lett.*, **58**, 2117–19.

Ismail, K., Nelson, S. F., Chu, J. O., and Meyerson, B. S. (1993). Electron transport properties of Si/SiGe heterostructures: Measurements and device implications. *Appl. Phys. Lett.*, **63**, 660–2.

Ito, T., and Okazaki, S. (2000). Pushing the limits of lithography. *Nature*, **406**, 1027–31.

Kahn, M. A., Chen, Q., Yang, J. W., Shur, M. S., Dermott, B. T., and Higgins, J. A. (1996). Microwave operation of GaN/AlGaN-doped channel heterostructure field effect transistors. *IEEE Electron Dev. Lett.*, **17**, 325–7.

Kittel, C., and Kroemer, H., (1980). *Thermal Physics*, second edition. San Francisco: W. H. Freeman and Co.

Kolnik, J., Oguzman, I. H., Brennan, K. F., Wang, R. and Ruden, P. P. (1997). Theory of electron initiated impact ionization in bulk zincblende and wurtzite GaN. *J. Appl. Phys.*, **81**, 726–33.

Kuech, T. F., Collins, R. T., Smith, D. L., and Mailhiot, C. (1990). Field-effect transistor structure based on strain-induced polarization charges. *J. Appl. Phys.*, **67**, 2650–2.

Kuo, J. M., Chang, T.-Y., and Lalevic, B. (1987). $Ga_{0.4}In_{0.6}As/Al_{0.48}In_{0.52}As$ pseudomorphic modulation-doped field-effect transistors. *IEEE Electron Dev. Lett.*, **EDL-8**, 380–2.

Lent, C. S., and Tougaw, P. D. (1997). A device architecture for computing with quantum dots. *Proc. IEEE*, **85**, 541–57.

Liu, W. (1999). *Fundamentals of III–V Devices, HBTs, MESFETs, and HFETs/HEMTs*. New York: Wiley-Interscience.

Martel, R., Schmidt, T., Shea, H. R., Hertel, T., and Avouris, Ph. (1998). Single and multi-wall carbon nanotube field-effect transistors. *Appl. Phys. Lett.*, **73**, 2447–9.

Mishra, U. K., Brown, A. S., Jelloian, L. M., Hackett, L. H., and Delaney, M. J. (1988). High-performance submicrometer AlInAs–GaInAs HEMTs. *IEEE Electron Dev. Lett.*, **9**, 41–3.

Mishra, U. K., Parikh, P., and Wu, Y-F. (2002). AlGaN-GaN HEMTs – An overview of device operation and applications. *Proc. IEEE*, **90**, 1022–31.

Morkoc, H., and Solomon, P. M. (1984). The HEMT: a superfast transistor. *IEEE Spectrum*, February, 28–35.

Nguyen, C., and Micovic, M. (2001). The state-of-the-art of GaAs and InP power devices and amplifiers. *IEEE Trans. Electron Dev.*, **48**, 472–8.

Nguyen, L. D., Brown, A. S., Thompson, M. A., and Jelloian, L. M. (1992). 50-nm self-aligned-gate pseudomorphic AlInAs/GaInAs high electron mobility transistors. *IEEE Trans. Electron Dev.*, **39**, 2007–14.

Nguyen, L. D., Schaff, W. J., Tasker, P. J., *et al.* (1988). Charge control, DC, and RF performance of a 0.35 μm pseudomorphic AlGaAs/InGaAs modulation doped field-effect transistor. *IEEE Trans. Electron Dev.*, **35**, 139–43.

Oguzman, I. H., Bellotti, E., Brennan, K. F., Kolnik, J., Wang, R., and Ruden, P. P. (1997). Theory of hole initiated impact ionization in bulk zincblende and wurtzite GaN. *J. Appl. Phys.*, **81**, 7827–34.

Packen, P. A. (1999). Pushing the limits. *Science*, **285**, 2079–80.

Pao, H. C., and Sah, C. T. (1966). Effects of diffusion current on characteristics of metal-oxide (insulator) – semiconductor transistors. *Solid-State Electron.*, **9**, 927–37.

Pierret, R. F. (1996). *Semiconductor Device Fundamentals*. Reading, MA: Addison-Wesley.

Porod, W. (1997). Quantum-dot devices and quantum-dot cellular automata. *J. Franklin Institute*, **334B**, 1147–75.

Regalado, A. (2000). DNA computing. *Technology Rev.*, May/June, 80–4.

Rosenberg, J. J., Benlami, M., Kirchner, P. D., Woodall, J. M., and Pettit, G. D. (1985). An InGaAs/GaAs pseudomorphic single quantum well HEMT. *IEEE Electron Device Lett.*, **EDL-6**, 491–3.

Selberherr, S. (1984). *Analysis and Simulation of Semiconductor Devices*. Vienna: Springer-Verlag.

Shichijo, H. (1981). A re-examination of practical scalability limits of n-channel and p-channel MOS devices for VLSI. *Proceedings IEDM*, 219–22.

Sriram, S., Messham, R. L., Smith, T. J., *et al.* (1997). RF performance of AlGaN/GaN MODFETs on high resistivity SiC substrates. *Materials Research Society Fall Meeting*, Boston, MA. Dec. 1–5.

Streetman, B. G., and Banerjee, S. (2000). *Solid-State Electronic Devices*, fifth edition. Upper Saddle River, NJ: Prentice Hall.

Sze, S. M. (1981). *Physics of Semiconductor Devices*, second edition. New York: John Wiley and Sons.

Taur, Y., Buchanan, D. A., Chen, W., *et al.* (1997). CMOS scaling into the nanometer regime. *Proc. IEEE*, **85**, 486–503.

Trew, R. J. (2002). SiC and GaN Transistors – Is there one winner for microwave power applications? *Proc. IEEE*, **90**, 1032–47.

Weitzel, C. E., Palmour, J. W., Carter, Jr., C. H., *et al.* (1996). Silicon carbide high-power devices. *IEEE Trans. Electron Dev.*, **43**, 1732–40.

Wu, Y.-F., Keller, B. P., Keller, S., *et al.* (1997). Short channel AlGaN/GaN MODFETs with 50-GHz f_T and 1.7 W/mm output-power at 10 GHz. *IEEE Electron Dev. Lett.*, **18**, 438–40.

Yao, Z., Kane, C. L., and Dekker, C. (2000). High-field electrical transport in single-wall carbon nanotubes. *Phys. Rev. Lett.*, **84**, 2941–4.

Index

absorption
  light, 33
  mechanisms, 239, 241, *242*
absorption coefficient, 32
absorption rate, 239, 241
acceptors, 16, 17
ACP (adjacent channel power), 277–8
active mode, 78, *79*, 80
adenine, 200
adjacent channel power (ACP), 277–8
algorithms, 207, 209, 210, 216
aliphatic molecules, 201, 202
aluminum
  as acceptor, 17
  for interconnects, 185
aluminum gallium arsenide, 109–12, 236, 288
  applications, 248
  energy gaps, 283
aluminum gallium indium phosphide, 234
aluminum gallium nitride, 283, 294
aluminum indium arsenide, 290
aluminum indium arsenide–gallium indium arsenide
    heterostructures, 289, 290, 293
aluminum nitride, 155
aluminum–silicon dioxide–silicon systems, in MIS
    systems, 128–9, 143
amplifiers, 223–5
  broadband, 224
  Class A, 281, 295
  Class AB, 281
  Class B, 281
  disadvantages, 224
  linearity, 277–8
  see also optical amplifiers; power amplifiers
AND gates, 205–6, 214
antibodies, 198
APDs *see* avalanche photodiodes (APDs)
applets, biological, 200
argon fluoride, 183
attenuation, in optical fibers, 221, 230
Auger generation–recombination, 30, 235
avalanche breakdown, 57–9, 267–8
  in bipolar junction transistors, 95–6
avalanche photodiodes (APDs), 260, 261, 265–70
  applications, 269
  avalanche breakdown, 267–8

breakdown field, 271
  gain, 268, 271–2
  materials, 269
  multiplication scheme, 266
  noise, 268–9, 271–2
  operation, 265–9
  separate absorption multiplication, 265
  superlattice/multiquantum well, 265
  thin multiplication region, 265

bacteria, 198
band bending, 26, 69–70
  heterointerfaces, 284–5
  MIS structures, 127–8, 130, 132, 134
band gaps *see* energy gaps
band-to-band Auger generation *see* impact
    ionization
base current, 80, 81, 84, 92
base narrowing, 87, 94–5
base resistance, 98
bases (semiconductors), 78
  n-type, 87
base stations, 226, 229
  adjacent channel power, 277–8
  hand-offs, 227
  power amplifiers, 275, 276
base transit time, 81, 99
  definition, 80
base transport factor, 86, 87
benzene, 202
bias
  forward, 47–51, 66, 71, 72–3
  metal–insulator–semiconductor systems under,
    133–44
  modes, 78–80
  and p–n junctions, 47–57
  reverse, 47–51, 57, 71, 72–3
  Schottky barriers, 72–3
  see also gate biasing
BiCMOS (bipolar CMOS), 280
bioagents, 198
biological computing, 197–201
biomolecules, applications, 200–1
bipolar junction transistors (BJTs), 78–99, 276
  applications, 275, 280
  avalanche breakdown, 95–6

bipolar junction transistors (BJTs) (*cont.*)
  base region, drift, 92–3
  base transport factor, 86, 87
  base width narrowing, 94–5
  biasing modes, 78–80
  capacitance, 98
  common base current gain, 86, 87
  common emitter current gain, 86, 87–8
  configurations, 78–80
  current determination, 82–5
    assumptions, 81–2
  doping, 92–3
  Ebers–Moll model, 88–90, 91–2
  emitter–collector voltage, 91–2
  emitter injection efficiency, 86
  equivalent circuits
    hybrid pi model, 97, 98
    low frequency, 97
  figures of merit, 98
  gain, 81
  linearity, 290
  maximum frequency of oscillation, 291
  operation, 78–92
    high frequency, 97–9
    maximum frequency, 99
  performance parameters, 86–8
  secondary effects, 92–6
  structure, 78
  see also heterostructure bipolar transistors
    (HBTs)
BJTs *see* bipolar junction transistors (BJTs)
Boltzmann's constant, 279
Boolean logic, 162
breakdown
  avalanche, 57–9
  avalanche photodiodes, 271
  p–n junction diodes, 60–1
  reverse, 57–60, *61*
  zener, 57, 59–60
breakdown voltage, 59, 295
built-in potential, *41*, 59
  definition, 40
  determination, 42–3, 45
  Schottky barriers, 70–1

capacitance
  bipolar junction transistors, 98
  MIS systems, 139
  p–n junctions, 61, 64–6
capacitance–voltage characteristics, MIS systems,
  140
carbon nanotubes, 195–7, 201
  conductance modulation, 196
  disadvantages, 196
  as emitters, 197
  as interconnects, 197
  thermal conductance, 196
carrier concentrations
  excess, 54, 62
  nonequilibrium, 244
  see also electron concentration; hole
    concentration
carrier extraction, p–n junctions, 61–3
carrier heating, 173–5
carrier injection, p–n junctions, 61–3
carrier transit time, 148–9
carrier tunneling, 59
carrier-velocity saturation, 169, 172
carriers
  dynamics, 23–36
  see also electrons; holes
CCDs (charge coupled devices), 261
CDMA *see* code division multiple access
  (CDMA)
cell splitting, 226
cellular telephones, 275, 276
cellular wires, 213–14
channel conductance, 122
channel length
  long channel theory, 169
  and scaling theory, 176–83
  short-channel effects, 169–76
channels
  doping, 171
  in JFETs, 101–4
  in MESFETs, 111
  in MODFETs, 111
  in MOSFETs, 144, 146–7, 151–5
charge control analysis, 148, 173
  p–n junctions, 61, 63–4
charge coupled devices (CCDs), 261
charge neutrality condition, 18–19
charge-sharing model, 171
chemical warfare, 198
chlorine, 131
CIE (Commission Internationale de l'Eclairage)
  curves, 234
classical computers, disadvantages, 197–8
closed systems, 7
CMOS circuits *see* complementary metal oxide
  semiconductor (CMOS) circuits
code division multiple access (CDMA), 277
  advantages, 227
  disadvantages, 227–8
collector–base junction depletion layer transit time,
  99
collector–base junctions, 80
  reverse-biased, 94, 95
collector–base voltage, 88, 94, 95, 96
collector capacitance, 98
collector current, 80, 81, 83, 84–5, 89, 91
  total, 88, 90

collector–emitter voltage, 97
collector junction charging time, 99
collectors (semiconductors), 78
Commission Internationale de l'Eclairage (CIE)
        curves, 234
common base configuration, 78, *79*, 95
common base current gain, 86, 87
common collector configuration, 78, *79*
common emitter configuration, 78, *79*, 80
common emitter current gain, 86, 87–8, 96
complementary metal oxide semiconductor
        (CMOS) circuits, 160–5
    advantages, 160
    bipolar, 280
    developments, 188–219
    inverters, 160–1, *162*
    logic gates, 161–5
    miniaturization, processing issues,
        183–6
    power dissipation, 160
    and scaling theory, 176–83
    silicon-based, 165
compositional grading, 92
compound semiconductors, 2–3
    advantages, 155
computation
    edge-driven, 214
    ground state, 214–15, 216, 217–18
computers
    conventional vs. tactile, 197–201
    limitations, 197–8
    Von Neumann, 197–8, 201
    see also teramac (tera multiple architecture
        computer)
computing
    biological, 197–201
    defect tolerant, 206–9
    molecular, 197–201
    quantum-dot, 199, 210–19
    semiconductor devices in, 1
    spin-based, 210
concentration gradient, 27
conductance modulation, carbon nanotube
        FETs, 196
conduction bands, 3, *4*, 5, 13–14, 58
    double heterostructure lasers, 249
    electron concentration, 241
    LEDs, 231–2
constant field scaling, 177–8
    limitations, 178
constant mobility model, 172, 173
constant voltage scaling, 179
contaminants, sources, 131
continuity equations, 33–6, 54, 247
    with no concentration gradient, 36
    one-dimensional, 51–2, 82

solution, 81–5
    in steady-state, 35
copper, for interconnects, 185, 197
Coulomb coupling, 210
Coulomb interaction, *213*, 214, 217
Coulomb potential energy, in quantum-dot cells,
        218–19
Coulomb repulsion, 211, *212*, 213
coupling losses, 247
critical field, 59
cross-talk, 260
crystalline solids
    classification, 2
    electrical conductivity, 2
    energy bands, 5
    translational symmetry, 3
current densities, 14, 52, 144
    high, 185
    ideal diodes, 53–6
    metal–semiconductor contacts, 72–3
    threshold, 247
current flow, p–n junctions, **47**
current–voltage characteristics
    FETs, 276–7
    JFETs, 103, *104*, *105*, *119*
    MESFETs, 107, *108*
    MOSFETs, 145–50
    p–i–n photodiodes, 261, *262*
    p–n junctions, 51, *61*
    Schottky barriers, 73
cutoff frequency
    definition, 98
    FETs, 123
    MESFETs, 157–8, 278–80
    MOSFETs, 158–9, 160
    transistors, 278
cutoff mode, 78–80
cutoff wavelength, 260–1
cytosine, 200

data storage, 228
DBRs (distributed Bragg reflectors), *253*, 254
de Broglie wavelength, 5–6, 59, 210, 284, 285
defect tolerant computing, 206–9
    definition, 207
2DEG (two-dimensional electron gas), 111–12
degenerate materials, 11, *12*
density of states, 8
    effective, 12
depletion, use of term, 134
depletion capacitance, 139
depletion regions, 40, 43–4, 62
    and capacitance, 64, 65
    definition, 38
    emitter–base junctions, 82–3
    gate controlled, 190

depletion regions (*cont.*)
   ideal diodes, 52–6
   in JFETs, 101, 102
   in LEDs, 230
   in MESFETs, 118–19
   in MOSFETs, 145
   nonideal diodes, 57, 60
   p–i–n photodiodes, 261–2
   Schottky barriers, 107–9
   and stored charge, 66–7
   under bias, 48
depletion region width, 115–17, 136–7, 141
   MIS capacitors, 143–4
   MOS capacitors, 140–1
   p–n junctions, 43, 45, 48, 50, 137
   Schottky barriers, 73, 75
detectors, 230
   selection criteria, 222–3
   signal to noise ratio, 222
   see also photodetectors; p–i–n photodetectors
diabetes, 200
DIBL (drain-induced barrier lowering), 171, 181
diffusion, 26–8
   in p–n junctions, 48–51
diffusion capacitance, 98
diffusion constant, 27, 28, 82, 85
diffusion current, 27, 72, 145, 150
diffusion length, 54–5
diode behavior
   ideal, 51–6
      deviations, 57–61
diode–diode logic, 205–6
diodes
   molecular, 202–6
   reverse breakdown, 57–60, *61*
   Schottky barrier, 74–5
   see also avalanche photodiodes (APDs); light
      emitting diodes (LEDs); p–i–n photodiodes;
      p–n junction diodes
direct gap semiconductors, 31
   applications, 233
   radiative efficiency, 235–6
   use of term, 232
dispersion
   in fiber optic communications, 222
   sources, 222
distributed Bragg reflectors (DBRs), *253*, 254
distribution functions, 242
DNA, 198
   structure, 200
DNA tiles, 200–1
donors, 16–17
dopants
   for glass, 221
   types of, 16
   see also acceptors

doping, 11, *12*
   bipolar junction transistors, 92–3
   channels, 171
   Schottky barriers, 73
   see also modulation doping
doping concentration
   fluctuations, 183
   increase, 177
   nonuniform, 92–3
doping schemes, p–n junctions, *41*, *44*
double heterostructure lasers, 248–50
drain conductance, 122
drain current
   JFETs, 113–18, 121–2
   MESFETs, 114, 118
   MOSFETs, 146–7, 148, 149–50, 151–5, 173
drain current overshoot, 191
drain-induced barrier lowering (DIBL), 171, 181
drain resistance, 120–1, 182–3
drains (semiconductors)
   in JFETs, 101–4
   in MOSFETs, 144–5
drain voltage, MOSFETs, 144–5
drift, 23–6, 27
   in bipolar junction transistors, 92–3
   definition, 23
   in p–n junctions, 47, 50–1
drift current density, 24
   total, 25
dual gate field effect transistors, 188, 191–2
   asymmetric, 192
   manufacture, 192
   symmetric, 192

Early Effect, 94–5
Ebers–Moll equations, 89, 90–1
Ebers–Moll model, 88–90, 91–2
EDFAs *see* erbium doped fiber amplifiers (EDFAs)
edge-driven computation, 214
effective mass, 24–5, 231
   electrons, 7, 12
Einstein coefficients, 241
Einstein relation, 28, 92
electrical conductivity, 5, 26
   crystalline solids, 2
   definition, 25, 144
   manipulation, 2
   n-type semiconductors, 24
   p-type semiconductors, 24
electric fields, in MOSFET channels, 169–70, 172–5
electromigration, 185
electron affinity, 68–9, 71
electron-beam lithography, 184
electron concentration, 15, 18, 34–5, 40–1
   decay, 36
   determination, 8

equilibrium, 16, 87
  excess, 62–4
  ideal diodes, 52–3, 54–6
  nonequilibrium, 244
  semiconductors, 5
  threshold, 247
electron drift current, 24, 25
electron energy, 13
electron–hole pairs
  generation, 50, 57, 95, 266
  generation/recombination, 28–30,
    33
electronic devices, materials, 155
electrons, 3–4, 13–14
  diffusion current, 27
  effective mass, 7, 12
  kinetics, 14
emitter–base junctions, 80
  depletion regions, 82–3
emitter–base voltage, 96
emitter capacitance, 98
emitter–collector delay time, 98–9
emitter–collector voltage, 96
  bipolar junction transistors, 91–2
emitter current, 80, 83, 84, 85
  and Early Effect, 94
  total, 88–9, 90
emitter injection efficiency, 86
emitter junction charging time, 99
emitters (semiconductors), 78
  carbon nanotubes, 197
  p-type, 87
emitters (telecommunications)
  device types, 230
  optical, 225
  see also transmitters
empirical scaling rules, 180–1
energy bands, 3, 17, 31, *32*
  crystalline solids, 5
  HBTs, 292
  parabolic, *6*
  p–n junctions, *41*, *48*
  quantum well lasers, 250, *251*
  semiconductors, *12*
  shape, 5–7
  see also band bending
energy gaps, 5, 42, 254
  forbidden, 3, *4*
  silicon, *233*
  wide, 155–6
enzymes, 198, 199, 200
equilibrium
  concept of, 7–8
  metal–insulator–semiconductor systems in,
    127–32
  p–n junctions in, 38–47

equilibrium carrier concentrations
  intrinsic, 14
  and intrinsic semiconductor materials,
    7–16
equilibrium concentration, 32–3, 34
equivalent circuits
  bipolar junction transistors, 97, *98*
  JFETs, 122–3
  MESFETs, 156–8
  MOSFETs, 158–9
  npn devices, *88*
erbium doped fiber amplifiers (EDFAs), 222, 230,
    259
  applications, 224–5
  disadvantages, 224, 258
  energy levels, 255, *256*
  gain, 258
  manufacturing costs, 225
  operation, 224, 255–8
  structure, 255
  threshold pump intensity, 257–8
EUV (extreme ultra-violet) lithography, 184
excess minority carrier charge, 81
external quantum efficiency, 235
extinction ratio, fiber optic communications,
    222
extraction efficiency, 235
  determination, 236
  improvements, 237
extreme ultra-violet (EUV) lithography, 184
extrinsic semiconductor materials, 16–19
  definition, 13, 16

fabrication
  economic issues, 186
  processing issues, 183–6
  see also manufacturing costs
fading, multi-path, 226
family-tree-like architecture, 208
fat-tree architecture, 208–9
FDMA (frequency division multiple access),
    227
feature size, 228
Fermi–Dirac distribution, 10–11, 15
Fermi–Dirac function, 11
Fermi levels, 9, 10, 11, *12*, *20*, 27
  determination, 18–19
  in equilibrium, 203–4
  in MIS systems, 127, 129, 130, 135, *136*
  and modulation doping, 109, 110
  in OLEDs, 237, *238*
  in p–n junctions, 38–9, *41*, 242–4
  in Schottky barriers, 68–9
  see also quasi-Fermi levels
fermions, 3
FETs *see* field effect transistors (FETs)

fiber optic communications, 220–3, 269
  advantages, 220
  capacity, 220, 221
  detectors, signal to noise ratio, 222
  dispersion, 222
  extinction ratio, 222
  link budget, 221–2
  maximum propagation distance, 221, 223
  noise, 222
  transmission bands, 221
  transmission power, 222
  see also optical fibers
Fick's Law, 27
field effect transistors (FETs), 101, 188
  applications, 275, 280
  current–voltage characteristics, 276–7
  cutoff frequency, 123
  materials, 155
  metal–insulator–semiconductor, 105
  microwave operation, 157
  nanoscale, 195–7
  output power, 276–7
  transistors, 275–81
  see also dual gate field effect transistors; junction
    field effect transistors (JFETs); metal oxide
    semiconductor field effect transistors
    (MOSFETs); metal semiconductor field
    effect transistors (MESFETs); modulation
    doped field effect transistors (MODFETs)
figures of merit, 98
FinFETs, 192
fixed oxide charge, 131–2
fluorine, 221
forbidden bands, 3, *4*
  see also energy gaps
free space mass, 7, 12
frequency division multiple access (FDMA),
    227
Fukui equation, 279

gain
  avalanche photodiodes, 268, 271–2
  in bipolar junction transistors, 81
  erbium doped fiber amplifiers, 258
  flatering, 224
  power, 98
  semiconductor lasers, 247
gallium arsenide, 15, 105, 109–12, 232, 294
  applications, 229, 233–4, 248, 269, 271,
    272
    HBTs, 293
    MODFETs, 286–90
  cutoff wavelength, 261
  energy band structure, 31, *32*
  energy gaps, 254, 283
  potential wells, 282–4

gallium arsenide–aluminum gallium arsenide
    heterostructures
  formation, 282
  HBTs, 293
  lasers, 248–9
  MODFETs, 281, 286–7, 288, 297
gallium arsenide antimonide, 293
gallium arsenide phosphide, 236
gallium indium arsenide, 269, 290
  electron mobility, 289
gallium indium nitride, 294
gallium nitride, 155, 283, 294
  advantages, 295
  applications, 1–2, 295
  substrates, 297
  thermal conductivity, 298
gallium nitride–aluminum gallium nitride
    heterostructures, 283, 296
gallium phosphide, 236
  applications, 233–4
gate biasing
  in JFETs, 101, 104
  in MESFETs, 107–9
gate capacitance, 123
gate lengths, MOSFETs, 169, 289–90
gate oxide charging, 173, 174–5
gates (logic) *see* logic gates
gates (semiconductors)
  in JFETs, 101–4
  materials, 182–3
  metal, 183
gate voltage, 150
Gaussian surfaces, 14
generalized scaling, 179–80
  advantages, 180
generation rates, 52, 82
generation–recombination events, 23, 28–33, 232–3,
    234, 235
  Auger, 30, 235
  band-to-band, 28, *29*
  band-to-bound, 28, *29*
  ideal diodes, 52
  mechanisms, 30
  quantitative behavior, 30–3
  radiative, 30
  thermal, 30
gene therapy, 200
germanium, 2
germanium diodes, 56–7
glass
  dopants, 221
  refractive index, 221
global wires, 186
graded index multi-mode optical fibers, 223
graded index separate confinement heterostructure
    (GRINSCH) lasers, 252

gradual channel approximation, 115
greenhouse gases, 1
GRINSCH (graded index separate confinement heterostructure) lasers, 252
ground state computation, 214–15, 216, 217–18
group III nitrides, 155, 156, 294, 297, 298
guanine, 200
Gummel number, 88

hafnium dioxide, 182
handsets, 229
HBTs *see* heterostructure bipolar transistors (HBTs)
heating
    carrier, 173–5
    integrated circuits, 155
    problems, 155–6
helium, charged particles, 4
HEMTs *see* high electron mobility transistors (HEMTs)
heterointerfaces, band bending, 284–5
heterojunctions, 192–3, 236
heterostructure bipolar transistors (HBTs), 275, 280, 285, 290–4
    advantages, 292–3, 294
    applications, 293
    energy bands, 292
    linearity, 294
    materials, 293
    noise, 294
    operating temperature, 293–4
    operation, 290–3
    structure, 290
heterostructures, 92, 109, *110*, 192–3, 281–5
    carrier flow, 285, *286*
    formation, 281–2
    lasers, 248–50
    pseudomorphic layers, 283, *287*
    Type I, 248, 283, *284*, 293, 295
    Type II, 283, *284*, 293
    Type III, 283, *284*
    see also gallium arsenide–aluminum gallium arsenide heterostructures
Hewlett-Packard Laboratories, 207
high density arrays, 210
high electron mobility transistors (HEMTs), 280, 286, 289
    structure, *111, 112*
highest occupied molecular orbital (HOMO), 204, *205*, 237, *238*
hole concentration, 12–13, 15, 18, 41–2
    equilibrium, 16
    excess, 62–4, 82, 83
    ideal diodes, 52–3, 54–6
    in MIS structures, 134–5
    nonequilibrium, 244

hole drift current, 24, 25
hole energy, 13
hole recombination, 80–1
holes
    definition, 13
    diffusion current, 27
    kinetics, 14
HOMO (highest occupied molecular orbital), 204, *205*, 237, *238*
homojunctions, 38
hot carrier injection, 175
hot-electron aging, 175
hot-electron emission, 174–5
hybrid pi model, 97, 98
hydrogen, charged particles, 4
hydrogen chloride, 131

ideal diode current, 89
immune system, pattern recognition, 198
impact ionization, 30, 58, 265–6
    rate, 173–4
impurity states, 28–30
indium aluminum nitride, 294
indium arsenide, 288
indium gallium arsenide, 287–8
indium gallium arsenide–indium phosphide heterostructures, 293
indium gallium arsenide phosphide, 269
indium nitride, 294
indium phosphide, 105, 269, 281, 293
    applications, 229
indium phosphide double heterostructure bipolar transistors (InP DHBTs), 293
insulator capacitance, 132
insulator charge, 132
insulators
    classification, 2
    electrical conductivity, 2
    energy bands, 5
    scaling, 185
    see also silicon on insulator (SOI) devices
integrated circuits
    heating, 155
    manufacturing costs, 207
    in RF applications, 229
Intel, 207
interconnects
    carbon nanotubes, 197
    materials, 185, 197
    resistance–capacitance delays, 185–6
    scaling, 185
interference, constructive, 245
intermodulation distortion, third order, 278
internal quantum efficiency, 235
intersymbol interference, 222
intrinsic levels, 14–15, 18–19, *20*, 135, 136

intrinsic semiconductor materials
  definition, 13
  and equilibrium carrier concentrations, 7–16
inversion, 135
inverted mode, 78
inverters, 160–1, *162*

JFETs *see* junction field effect transistors
      (JFETs)
junction capacitance, 65–6
junction field effect transistors (JFETs), 101,
      155
  channel conductance, 122
  circuit design, 120–1
  current–voltage characteristics, 103, *104*, *105*,
      119
  depletion regions, 101, 102
  disadvantages, 104
  drain conductance, 122
  drain current, 113–18, 121–2
  drain resistance, 120–1
  equivalent circuits
    high frequency, 123
    low frequency, 122, *123*
  gate biasing, 101, 104
  operation, 101–4
  pinch-off point, 103, 104, *105*, *106*
  pinch-off voltage, 115–17, 120, 123–4
  quantitative description, 112–18
  saturation current, 103, 118, 120
  saturation voltage, 103–4, 117
  small signal model, 121–3
  source resistance, 120–1
  structure, 101, *102*
  transconductance, 122, 123–4
junctions, 38–75
  definition, 38
  see also collector–base junctions; emitter–base
      junctions; p–n junctions

kink effect, 191
Kirchoff's Current Law, 52
Kirchoff's Voltage Law, 136, 154
k relationship, vs. parabolic energy, 6
krypton fluoride, 183
k-space, electron motion in, 6

lasers
  applications, 230
  development, 1
  double heterostructure, 248–50
  GRINSCH, 252
  operation, 244–7
  resonance condition, 245–6
  resonant cavities, 245
  threshold conditions, 247

  see also quantum well lasers; semiconductor
      lasers; vertical cavity surface emitting lasers
      (VCSELs)
laterally diffused metal oxide semiconductor
      (LDMOS), 280
Law of Mass Action, 16, 18, 19, 42, 56
LDD MOSFETs (lightly doped drain metal oxide
      semiconductor field effect transistors), 175
LDMOS (laterally diffused metal oxide
      semiconductor), 280
light absorption, in semiconductors, 33
light emitting diodes (LEDs), 230–8
  applications, 1, 225, 230
  blue, 1, 234
  blue-green, 1
  commercial issues, 234
  construction issues, 234
  depletion regions, 230
  edge emitters, 234
  efficiency, 1
  external quantum efficiency, 235
  extraction efficiency, 235, 236
  green, 234
  heterojunction, 236
  high brightness, 230
  internal quantum efficiency, 235
  for lighting, 230, 234
  luminous performance, 234
  materials, 233–4
  operation, 230–3
  optical yield, 234
  orange, 234
  package design, 236–7
  red, 234
  structure, 234, *235*
  yellow, 234
  see also organic light emitting diodes (OLEDs)
lightly doped drain metal oxide semiconductor field
      effect transistors (LDD MOSFETs), 175
lightwave telecommunications systems
  amplifiers, 223–5
  classification, 220
  detectors, 222–3
  optoelectronic devices, 230
  repeaters, 223–5
  transmitters, 222–3
  see also fiber optic communications
link budget, 221–2
lithium, charged particles, 4
lithography
  and miniaturization, 183–5
  optical, 183–4
logic gates, 161–5, 205–6
  AND gates, 205–6, 214
  OR gates, 205–6, 214
  see also NAND gates; NOR gates

losses
  coupling, 247
  propagation, 226
  scattering, 247
  semiconductor lasers, 246–7
lowest unoccupied molecular orbital (LUMO),
      203–5, 237, *238*
lumens, 234
luminous performance, light emitting diodes, 234
LUMO (lowest unoccupied molecular orbital),
      203–5, 237, *238*

manufacturing costs, 131
  optical amplifiers, 225
maximum frequency of oscillation, 279, 280
  BJTs, 291
  definition, 98, 291
  MESFETs, 157, 158
maximum operating frequency
  MODFETs, 290
  transistors, 278–9
Maxwell–Boltzmann distribution, 11, 15
Maxwell's equations, 44
MESFETs *see* metal semiconductor field effect
      transistors (MESFETs)
metal–insulator–semiconductor field effect
      transistors (MISFETs), 105
metal–insulator–semiconductor (MIS) capacitors,
      p-type, 143–4
metal–insulator–semiconductor (MIS) systems,
      127–68
  band diagrams, 127–8, 129–30
  capacitance, 139
  capacitance–voltage characteristics, 140
  in equilibrium, 127–32
  ideal, 127–8, 133–7
  materials, 128–9, 155
  realistic, 128–30, 138–9
  with realistic surfaces, 141–2
  structure, 127, *128*
  threshold voltages, 135, 137–8, 142–3
  under bias, 133–44
  voltages, *137*
metal oxide semiconductor field effect transistors
      (MOSFETs), 101, 127
  applications, 1, 159
  current–voltage characteristics, 145–50
  cutoff frequency, 158–9, 160
  depletion regions, 145
  drain current, 146–7, 148, 149–50, 151–5, 173
  dual gate, 191–2
  equivalent circuits, 158–9
  gate lengths, 169, 289–90
  high electric fields in channel, 169–70, 172–5
  inversion layer, *146*
  lightly doped drain, 175

long-channel, 169
  miniaturization, 169, 176–7
    processing issues, 183
  n-channel, 144, 146–7, 161
    circuit, 153–5
    drain current, 152–5
    operation, 151–2
  operation, 144–55
    small signal, 158–9
  p-channel, 161
  pinch-off point, 145, 148, 150
  saturation current, 145
  scaling theory, 176–83
  short-channel effects, 169–76
  source–drain separation distance, 175–6
  structure, 144
  subthreshold current, 150–1
  threshold voltage, 148, 152
  transconductance, 159, 194
  two-dimensional potential profiles, 169,
      170–1
metal oxide semiconductor (MOS) capacitors,
      depletion region width, 140–1
metals
  classification, 2
  electrical conductivity, 2, 5
  energy bands, 5
  see also semimetals
metal–semiconductor contacts, 68
  band diagrams, 69–70
  built-in potential, 70
  current densities, 72–3
  see also ohmic contacts; Schottky barriers
metal semiconductor field effect transistors
      (MESFETs), 101, 286
  applications, 104–5
  current–voltage characteristics, 107, *108*
  cutoff frequency, 157–8, 278–80
  depletion regions, 118–19
  drain current, 114, 118
  equivalent circuits, 156–8
  gate biasing, 107–9
  materials, 155
  maximum frequency of oscillation, 157, 158
  operation, 104–9
    small signal, 155–8
  pinch-off point, *108*, 109
  pinch-off voltage, 118–19
  quantitative description, 112–18
  structure, 105, *106*
  transconductance, 279–80
metal–semiconductor–metal (MSM) photodetectors,
      261
  structure, 270
methylene, 202
microprocessors, defect tolerant, 206–7

miniaturization
  constraints, 181–3
  economic issues, 186
  MOSFETs, 169, 176–7
  processing issues, 183–6
  short-channel effects, 169–76
  trends, 228
  see also scaling
minority carrier injection, 244
MISFETs (metal–insulator–semiconductor field
    effect transistors), 105
MIS (metal–insulator–semiconductor) capacitors,
    p-type, 143–4
MIS systems *see* metal–insulator–semiconductor
    (MIS) systems
MMICs (monolithic microwave integrated circuits),
    229
mobile cellular telecommunications systems, 220,
    225–8
  components, 226
  device types, 228–9
  growth, 225
  hand-offs, 227
  issues, 226–7
  service quality, 226–7
  structure, 226
  transmission bands, 226
mobile telephone switching offices (MTSOs),
    226
mobile units, 226
mobility, 24–5, 28
  reduction, 171
modulation doped field effect transistors
    (MODFETs), 155, 275
  advantages, 286
  applications, 229, 280
  definition, 109
  linearity, 290, 294
  materials, 280–1, 286–90, 297
  maximum operating frequency, 290
  noise, 294
  operation, 111–12
  output power, 295
modulation doping
  advantages, 110
  principles, 286
  processes, 109–11
modulators, 230
molecular computing, 197–201
molecular diodes, 202–6
molecular materials, advantages, 199–200
molecules
  in computer systems, 198, 199–206
  types of, 201–2
moletronics, 201–6
  use of term, 201

monolithic microwave integrated circuits (MMICs),
    229
Moore's Law, 183, 185
Moore's Second Law, 186
MOSFETs *see* metal oxide semiconductor field
    effect transistors (MOSFETs)
MOS (metal oxide semiconductor) capacitors,
    140–1
MSM photodetectors *see*
    metal–semiconductor–metal (MSM)
    photodetectors
MTSOs (mobile telephone switching offices),
    226
multi-mode optical fibers, 223

NAND gates, 161–2
  biomolecular, 201
  operation, 163–5
nanotubes *see* carbon nanotubes
near–far problem, 277
noise
  avalanche photodiodes, 268–9, 271–2
  excess, 268, 271–2
  in fiber optic communications, 222
  power amplifiers, 294
  see also signal to noise ratio (SNR)
nondegenerate materials, 11, *12*, 15
NOR gates, 161–2
  biomolecular, 201
  operation, 162–3
normal valleys, 194
npn devices, equivalent circuits, *88*
n-type semiconductors, 16, 18–19, *20*
  electrical conductivity, 24
  work function, 69
nucleic acids, 200

ohmic contacts, 68
  band diagrams, 74
  formation, 73–4
OLEDs *see* organic light emitting diodes
    (OLEDs)
optical amplifiers
  applications, 230
  see also erbium doped fiber amplifiers (EDFAs);
      semiconductor optical amplifiers (SOAs)
optical emitters, 225
optical fibers
  attenuation, 221, 230
  materials, 220–1
  multi-mode, 223
  single-mode, 223
  structure, 220, *221*
  types of, 223
optical lithography, 183–4
optical power, 241

optical radiation
 mechanisms, 238
 see also spontaneous emission; stimulated
  emission
optical telecommunications systems, 220
optical yield, light emitting diodes, 234
optoelectronic devices, 230–74
 materials, 155
optoelectronic telecommunications systems,
  220
organic light emitting diodes (OLEDs)
 advantages, 237
 limitations, 237–8
 operation, 237
 types of, 237
OR gates, 205–6, 214
oscillation condition, 245
oxide charge, 130, 132
 factors affecting, 131
oxide–semiconductor interface
 dangling bonds, 131
 hole concentration, 134–5
oxide thickness, scaling, 177–8, 181–2
oxide trapped charge, 131

packing density limit, 210
PAE *see* power added efficiency (PAE)
parabolic energy, vs. *k* relationship, 6
parallel valleys, 194
parameter fluctuations, 182
parasitic bipolar effect, 173, 174
particle flux
 p–n junctions, **47**
 Schottky barriers, 72
particles, classical, 59–60
pattern recognition, 198, 199
Pauli Principle, 3–4, 8, *9*
perfectly operating chip paradigm, 206–7
periodic potentials, 3
PHEMTs *see* pseudomorphic high electron mobility
  transistors (PHEMTs)
phenylene, 202
phonons, definition, 30
phosphorus, as donor, 16–17
photoconductors, 261
photodetectors, 230
 see also metal–semiconductor–metal (MSM)
  photodetectors; p–i–n photodetectors
photometry, 234
photons, 236
 definition, 30
 emission, 238–9, 249–50
phototransistors, 261
piezoelectric polarization, 295–6
pinch-off point
 in JFETs, 103, 104, *105*, *106*

in MESFETs, *108*, 109
in MOSFETs, 145, 148, 150
pinch-off voltage
 in JFETs, 115–17, 120, 123–4
 in MESFETs, 118–19
p–i–n photodetectors, 260–5
 applications, 260
 cutoff wavelength, 260–1
 frequency response, 265
p–i–n photodiodes, 260–5
 bandwidth, 261–2
 current–voltage characteristics, 261, *262*
 depletion regions, 261–2
 electric field strength, 262, *263*
 illumination, 262, *263*
 operation, 261, 264
 quantum efficiency, 261, 264–5
 response speed, 265
Planck's constant, 6, 241, 254
p–n junction diodes, 203
 breakdown, 60–1
 capacitance, 67–8
 forward biased, 230
 germanium, 56–7
 maximum electric field, 67–8
 minority carrier injection, 244
 stored charge determination, 66–7
p–n junctions
 built-in potential, 40, *41*, 42–3, 45
 capacitance, 61, 64–6
 carrier extraction, 61–3
 carrier injection, 61–3
 charge control analysis, 61, 63–4
 current flow, **47**
 current–voltage characteristics, 51, *61*
 depletion regions, 38, 40, 43–5, 48, 50, 137
 diffusion, 47, 48–51
 doping schemes, *41*, *44*
 drift, 47, 50–1
 energy bands, *41*, *48*
 in equilibrium, 38–47
 Fermi levels, 38–9, *41*, 242–4
 in JFETs, 101
 particle flux, **47**
 potential diagrams, *41*
 under bias, 47–57
Poisson's equation, 43, 179–80, 271
polyphenylenes, 201–2
polysilicon, 128–9, 182–3
polytypes, 297
population inversion, 244, 246, 255, 257, 258, 259
 determination, 241–2
potential profiles, two-dimensional, 169, 170–1
potential wells, 282–4
power added efficiency (PAE), 275, 281
 definition, 277

power amplifiers
    device types, 229, 280, 286
    HBTs in, 293
    noise, 294
    operating temperature, 293–4
    optimization, 275–6
    output power, 295
    performance parameters, 275
    requirements, 228
    in wireless systems, 229, 275
power gain, 98
probability distribution functions, 8–11
    equilibrium, 8, 10
    Fermi–Dirac distribution, 10–11, 15
propagation loss, 226
protons, 3–4
proximity effect, 184
pseudomorphic high electron mobility transistors
        (PHEMTs), 280, 290, 294
    frequency performance, 288
    materials, 288–9
p-type semiconductors, 16, 18–19, *20*
    continuity equation, 35
    electrical conductivity, 24
pump photon fluxes, 257–8
punchthrough, 94–5, 176

QCA *see* quantum-dot cellular automata (QCA)
quantum dash structures, 217
quantum-dot arrays, two-dimensional, 210
quantum-dot cells, 211–14
    Coulomb potential energy, 218–19
quantum-dot cellular automata (QCA), 210–19
    architecture, 210–14
    arrays, 214–18
    limitations, 217–18
    operation, 214–18
    structure, 214, *216*
quantum-dot computing, 199, 210–19
quantum-dots, 211
quantum efficiency, 269
    external, 235
    internal, 235
    p–i–n photodiodes, 261, 264–5
quantum mechanical tunneling, 60, 174–5
quantum mechanics, 59–60, 201
quantum well lasers
    energy bands, 250, *251*
    limitations, 250–1
    multiple, 251–2
    operation, 250
    photon energy, 254–5
    single, 250–1
quantum wells, two-dimensional, 210
quantum wires, 210–11
quasi-constant voltage scaling, 179

quasi-Fermi levels, 243, 244
    concept of, 242
quaternary compound semiconductors, 2–3

radiation hardness, 189
radiative generation rate, 31, 34, 35
radiative generation–recombination, 30, 232
radiometric power efficiency, 234–6
recombination rates, 52
reduced mass, 12
refractive index, 222, 259
    glass, 221
    semiconductors, 237
repeaters, 223–5
    disadvantages, 224
    operation, 223
resistance–capacitance delays, 185–6
resistivity, 25–6
resonant amplification, 245
resonant cavities, lasers, 245
ruthenium–tantalum alloys, 183

saturation current
    JFETs, 103, 118, 120
    MOSFETs, 145
saturation drift velocity, 172
saturation mode, 78
saturation voltage, JFETs, 103–4, 117
scalability, 201
scaling, 185
    constant field, 177–8
    constant voltage, 179
    constraints, 181–3
    empirical rules, 180–1
    generalized, 179–80
    goals, 177
    quasi-constant voltage, 179
    theory, 176–83
    see also miniaturization
scanning tunneling microscopy (STM), 184–5
scattering
    events, 251
    losses, 247
Schottky barrier diodes, 74–5
Schottky barrier height, 70, 71, 194
Schottky barriers, 68–74, 155
    bias, 72–3
    built-in potential, 70–1
    current–voltage characteristics, 73
    definition, 68
    depletion regions, 73, 75, 107–9
    doping, 73
    Fermi levels, 68–9
    formation, 105–7, 296
    particle flux, 72
    reverse biased, 104, 118–19

Schottky contacts, 105, 119
SCISCM (single carrier injection, single carrier
        multiplication), 266–7, 268, 269
self-assembled structures, 200, 201
semiconductor devices
    applications
        commercial, 1
        in computing, 1
        in telecommunications, 1
    developments, 188–95
    fundamentals, 1–22
    heating problems, 155–6
    see also optoelectronic devices; silicon on
        insulator (SOI) devices
semiconductor lasers, 225
    gain, 247
    heterostructure, 248–50
    losses, 246–7
    types of, 248–54
semiconductor materials
    developments, 1–2
    see also extrinsic semiconductor materials;
        intrinsic semiconductor materials
semiconductor optical amplifiers (SOAs), 230,
        258–60
    advantages, 258
    applications, 224
    cross-talk, 260
    disadvantages, 259–60
    manufacturing costs, 225
    operation, 224, 258–9
    structure, 225, 258, *259*
semiconductors
    carrier dynamics, 23–36
    classification, 2
    definition, 2–7
    electrical conductivity, 2, 5
    electron concentration, 5, 8
    energy bands, 5, *12*
    identification, 3
    indirect gap, 31
    light absorption, 33
    quaternary compound, 2–3
    see also compound semiconductors; direct gap
        semiconductors; n-type semiconductors;
        p-type semiconductors; wide band gap
        semiconductors
semimetals
    classification, 2
    electrical conductivity, 2
    energy bands, 5
separate absorption multiplication (SAM) avalanche
        photodiodes, 265
shadowing, 226
Shockley equation, 56, 57
Shockley–Read–Hall (SRH) recombination, 235

short-channel effects, 191, 192
    MOSFETs, 169–76
signal to noise ratio (SNR), 268
    definition, 222
    detectors, 222
    transistors, 279
silica glass, in optical fibers, 221
silicon, 2, 105, 294
    acceptors, 17
    advantages, 182
    in CMOS, 165
    donors, 16–17
    electron mobility, 280
    energy band structure, 31
    energy gaps, 233
silicon carbide, 155, 156, 294
    advantages, 298
    applications, 1–2, 295
    crystal growth, 297–8
    substrates, 297
    thermal conductivity, 298
silicon germanide heterostructures, 293
silicon germanide–silicon heterostructures, 188,
        192–4
silicon germanide technology, advantages, 229, 293
silicon on insulator (SOI) devices, 188–91
    advantages, 189
    applications, 188–9
    fully depleted, 190–1
    partially depleted, 190–1, 194–5
    radiation hardness, 189
    structure, 189
silicon–silicon dioxide interface
    dangling bonds, 131–2
    MOSFETs, 146
    oxide thickness, 181–2
    states, *132*
single carrier injection, single carrier multiplication
        (SCISCM), 266–7, 268, 269
single-mode optical fibers, 223
small signal input conductance, 98
SNR *see* signal to noise ratio (SNR)
SOAs *see* semiconductor optical amplifiers (SOAs)
sodium ions, 131
soft errors, 188
SOI devices *see* silicon on insulator (SOI) devices
source–drain separation distance, MOSFETs, 170,
        175–6
source resistance, 120–1, 182–3
spatial quantization, 210
spontaneous emission, 238
    mechanisms, 240
spontaneous emission rate, 240
square law, 118, 120
SRH (Shockley–Read–Hall) recombination, 235
steady-state conditions, 23

steady-state systems, 7
step-index multi-mode optical fibers, 223
stimulated emission, 238–44, 249–50
    induction, 238–9
    mechanisms, 239–40, 241, *242*
    rate, 239–40, 241–2, 243–4
STM (scanning tunneling microscopy), 184–5
strong inversion, 135
subthreshold current, MOSFETs, 150–1

tactile computers, 198–9
TDMA (time division multiple access), 227
telecommunications systems, 220–9
    semiconductor devices in, 1
    see also lightwave telecommunications systems;
        mobile cellular telecommunications systems;
        wireless systems
teramac (tera multiple architecture computer), 207–9
    architecture, 208–9
    defects, 209
terrorist bioattacks, 198
thermal conductance, carbon nanotubes, 196
thermal generation–recombination, 30
thermionic emission currents, 72
threshold current density, 247, 253–4
threshold pump intensity, 257–8
threshold voltages
    MIS systems, 135, 137–8, 142–3
    MOSFETs, 148, 152
thymine, 200
time division multiple access (TDMA), 227
toxins, 198
transconductance, 97
    JFETs, 122, 123–4
    MESFETs, 279–80
    MOSFETs, 159, 194
transistors
    cutoff frequency, 278
    figures of merit, 275–81
    linearity, 277–8
    maximum operating frequency, 278–9
    output power, 276–7
    signal to noise ratio, 279
    for wireless systems, 275–99
        selection criteria, 280–1
    see also bipolar junction transistors (BJTs); field
        effect transistors (FETs); high electron
        mobility transistors (HEMTs)
translational symmetry, 3
translations, linear, 3
transmission bands
    fiber optic communications, 221
    mobile cellular telecommunications systems, 226
transmitters
    selection criteria, 222–3
    see also emitters (telecommunications)

tunneling, 285
    carrier, 59
    quantum mechanical, 60, 174–5
    zener, 60
two-dimensional electron gas (2 DEG), 111–12
two-port networks, 121–3, 279

University of California, Los Angeles (UCLA),
        207

valence bands, 3, *4*, 5, 7, 12–14
    double heterostructure lasers, 249
    electron concentration, 241
    LEDs, 231–2
VCSELs *see* vertical cavity surface emitting lasers
        (VCSELs)
velocity–field relationships, 172
vertical cavity surface emitting lasers (VCSELs)
    advantages, 253
    disadvantages, 253–4
    operation, 252–3
viruses, 198
voice quality, 226
Von Neumann computers, 197–8, 201

wavelength-division multiplexing (WDM), 224
wave-particle duality, 5–6
wavevectors, 6
WDM (wavelength-division multiplexing), 224
wide band gap semiconductors, 280, 291–2,
        294–8
    applications, 295
    definition, 294
    future trends, 298
    materials, 294–5
wireless systems
    device attributes, 228
    device types, 228–9
    power amplifiers, 229, 275
    transistors for, 275–99
        selection criteria, 280–1
    see also mobile cellular telecommunications
        systems
wires
    cellular, 213–14
    global, 186
    local, 185–6
    quantum, 210–11
    see also interconnects
work function difference, 132
work functions, 68–9, 71, 128, 130

X-ray lithography, 184

zener breakdown, 57, 59–60
zener tunneling, 60